Electron-Molecule and Photon-Molecule Collisions

Electron-Molecule and Photon-Molecule Collisions

Edited by

Thomas Rescigno

Lawrence Livermore Laboratory
University of California
Livermore, California

Vincent McKoy

California Institute of Technology
Pasadena, California

and

Barry Schneider

Los Alamos Scientific Laboratory
University of California
Los Alamos, New Mexico

Plenum Press · New York and London

Library of Congress Cataloging in Publication Data

Asilomar Conference on Electron–molecule and Photon–molecule Collisions, 1st, Pacific Grove, Calif., 1978.
Electron–molecule and photon–molecule collisions.

1. Electron–molecule scattering – Congresses. 2. Photoionization – Congresses.
I. Rescigno, Thomas. II. McKoy, Vincent. III. Schneider, Barry, 1940-
IV. Title.
QC793.5.E628A79 1978 539.7'54 79-15211
ISBN 978-1-4684-6990-5 ISBN 978-1-4684-6988-2 (eBook)
DOI 10.1007/978-1-4684-6988-2

Proceedings of the First Asilomar Conference on Electron–Molecule and Photon–Molecule Collisions, held in Pacific Grove, California, August 1-4, 1978

A Division of Plenum Publishing Corporation
227 West 17th Street, New York, N.Y. 10011

Participants

Conference Chairmen

T. N. Rescigno, Lawrence Livermore Laboratory, Livermore, California

V. McKoy, California Institute of Technology, Pasadena, California

B. Schneider, Los Alamos Scientific Laboratory, Los Alamos, New Mexico

Participants

N. Bardsley	University of Pittsburgh
C. Bender	Lawrence Livermore Laboratory
C. Bottcher	Oak Ridge National Laboratory
B. Buckeley	Queens University of Belfast, Northern Ireland
S. Choi	University of California, Riverside
E. Davidson	University of Washington
J. Dehmer	Argonne National Laboratory
D. Dill	Boston University
F. Flannery	Georgia Institute of Technology
A. Fliflet	California Institute of Technology
G. Gallup	University of Nebraska at Lincoln
J. Hay	Los Alamos Scientific Laboratory
E. Hayes	National Science Foundation
A. Hazi	Lawrence Livermore Laboratory
E. Heller	University of California, Los Angeles
R. Henry	Louisiana State University at Baton Rouge
A. Herzenberg	Yale University
U. Kaldor	Tel Aviv University, Israel
H. Kelly	University of Virginia

Vo Ky Lan	Observatoire de Paris (Meudon)
N. Lane	Rice University
P. W. Langhoff	Indiana University
M. Le Dourneuf	Observatoire de Paris (Meudon)
J. Macek	University of Nebraska at Lincoln
C. McCurdy	University of California, Berkeley
V. McKoy	California Institute of Technology
D. L. Moore	University of London, University College
M. Morrison	University of Oklahoma
R. Nesbet	International Business Machines Corporation
S. V. O'Neil	Joint Institute of Laboratory Astrophysics, University of Colorado
R. Poe	University of California, Riverside
W. Reinhardt	University of Colorado
T. N. Rescigno	Lawrence Livermore Laboratory
D. Robb	Los Alamos Scientific Laboratory
J. Rumble	University of Colorado
B. Schneider	Los Alamos Scientific Laboratory
H. Taylor	University of Southern California
K. J. Taylor	University of Colorado
A. Temkin	National Aeronautics and Space Administration
D. Truhlar	University of Minnesota
D. Watson	California Institute of Technology

Preface

The First Asilomar Conference on Electron- and Photon-Molecule Collisions was held August 1-4, 1978 in Pacific Grove, California. This meeting brought together forty scientists who are actively involved in theoretical studies of electron scattering by, and photoionization of, small molecules. In this volume, are collected the contributions of the invited speakers, as well as the roundtable and evening discussions condensed from taped recordings of the entire proceedings. The subject matter reflects current activity in the field and describes many of the techniques that are being developed and applied to molecular collision problems.

We would like to thank the Air Force Office of Scientific Research (AFOSR) and the Office of Naval Research (ONR) for providing the financial support that made this conference possible. Special thanks are due to Dr. Robert Junker of ONR and Dr. Ralph Kelley of AFOSR for the interest and encouragement they provided in all phases of this meeting. We also thank all the participants whose efforts and contributions made this conference a success. Finally, we thank Ms. Charlotte MacNaughton and Ms. Sara Jackson for the many hours they spent transcribing tapes and preparing this volume for publication.

T. Rescigno, V. McKoy, B. Schneider

Contents

Introduction

Present and future efforts in gas-laser development, laser fusion, isotope separation, combustion, atmospheric physics, as well as other energy-related fields, rely on a quantitative understanding of the electron and photon scattering processes for a variety of small molecules. The need for quantitatively accurate electron- and photon-molecule scattering cross sections is putting stringent demands on present theoretical methods. As this need becomes more pressing, it is important to assess the scope of those methods which are presently being used in the study of electron- and photon-molecule collisions, to identify the key advantages and limitations of the various methods and to see which techniques are best suited for a quantitative study of the important problems in molecular scattering.

A variety of techniques are currently being developed and applied to study these collision problems. These include both numerical close-coupling methods traditionally associated with atomic collision problems as well as discrete basis set expansion approaches based on techniques normally used in molecular electronic structure problems. While no definitive technique has yet emerged, a number of different approaches have been developed which appear to offer significant promise and it is clear that the outlook is for a continued increase in activities related to such techniques. It is clear that for significant progress to continue, it is essential that there be a clear understanding on the part of the various investigators of the methods which are presently in use, of those which are being developed and of the potential advantages to be gained from a synthesis of the available approaches.

With this goal in mind, a small number of atomic and molecular physicists, actively studying electron-molecule collisions and photoionization processes from different theoretical approaches, met for an intensive three day workshop. The invited papers and discussions collects in a single source material which would normally be found in a variety of technical journals. Many of these techniques are still being developed and have not yet found their way into text books.

APPLICATION OF THE CLOSE COUPLING METHOD TO ELECTRON-MOLECULE SCATTERING

D. L. Moores

Department of Physics and Astronomy
University College London
London, England

I. INTRODUCTION

The close coupling method has been applied to the scattering of electrons by atomic and molecular systems for a long time now and the basic idea is very well known. The total wave function for an (N+1)-electron system consisting of an N-electron atom, ion, or molecule plus an additional electron is expanded in antisymmetrized products of N-electron target eignefunctions and a set of unknown functions representing the added electron. The unknown functions are calculated by solving a set of coupled integro-differential (ID) equations which are derived from a variational principle: the asymptotic forms of these functions give the reactance and scattering matrices and thence the cross sections. If it were possible to retain an infinite number of terms in the expansion (including an integral over the continuum states of the system) the close coupling method would yield an exact solution to the problem. In practice, of course, a finite number of terms must be retained. The accuracy of the method should improve as this number is increased and in fortunate cases some kind of convergence may be obtained. However, in many cases the convergence is very slow and so many equations have to be solved that the method becomes intractable. If the method is to be extended, we need to develop techniques for dealing, in a fast and efficient manner, with large sets of coupled integro-differential equations. If this cannot be done then we must conclude that the method has been carried to its limits, and should seek alternative formulations of the problem.

The method described below is one originally developed by M. J. Seaton for atomic work which has been modified and adapted by Martin Crees and myself for electron-molecule scattering. It is the method used in the program IMPACT,[1] which has been applied with great success to the solution of the large sets of coupled integro-differential equations which result from application of the close coupling method to electron-atom and electron-ion problems. In brief, the coupled ID equations are converted to a set of linear algebraic equations which are solved by matrix inversion. The limitations of the method are then determined by the size of matrix that can be inverted and the speed with which this can be done. The philosophy underlying the method is basically that since in many cases rather severe physical approximations are being made, it makes no sense to solve the resulting approximate problem to high numerical precision. More is gained by including additional physics but working to less numerical accuracy. It is better to solve a 30-channel problem to a few per cent than a 5-channel problem precisely. Having said that, it is important of course that numerical stability be maintained in any procedure adopted.

A serious limitation of some of the traditional methods of solving the ID equation arises when exchange is included. The exchange kernels are handled by converting the ID equations into another set, usually much larger in number, of coupled differential equations. Each multipole component of each exchange term yields its own differential equation, which causes the number of equations to be solved to become very large. Another way is to iterate on the exchange terms, but convergence can be slow, or not obtainable at all. In the present method, the inclusion of exchange increases the amount of labour only very marginally; in particular, the size of matrix to be inverted is not altered, an important advantage.

II. FORMULATION

In order to demonstrate the method, in the interests of clarity and of simplicity of presentation we shall consider the case of electron scattering by a homonuclear diatomic molecule, in the molecular frame, and with only the ground electronic state included in the close coupling expansion (the static-exchange approximation). The close coupling expansion may then be written

$$\Psi(1,2 \cdots, N+1) = A\Phi_0(1, 2 \cdots N)F(N+1) \tag{1}$$

For diatomic molecules we choose to work in the prolate spheroidal coordinate system

$$\xi = \frac{r_a + r_b}{R}, \quad \eta = \frac{r_a - r_b}{R} \tag{2}$$

where R is the internuclear separation. The advantages of doing this are first, that the multipole expansion of the potential converges more rapidly in this system (indeed, the nuclear part is given by a single term) and secondly, the potential varies slowly with the radial coordinate ξ, which is important for the success of the method.

We write

$$F(\kappa, \xi, \eta, \phi) = \sum_{\ell,\Lambda} \frac{f_i^{\Lambda}(\xi)}{(\xi^2-1)^{\frac{1}{2}}} e^{i\Lambda\phi} T_{\ell\Lambda}(\kappa, \eta) \tag{3}$$

where $\kappa = \frac{1}{2}Rk$, k^2 being the electron energy in Rydbergs, and the $T_{\ell\Lambda}(\kappa,\eta)$ are energy dependent spheroidal angular functions. For homonuclear molecules we choose $T_{\ell\Lambda}(\kappa,\eta) = S_{\ell\Lambda}(\kappa,\eta)$ where $S_{\ell\Lambda}(\kappa,\eta)$ are solutions of

$$\left(\frac{\partial}{\partial\eta}(1-\eta^2)\frac{\partial}{\partial\eta} - \frac{\Lambda^2}{1-\eta^2} + \lambda_{\ell\Lambda}(\kappa) - \kappa^2\right) S_{\ell\Lambda}(\kappa,\eta) = 0 \tag{4}$$

The properties of the functions $S_{\ell\Lambda}(\kappa,\eta)$ and the eigenvalues $\lambda_{\ell\Lambda}(\kappa)$ are well known.[2] The radial functions $f_{\ell}^{\Lambda}(\kappa)$ then satisfy the coupled ID equations

$$\left(\frac{d^2}{d\xi^2} + \frac{1-\Lambda^2}{(\xi^2-1)} - \frac{\lambda_{\ell\Lambda}(\kappa)}{(\xi^2-1)} + \frac{\kappa^2\xi^2}{\xi^2-1}\right) f_i^{\Lambda}(\xi) = \sum_{\ell'}\left(V_{\ell\ell'}^{\Lambda}(\xi) + W_{\ell\ell'}^{\Lambda}(\xi)\right) \times f_{\ell'}^{\Lambda}(\xi) \tag{5}$$

The system is diagonal in the quantum number Λ which will be omitted in what follows. The exchange terms $V_{\ell\ell'}^{\Lambda}(\xi)$ will be discussed later. The direct potentials $V_{\ell\ell'}^{\Lambda}(\xi)$ are given by

$$V_{\ell\ell'}^{\Lambda}(\xi) = \frac{R^2}{2(\xi^2-1)} \int_{-1}^{1} S_{\ell\Lambda}(\kappa,\eta)(\xi^2-\eta^2)\, V_{static}(\xi,\eta) \times S_{\ell'\Lambda}(\kappa,\eta)\, d\eta \tag{6}$$

where

$$V_{static}(\xi_{N+1}, \eta_{N+1}) = \Phi_o(1,2\cdots N)\,|V_{N+1}|\,\Phi_o(1,2\cdots N) \tag{7}$$

and

$$V_{N+1} = \sum_{i=1}^{N} \frac{1}{r_{i,N+1}} - \frac{Z}{r_{a,N+1}} - \frac{Z}{r_{b,N+1}} \tag{8}$$

The first member of the right-hand side of equation (8) is the electronic and the remaining two terms the nuclear part of the potential. We may thus write

$$V_{static} = V_{el} + V_{nuc} \tag{9}$$

To illustrate the method, we consider first the no-exchange case. We use

$$\frac{1}{r_{ij}} = \frac{2}{R} \sum_{t,\nu} D_t^\nu Q_t^\nu(\xi_>)P_t^\nu(\xi_<)P_t^\nu(\eta_i)P_t^\nu(\eta_j) \cos \nu(\phi_i-\phi_{j'}) \tag{10}$$

where $\xi_>$ is the greater and $\xi_<$ the lesser of ξ_i and ξ_j and where P_t^ν and Q_t^ν are the Legendre functions of the first and second kinds, respectively. Using this expansion, the electronic part of the static potential can be expressed in the form

$$V_{el}(\xi,\eta) = \sum_{t=0}^{\infty} V_t(\xi)\, P_t(\eta) \tag{11}$$

The nuclear part is given by

$$V_{nuc}(\xi,\eta) = -\frac{4Z\xi}{R(\xi^2-\eta^2)} \tag{12}$$

The coupled equations then reduce to

$$\left(\frac{d^2}{d\xi^2} + \frac{1-\Lambda^2}{(\xi^2-1)^2} + \frac{\kappa^2\xi^2}{\xi^2-1} + \frac{2RZ\xi}{\xi^2-1} - \frac{\lambda_{\ell\Lambda}(\kappa)}{\xi^2-1}\right) f_\ell^\Lambda(\xi) \tag{13}$$

$$= \sum_{\ell' t} \left(C_0^\Lambda(\kappa,\ell,\ell',t)\ \xi^2 + C_2^\Lambda(\kappa,\ell,\ell',t)\right) \frac{V_t(\xi)}{\xi^2-1} f_{\ell'}^\Lambda(\xi)$$

where $C_0^\Lambda(\kappa,\ell,\ell',t)$ and $C_2^\Lambda(\kappa,\ell,\ell',t)$ are numerical coefficients. Note that the nuclear part of the potential gives rise to a single diagonal term $2RZ\xi/(\xi^2 - 1)$. Near the origin, $\xi \to 1$ and we write $\xi = 1+\varepsilon$ Expanding all terms in powers of ε Eq. (13) becomes

$$\left(\frac{d^2}{d\varepsilon^2} + \frac{1-\Lambda^2}{4\varepsilon^2}\right) f_\ell^\Lambda = \sum_{\ell'} \left(\frac{a_{\ell\ell'}^\Lambda}{\varepsilon} + b_{\ell\ell'}^\Lambda + O(\varepsilon)\right) f_\ell^\Lambda \tag{14}$$

so that

$$f_\ell^\Lambda(\varepsilon) \underset{\varepsilon\to 0}{\sim} A_\ell\, \varepsilon^{\frac{\Lambda+1}{2}} (1 + O(\varepsilon)) \tag{15 }$$

In the asymptotic region, ξ large, we have

$$V_t(\xi) \sim \frac{1}{\xi^{t+1}} (V_t^o + \frac{V_t'}{\xi^2} + \cdots) \tag{16}$$

and thus

$$\left(\frac{d^2}{d\xi^2} + \kappa^2\right) f_\ell^\Lambda(\xi) = \sum_{\ell'} \sum_{n=2}^{\infty} \frac{B_n(\ell,\ell')}{\xi^n} f_{\ell'}^\Lambda(\xi) \tag{17}$$

In (14) and (17) B_n, $a^{\Lambda}_{\ell\ell'}$, and $b^{\Lambda}_{\ell\ell}$ are numerical constants.

III. SOLUTION OF THE EQUATIONS

We divide the range of integration into N_p-1 intervals defined by N_p mesh points, not equally spaced, $\xi_1 = 1, \xi_2, \xi_3 \cdots \xi_{N_p}$. We take as unknowns the values of the functions $f_\ell(\xi_t)$ at the mesh points. The number of unknowns is thus $N_p \times N_c$ where N_c is the number of channels retained in (13), all assumed open. We then use interpolation and other finite difference formulae to express the value of each function, and its second derivative, at each mesh point, in terms of the values at neighboring points. We thus express each unknown as a linear combination of the others, to obtain a set of linear equations. The range of integration is divided into three regions, which may be referred to as near the origin, the main integration region and the asymptotic region.

A. Near the Origin

This region is defined by the first c points $\xi_1, \xi_2 \cdots \xi_c$ where c is of the order of 3 or 4. Here we have

$$\xi_t = 1 + \varepsilon_t \qquad t = 1, \cdots c \tag{18}$$

We let

$$f_\ell = \varepsilon^{\frac{1+\Lambda}{2}} (A_\ell + \varepsilon G_\ell(\varepsilon)) \tag{19}$$

where A_ℓ is an arbitrary constant and $G_\ell(\varepsilon)$ is finite as $\varepsilon \to 0$. In fact

$$\lim_{\varepsilon\to 0} G_\ell(\varepsilon) = -\frac{1}{\Lambda+1} \sum_{\ell'} a_{\ell\ell'} A_{\ell'} \quad . \tag{20}$$

$G_\ell(\varepsilon)$ is obtained by assuming it to have the form of a polynomial of degree c-1, whose coefficients are given by a c-point Lagrangian interpolation formula.

For a function f(x) tabulated at $x_1 \cdots x_c$, the value at an arbitrary point in this range is given by the Lagrange formula by

$$f(x) = \sum_{t=1}^{c} L_c(x,x_t)\; F(x_t) \tag{21}$$

where

$$L_c(x,x_t) = \prod_{t=1}^{c} \frac{(x - x_t)}{(x_t - x_t)} \tag{22}$$

Equation (21) may be rearranged in the form of a polynomial of degree c-1 in x whose coefficients involve the values of f at the tabulated points, $f(x_t)$. Applying this formula to $G_\ell(\varepsilon)$, we may write

$$G_\ell(\varepsilon) = \sum_{t=1}^{c} \varepsilon^{t-1} \sum_{t'=1}^{c} FI(t, t')\, G_\ell(\varepsilon_{t'}) \tag{23}$$

where FI(t, t') is an array of coefficients whose values depend on c and the choice of mesh points ε_t.

The second derivatives $G''_\ell(t)$ required in the coupled equations may be obtained by differentiating (23) twice. Substitution into (13) at each point in the range 1 to c then yields the first c linear equations for each value of ℓ.

B. The Main Integration Region:

The functions can only be represented by (19) over a restricted range near the origin. Away from the origin the unknowns are the functions $f_\ell(\xi)$ themselves. We use a generalized Numerov formula derived from Taylor series expansions. For a function f tabulated at points x_{-1}, x_0, x_1 such that

$$x_0 - x_{-1} = \alpha, \quad x_1 - x_0 = \beta$$

we may write down the Taylor expansions

$$f_{-1} = f_o - \alpha f'_o + \frac{\alpha^2}{2} f''_o$$

$$f''_{-1} = f''_o - \alpha f'''_o + \frac{\alpha^2}{2} f^{iv}_o$$

$$f_1 = f_o + \beta f'_o + \frac{\beta^2}{2} f''_o$$

$$f''_1 = f''_o + \beta f'''_o + \frac{\beta^2}{2} f^{iv}_o \tag{24}$$

where the superscripts denote derivatives of different orders. We form linear combinations so as to eliminate f'_0, f''_0 and f_0^{iv}:

$$(\beta f_{-1} - (\alpha+\beta) f_o + \alpha f_1) - (T_{-1}(\alpha,\beta) f''_{-1} + T_o(\alpha,\beta) f''_o + T_1(\alpha,\beta) f''_1) = P(\alpha,\beta) f^{v}_o + Q(\alpha,\beta) f^{vi}_o \tag{25}$$

where T_1, P and Q are algebraic functions of α and β.

Neglecting the right-hand side gives a working formula. The generalized Numerov formula expresses the value of a function and its second derivative at a mesh point in terms of their values at the adjacent points. The second derivatives are related to the function values themselves via the coupled equations. Hence, substitution of the coupled equations into the Numerov formula gives a linear equation in the unknowns.

C. The Asymptotic Region:

The last point ξ_{N_p} is chosen such that for $\xi > \xi_{N_p}$, the coupled equations have attained their asymptotic form (17). Two extra points $N_p + 1$, $N_p + 2$ are then chosen such that $\xi_{N_{p+2}} - \xi_{N_{p+1}} = \xi_{N_{p+1}} - \xi_{N_p} = \xi_{N_p} - \xi_{N_{p-1}}$. The system of equations (17) has solutions $S_{\ell\ell'}(\xi)$ and $C_{\ell\ell'}(\xi)$ with asymptotic form which may be obtained by the methods of Burke and Schey,[3] Norcross[4] and Norcross and Seaton.[5] Imposing the boundary conditions for ξ large

$$f_\ell(\xi) \sim \sum_{\ell'=1}^{N_c} S_{\ell\ell'}(\xi)\, \alpha_{\ell'} + C_{\ell\ell'}(\xi)\, \beta_{\ell'} \tag{26}$$

we introduce $(2\,N_c)$ additional unknowns $\alpha_{\ell'}$, $\beta_{\ell'}$.

One additional boundary condition remains to be imposed; the choice of the set of arbitrary constants A_ℓ. This may be done in N_c ways, thus giving all columns of the full matrix solutions.

Summarizing, we then have for our set of unknowns (1) the solutions near the origin $G_\ell(\varepsilon_t)$, $\ell=1, \cdots N_c$, $t = 1 \cdots c$; (2) the solutions in the main integration region $f_\ell(\xi_t)$, $\ell = 1, \cdots N_c$, $t = c \cdots N_p$, and (3) the asymptotic coefficients α_ℓ, β_ℓ, $= 1 \cdots N_c$. The number of unknowns is thus $N = N_c(N_p + 2)$. Labelling them $F_1 \cdots F_N$ the coupled equations may then be written

$$\underline{C}\,\underline{F} = \underline{X}\,\underline{A} \tag{27}$$

where $\underline{C}$, $\underline{F}$, $\underline{X}$ and $\underline{A}$ are respectively matrices with dimensions (NxN), $(N x N_c)$, $(N x N_c)$ and $N_c x N_c$). The solution yields among other things the constants $\alpha_{\ell\ell'}$ and $\beta_{\ell\ell'}$. The reactance matrix is then obtained from $\underline{\beta}\,\underline{\alpha}^{-1}$. To ensure good linear independence we rearrange terms to obtain

$$(\underline{C}, -\underline{X}) \left(\frac{\underline{F}}{\underline{A}}\right) = 0 \tag{28}$$

or

$$\underline{C}'\,\underline{F}' = 0 \tag{29}$$

This system is solved by Gauss elimination with complete pivoting using a subroutine adapted from the IBM SSP.[6]

IV. THE EXCHANGE TERMS

The insertion of electron exchange introduces terms of the type

$$\int_1^{\infty} U(\xi') \; f_{\ell}(\xi') \; d\xi' \; P(\xi) \tag{30}$$

where $P(\xi)$ are target wavefunctions. The unknown functions $f_{\ell}(\xi)$ appear in the integrand. We consider

$$I = \int_1^{\xi_{N_p}} U(\xi) \; f_{\ell}(\xi) \; d\xi \tag{31}$$

Near the origin, f_{ℓ} is given by

$$f_{\ell} = \varepsilon^{\frac{1+\Lambda}{2}} \left(A_{\ell} + \varepsilon \, G_{\ell}(\varepsilon)\right) \tag{32}$$

where

$$G_{\ell}(\varepsilon) = \sum_{t=1}^{c} \varepsilon^{t-1} \sum_{t'=1}^{c} FI(t, t') \; G_{\ell}(\varepsilon_{t'}) \tag{33}$$

This is substituted into the integrand, and the integration carried out term by term. The result is an expression linear in the $G(\varepsilon_{t'})$.

In the main integration region $c + 1 \leq t \leq N_p$ we divide each interval into four equal sub-intervals. The sub-tabular points are thus given by

$$e_i^t = \xi_{t-1} + i\Delta \; ; \quad i = 1, 2, 3 \tag{34}$$

where

$$\Delta = \frac{1}{4} (\xi_t - \xi_{t-1}) \tag{35}$$

The values of the functions at the sub-tabular points are obtained in terms of their values at ξ_{t-2}, ξ_{t-1}, ξ_t, and ξ_{t+1} again using the Lagrange formula

$$f(e_i^t) = \sum_{\sigma=1}^{4} FLAG(t, i, \sigma) \; f(\xi_{t-3+\sigma}) \tag{36}$$

The integral from ξ_{t-1} to ξ_t is obtained from a 5-point integration formula

$$\int_{\xi_{t-1}}^{\xi_t} U(\xi)\ f(\xi)\ d\xi = \frac{\Delta}{720} \sum_{i=1}^{5} SH(n,i)\ U(e^t_{i-1})\ f(e^t_{i-1}) \qquad (37)$$

The error in (37) is $\frac{\Delta}{1260}\ \delta^6(Uf)$. The quantities FLAG and SH are arrays of coefficients.

Substitution of (37) and (36) into (31) gives the integral as a linear combination of the values of the functions at the mesh points. No new unknowns are introduced when exchange is included and hence the dimension of the matrix $\underline{C}$ is unaltered.

V. ORTHOGONALITY CONSTRAINTS

In a practical calculation, the functions F(N+1) in equation (1) are constrained to be orthogonal to the target orbitals. This introduces additional Lagrange multiplier terms which have the form of integral operators similar to those discussed in the previous section. Each Lagrange multiplier will then constitute an additional unknown and each constraint will introduce an additional equation. Thus $N = N_c(N_p + 2) + N_L$ where N_L is the number of different Lagrange multipliers. For scattering by an open-shell molecule, so-called "bound channels" must also be included, further increasing the number of unknowns.

The success of the method depends very much on the choice of the mesh points ξ_t. If the number N_p becomes too large, then we have to invert huge matrices and the method becomes impracticable. On the other hand, if N_p is too small then the integration can become unstable. Systematic methods for choosing the points have been discussed in detail by Seaton.[6] In practice, for $e^- + N_2$ scattering, the value of N_p for energies of a few volts is 30 with $\xi_{N_p} \simeq 5.0$. For 5 ℓ-channels the typical running time on the CDC 7600 machine is then 8 seconds per energy point, for a static-exchange run.

VI. CONCLUDING REMARKS

The chief advantages of the method are its speed, enabling large numbers of channels to be included, and the fact that the inclusion of exchange merely leads to extra algebraic complexity before the programming stage, rather than a large increase in computer time. Present disadvantages stem from the fact that one requires slowly-varying potential functions and is hence obliged to work in a two-center co-ordinate system. This currently restricts the method to diatomic molecules. In addition, it relies at present upon target wavefunctions computed elsewhere. Since few accurate two-center calculations exist, single-center functions must be transformed into prolate spheroidal co-ordinates.

REFERENCES

1. M. A. Crees, M. J. Seaton, and P. M. H. Wilson, Comp. Phys. Comm. 15, 23 (1978).
2. C. Flammer, Spheroidal Wave Functions (Stanford University Press, Stanford, California, 1957).
3. P. G. Burke, and H. Schey, Phys. Rev. 126, 147 (1962).
4. D. W. Norcross, Comp. Phys. Comm. 1, 88 (1969).
5. D. W. Norcross and M. J. Seaton, J. Phys. B: Atom. Molec. Phys. 6, 614 (1973).
6. M. J. Seaton, J. Phys. B: Atom. Molec. Phys. 7, 1817 (1974).

DISCUSSION

Schneider: Have you experimented at all with iterative methods of solving linear equations? If those matrices are diagonally dominant, which they might well be, you can solve the problems by iterative techniques, which go as N^2 as opposed to N^3. Furthermore, with iterative methods, all you have to be able to do is dot products and you do not need to have the whole matrix in core. This feature could make the solution of these equations very efficient.

Dill: What are the operational difficulties involved in using this method for more complicated systems such as polyatomic molecules?

Moores: Well, this system of spheroidal coordinates is most appropriate for diatomic molecules. For a polyatomic, one would use a one-center, or a multi-center expansion. You want the wavefunctions to behave smoothly. In our experience with molecules, one-center expansions give problems with numerical stability because everything is varying too rapidly.

Schneider: It's probably only the nuclear parts, and not the electronic parts, that cause trouble.

Moores: That's right.

Bardsley: Can you give us some idea as to how many angular momentum channels you typically need, say in the simplest system - $e^- + H_2$?

Moores: About four or five.

Bardsley: Taking those numbers for each electronic channel, how far beyond static-exchange do you think you can go?

Moores: Well, that depends very much on the system, but in this case we could probably handle about four electronic channels in the expansion.

Poe: How easily can these methods be adapted to treat vibrational channels?

Moores: Well, we would have to think carefully before trying to apply the method to treat vibrational effects; that is, to treat them non-adiabatically. I haven't thought about it very deeply, but I can foresee problems.

McCurdy: If you are considering a static-exchange problem in a system with several occupied orbitals, there are nodes in the radial wavefunctions, at least for the lower partial waves, that come from the constraint that the scattering solution be orthogonal to the bound solutions. It seems like that is going to make the inner region very complicated. How do you cope with that numerically?

Moores: Well, I didn't emphasize that aspect of the problem, but orthogonality is imposed by the inclusion of Lagrange multipliers which constitute additional unknowns.

Henry: One last question. Can you get wavefunctions out of this that are of reasonable quality that you could use to calculate, say, a photoionization cross section?

Moores: Yes, we do solve for the wavefunction.

Henry: But they are of reasonable quality, even though they are fairly sparse?

Moores: Yes. Wavefunctions computed by this method have been used to do atomic photoionization successfully. Even though they are sparse, they can be interpolated with reasonable accuracy.

THE COUPLED-CHANNELS INTEGRAL-EQUATIONS METHOD IN THE THEORY OF LOW-ENERGY ELECTRON-MOLECULE SCATTERING

Michael A. Morrison

Department of Physics and Astronomy
University of Oklahoma
Norman, Oklahoma 73019

I. INTRODUCTION

The part of this workshop that is devoted to electron-molecule collisions is concerned with two somewhat complimentary classes of approaches to this problem: L^2-variational methods, such as the R-matrix and T-matrix methods, and eigenfunction-expansion methods, such as the close-coupling or coupled-channel method. The former approaches are discussed in accompanying articles by Schneider and by Fliflet. In this paper, we shall be concerned with the *coupled-channels* method for low-energy electron-molecule scattering.[1-4] Various aspects of the physics of electron-molecule collisions have been discussed in detail elsewhere. Thus the present review will emphasize techniques for solving the problem. We first describe how to formulate and carry out such a calculation (Secs. II and III) and then suggest where likely pitfalls lie and ways to avoid them (Sec. IV). Since most of the applications to date of the coupled-channel method have employed approximate local potentials, the development in this paper will deal primarily with the solution of *differential* equations via an integral-equations algorithm. The case of *integrodifferential* equations (which arise when electron exchange effects are rigorously taken into account) will be discussed briefly in Sec. V.

The present application of integral equations methods to electron-molecule collision problems was stimulated by early work by Sams and Kouri[5] and others.[1] Similar methodology has been applied to a variety of problems in heavy-particle and reactive scattering.[6]

II. THEORETICAL DESCRIPTION OF THE SCATTERING EVENT

A. The Eigenfunction Expansion Approach

The problem at hand is the scattering of low-energy electrons by a linear molecule in the ground electronic state, e.g., e-$N_2(X^1\Sigma_g^+)$. We shall consider a closed-shell target with N_e-electrons. By "low energy" we mean incident energies $E_{inc.} \lesssim 1$ Rydberg (13.6 eV). Thus we must solve the non-relativistic time-independent Schroedinger equation for the electron-molecule system,

$$(\mathcal{H}-E)\Psi_E(x,\tau,\vec{R}) = 0 \ , \quad (2.1)$$

where x denotes the coordinates (spatial $\vec{r}$ and spin σ) of the scattering electron, τ the target electronic coordinates and $\vec{R}$ the target nuclear coordinates. The system Hamiltonian is

$$\mathcal{H} = \mathcal{H}_{mol}(\tau,\vec{R}) + T_e(\vec{r}) + V_{e\text{-}mol}(x,\tau,\vec{R}) \ , \quad (2.2)$$

where the electron-molecule interaction potential $V_{e\text{-}mol}$ includes electrostatic contributions. The potential energy may also include approximate exchange and polarization terms, depending on how these effects are handled in the scattering formulation. $T_e=(-\hbar^2/2m_e)\nabla^2_{\vec{r}}$ is the kinetic energy operator of the scattering electron, and $\mathcal{H}_{mol}$ is the target Hamiltonian with eigenfunctions Φ_γ and eigenvalues ε_γ, i.e.,

$$\mathcal{H}_{mol}\Phi_\gamma(\tau,R) = \varepsilon_\gamma\Phi_\gamma(\tau,\vec{R}) \ , \quad (2.3)$$

where γ represents a set of quantum numbers which label a particular molecular state (i.e., electronic, vibrational and rotational labels). The total energy E in (2.1) is simply related to the asymptotic asymptotic kinetic energy of the incident electron, with wavenumber k_γ, by

$$E = \varepsilon_\gamma + \frac{\hbar^2 k_\gamma^2}{2m_e} \ . \quad (2.4)$$

In eigenfunction-expansion methods,[7] the system wavefunction Ψ_E is expanded in a complete set of target states, which are usually chosen to be the Born-Oppenheimer eigenfunctions of $\mathcal{H}_{mol}$. The coefficients of this expansion are the scattering functions. The equations for these functions, which obtain from substitution of the eigenfunction expansion into (2.1), are coupled in the target wavefunctions. The scattering functions are further expanded in some appropriate angular basis set. Summarizing these expansions, we can write

$$\Psi_E(x,\tau,\vec{R}) = A\sum_\nu \frac{1}{r} f_\nu(r)\Psi_\nu(\hat{r},\sigma,\tau,\vec{R}) \qquad (2.5)$$

where ψ_ν is a basis set complete in the indicated spatial and spin coordinates and A is the antisymmetrizer required to enforce the proper behavior of Ψ_E under pairwise electron interchange. Substitution of (2.5) into the time-independent Schroedinger equation (2.1) leads[4] to a set of coupled integrodifferential equations for the radial scattering functions f_ν. The coupling manifested in these equations arises from a variety of sources; e.g., the angular coupling due to the non-spherical nature of the interaction potential.

By truncating the eigenfunction and angular expansions, one obtains a finite set of coupled equations; this is the "close-coupling" or "coupled-channels" approximation. This set of equations is numerically integrated to the asymptotic region, where the interaction potential is typically of the order 10^{-4} - 10^{-6} Hartree, and the K-matrix is extracted. From this matrix, the desired cross sections are easily calculated.

B. Options in Formulating the Collision Problem

In implementing an eigenfunction expansion procedure, one immediately faces several important alternatives at the fundamental level of problem formulation. First, coordinates appropriate to the particular problem under consideration must be selected. The use of a two-center coordinate system is discussed in the accompanying paper by Moores. In this approach, one introduces prolate spheroidal coordinates with the foci of the ellipses located at the nuclear centers. The angular basis functions are spheroidal harmonics.[8] By contrast, in a single-center coordinate system, the familiar spherical polar coordinates are used. The origin is usually placed at the center-of-mass of the molecule. In this case, the angular basis functions are constructed from simple spherical harmonics.

The two-center system is certainly suitable for electron collisions with diatomic molecules. And, as the work of Crees and Moores[9] and others has shown, in such problems the averaged Coulomb potential energy due to the electron-nucleus interaction converges quite rapidly in prolate spheroidal coordinates. However, for collisions with linear or nonlinear polyatomic targets, it is probable that single-center coordinates are more suitable. Moreover, if the single-center system is used, considerable care must go into the treatment of the static potential energy terms (see Sec. III).

A second fundamental question is whether to use a "lab frame" or a "body-frame" representation. In the laboratory (space-fixed) reference frame,[10] the symmetry axis of the molecule is oriented at an angle $\hat{R}$ with respect to coordinate axes fixed in space. The scattering electron is (usually) taken to be incident along the $\hat{z}$ axis. Strictly speaking, space-fixed and body-fixed formulations are equivalent. However, in referring to a "body frame," we usually mean "with the fixed-nuclei[11] approximation." In the body-fixed reference frame for electron-molecule collisions,[12] the orientation of the nuclear axis is fixed along the polar coordinate axis. Thus the molecule does not rotate in the body-frame so defined. Cross sections for elastic scattering may be obtained by averaging the body-frame scattering amplitude over all molecular orientations. The principal assumption used in this fixed-nuclei-body-frame representation is that the collision time is short compared to the period of molecular rotation. Rotational excitation cross sections can be obtained by invoking the adiabatic nuclei approximation,[13] which is valid except near threshold. This approximation can also be applied to the vibrational motion of the nuclei,[14] and the internuclear separation is often held fixed (i.e., the molecule is rigid).

The relationship between the body- and lab-frame theories is clearly seen in the frame-transformation theory of Chang and Fano.[15] The body-frame formulation is appropriate for small r, corresponding to the scattering electron being near the target, where this electron's orbital motion is strongly coupled to the internuclear axis $\hat{R}$. In this region, the projection of the electron's orbital angular momentum $\vec{\ell}$ along the internuclear axis, $\Lambda = \pm|\vec{\ell} \cdot \hat{R}|$, is a constant of the motion. By contrast, the lab-frame theory is appropriate for large r, i.e., far from the target, where the interaction potential is comparatively weak and it is necessary, in general, to take explicit account of the rotational (and vibrational) Hamiltonians. In the lab frame the rotational angular momentum of the target and the projection of the electron's orbital angular momentum along $\hat{R}$ are constants of the motion in the asymptotic region.

The choice of which formulation to use in a given calculation is determined primarily by the system and range of scattering energies of interest. For example, a "pure" lab-frame formulation is intractable for treating low-energy collisions of electrons with moderately- to highly-aspherical targets (e.g., N_2 to CO_2) owing to the enormous number of coupled channels which must be included to converge the calculation. On the other hand, if one seeks rotational excitation cross sections near threshold, the body-frame-fixed-nucleus theory is invalid once the criterion

$$|k_i - k_f|\, R_e << 1 \tag{2.6}$$

is violated.[16] [In (2.6), k_i (k_f) is the initial (final) wavenumber of the scattering electron, and R_e is the effective radius of the interaction region.]

The angular basis for expansion of the scattering functions in the body frame is especially simple, being composed of spherical harmonics. In the lab frame, it is necessary to account for the orientation angles of the internuclear axis. The most convenient angular basis in this case is the coupled angular momentum representation of Arthurs and Dalgarno.[17]

The third option involves truncation of the aforementioned expansion of ψ_E in target electronic eigenfunctions. Retaining only the ground electronic state of the molecule in this expansion leads to enormous simplifications in the resultant coupled equations. This approximation is predicated on two assumptions: (1) either there are no open electronic channels in the range of collision energies under consideration or coupling to such channels as are open is unimportant to the calculation of the desired cross sections; (2) the effects of virtual excitation of closed electronic channels is negligible. The last assumption is, in fact, false for low-energy collisions.[10] The effects of virtual excitation are usually included by means of an induced polarization potential[18] or psuedostates.[19]

The approximation that only the ground electronic state need be retained in the eigenfunction expansion leads to an appealingly simple physical picture of the nature of the static component of the electron-target interaction. In this picture, the electron "scatters" from the averaged static potential energy, which is the matrix element

$$V_{st}(r,\theta) = \langle \Phi_0 | V_{int} | \Phi_0 \rangle , \tag{2.7}$$

where the integration is over the coordinates of all target electrons. V_{int} is the sum of the Coulomb interactions, i.e.,

$$V_{int} = -\sum_{\alpha} \frac{Z_\alpha}{|\vec{r} - \vec{R}_\alpha|} + \sum_{i} \frac{1}{|\vec{r} - \vec{r}_i|} , \tag{2.8}$$

where the first sum runs over the nuclei (charge Z_α) and the second over the N_e molecular electrons.

The theory to be discussed herein is a single-center, body-frame formulation in which the fixed-nuclei approximation is made and the molecule is treated as rigid. Only the ground electronic state of the target $\Phi_0(\tau;R)$ is explicitly included. We further make the physical assumption that the electron-molecule interaction can reasonably be approximated by a local potential energy function, the components of which will be discussed in the next section. This fairly simple problem provides a convenient framework for the discussion of numerical procedures that is to follow.

C. Options in Determination of the Interaction Potential

Having decided how to formulate the theoretical collision problem, we turn to the definition of the electron-molecule interaction potential energy, $V_{e\text{-}mol}$ in Eq. (2.2). We must first decide what types of interactions should be included, e.g., static, exchange, polarization, correlation. The need for accurately representing the electrostatic interactions and for incorporating the effects of exchange and polarization in low-energy electron collisions is well known. The question of the importance of short-range correlation in such problems is largely unexplored. The second issue related to the interaction potential is how each type of interaction should be represented: e.g., how accurately must each contribution to $V_{e\text{-}mol}$ be calculated, and what physical assumptions can reasonably be introduced?

The average electrostatic (Coulomb) interaction term reflects the symmetry of the target. In a single-center formulation, it is conveniently expanded in Legendre polynomials,[3] with only even-order terms appearing for targets belonging to the point group $D_{\infty h}$. For Σ target states, we have

$$V_{st}(r,\theta) = \sum_{\lambda=0}^{\lambda_{max}^{st}} v_{\lambda}^{st}(r) P_{\lambda}(\cos\theta) \ . \tag{2.9}$$

Each expansion coefficient can be written as

$$v_{\lambda}^{st}(r) = v_{\lambda}^{el}(r) + v_{\lambda}^{nuc}(r) \ , \tag{2.10}$$

where $v_{\lambda}^{el}(r)$ arises from the electron-electron interactions in (2.8) and $v_{\lambda}^{nuc}(r)$ from the electron-nucleus terms. As λ increases, the nuclear contribution dominates v_{λ}^{st}.

Methods for rapid calculation of approximations to the static potential (such as the INDO method) have been studied by Truhlar and others[20] and are discussed in an accompanying article. As the extent of the validity of these methods is uncertain at present, we calculate v_{λ}^{st} directly from the near-Hartree-Fock electronic wavefunction of the molecule. This procedure is described in Sec. III. In regard to V_{st}, other important questions relate to computational details such as choice of quadrature scheme, integration mesh, etc. These matters have been fully discussed in descriptions of particular applications.[1-4]

At present it is not so obvious how best to incorporate the effects of induced polarization, i.e., distortion of the electron charge cloud of the target by the electric field of the scattering electron. Including these effects directly in a coupled-channels procedure by including closed electronic channels is not feasible with present-day computers. Alternative approaches include the use of psuedostates[19] and an optical potential.[21] But the most widely-used procedure is to simply add to the static term a semi-empirical adiabatic polarization potential of the form[10,18]

$$V_{pol}(r,\theta) = \left[-\frac{\alpha_0}{2r^4} - \frac{\alpha_2}{2r^4} P_2(\cos\theta)\right] C(r) , \qquad (2.11)$$

where α_0 (α_2) is the spherical (non-spherical) polarizability of the target and $C(r)$ is a cutoff function, which is usually taken to be spherical and of the form

$$C(r) = 1-\exp\left[(r/r_c)^6\right] \qquad (2.12)$$

with r_c, the cutoff radius, a parameter in the theory. This function is introduced to account for the fact that the assumption that the motion of the scattering electron can be treated adiabatically is false at small r, where the polarization potential should depend on the electron's velocity as well as its position.[22] The form (2.12) introduces a rather sharp cutoff at $r \simeq r_c$. The central concern in implementing (2.11) is how to determine r_c.

Two additional questions regarding V_{pol} remain largely unexplored for electron-molecule collisions: (1) is it necessary to take into account deviations of V_{pol} from the asymptotic form $-(\alpha_0/2r^4) - (\alpha_2/2r^4)P_2(\cos\theta)$, still working within the adiabatic theory? (2) how can non-adiabatic (distortion) effects be introduced into this approach? The former question has been examined in a series of ab initio calculations[23,43] of an adiabatic polarization potential at the near-Hartree-Fock level of accuracy for e-N_2 and e-CO_2. The results showed substantial deviations from the asymptotic form, even well beyond the region of the molecular charge cloud. Because of these deviations, use of a simple single-parameter spherical cutoff function such as (2.12) in scattering calculations was not feasible, suggesting that the convenience, simplicity, and apparent flexibility of the semi-empirical adiabatic polarization potential (2.11) may be fortuitous. Further study of these questions would seem to be in order.

Perhaps the most troubling component of the electron-molecule interaction is that which arises from electron exchange, i.e., from the requirement that the system wavefunction Ψ_E must be antisymmetric under interchange of the scattering electron and any of the bound electrons of the molecule. On the one hand, it is formally possible to enforce this requirement rigorously, a procedure which leads to coupled integrodifferential equations including non-local "exchange-potential" terms. The numerical solution of these equations rapidly becomes intractable as the complexity of the target increases beyond, say, H_2 or N_2. Both the number of target electrons and the degree of asphericity of the molecule are responsible for this difficulty.

This predicament has motivated study of the use of local approximations to represent the exchange potential. At present, the question of the "best" such model potential remains open. Two classes of approximations, one based on a free-electron-gas model[4] and the other on a semiclassical approximation,[24] show considerable promise and have been the focus of recent attention.

An alternative to the use of model exchange potentials is provided by an orthogonalization procedure of Burke and Chandra[25] in which the non-local exchange terms in the coupled scattering equations are set to zero. However, the scattering function is forced to be orthogonal to all bound molecular orbitals of the same symmetry as the scattering function through the use of Lagrange Undetermined Multipliers. This procedure leads to sets of coupled <u>inhomogeneous</u> differential equations <u>except</u> for symmetries for which there are no bound orbitals to which the scattering function can be orthogonalized. In such instances, one is left with the <u>static</u> equations uncorrected for exchange. This difficulty arises, for example, in the important resonant Π_g symmetry in e-N_2 scattering.

In the present application, we construct $V_{e\text{-mol}}$ from a highly accurate near-Hartree-Fock static potential, a local exchange potential based on the free-electron-gas model,[4,44,45] and the semi-empirical adiabatic polarization potential of Eqs. (2.11) and (2.12).

III. THE COUPLED-CHANNELS PROCEDURE

In this section, we shall describe the solution of the scattering problem by the coupled-channels procedure, using results from previous studies of e-N_2 and e-CO_2 scattering for purposes of illustration.[2-4] Our philosophy at each stage of these calculations has been to try to treat all numerical aspects of the problem as accurately as feasible. The hope in so doing is that the results will reliably reflect strengths and weaknesses in the physical assumptions of the theory rather than defects in the numerical or computational analysis. We shall assume "production mode" in this section, i.e., that all necessary convergence studies and other numerical tests have been completed. These matters will be discussed in Sec. IV.

A. <u>The Coupled Differential Equations</u>

In the single-center body-frame-fixed-nuclei formulation,[4,12] the equation for the scattering function is reduced to a set of coupled homogeneous radial equations by expanding the function in the angular basis of spherical harmonics $Y_\ell^\Lambda(\theta,\phi)$. This expansion is truncated at some number of channels N, which must be determined by convergence studies. The resulting set of N coupled equations for scattering energy k^2 (in Rydbergs) has the form

$$\left[\frac{d^2}{dr^2}\underset{\sim}{1} - \frac{1}{r^2}\underset{\sim}{L} + k^2\underset{\sim}{1}\right]\underset{\sim}{f}(r) = 2\underset{\sim}{V}(r)\underset{\sim}{f}(r) \quad , \tag{3.1}$$

where $\underset{\sim}{f}$ is the NxN solution matrix, the columns of which correspond to the N linearly-independent solution vectors that are regular at the origin. Thus the column index of the element $f_{\ell\ell_0}(r)$ can be chosen to correspond to the initial channel. The row index ℓ labels the partial-wave components of each solution vector. The solution matrix depends on the good quantum number Λ, which corresponds to the projection of the electron's orbital angular momentum along $\hat{R}$, but since the equations are not coupled in Λ we have omitted this label. (This simplification obtains only if the fixed-nuclei approximation is invoked in the body frame.) In Eq. (3.1), $\underset{\sim}{L}$ is a constant diagonal NxN matrix with elements $\delta_{\ell\ell'}\ell(\ell+1)$, $\underset{\sim}{1}$ is the NxN unit matrix, and V(r) is the potential matrix with $\ell\ell'$ element

$$V_{\ell\ell'}(r) = \langle Y_\ell^\Lambda | V(\vec{r}) | Y_{\ell'}^\Lambda \rangle \quad , \tag{3.2}$$

where $V(\vec{r})$ includes V_{st} of Eq. (2.7), V_{pol} of (2.11) and the local exchange potential term. We solve these equations subject to the usual boundary condition of regularity at the origin,

$$f_{\ell\ell_0}(r=0) = 0 \quad , \tag{3.3}$$

and K-matrix asymptotic boundary conditions, i.e.,

$$f_{\ell\ell_0}(r) \underset{r\to\infty}{\sim} \sin\left(kr - \ell_0\frac{\pi}{2}\right)\delta_{\ell\ell_0} + K_{\ell\ell_0}^{(\Lambda)} \cos\left(kr - \ell\frac{\pi}{2}\right). \tag{3.4}$$

B. Outline of the Computational Procedure

The coupled-channels procedure for electron-molecule scattering is outlined in the flow chart of Figure 1. The seven steps shown will be described briefly here, with the exception of the integral equations algorithm, which is the subject of Sec. III.C.

In order to calculate the static and exchange contributions to the potential matrix $\underset{\sim}{V}$ in (3.1), we require the charge density of the molecule, $\rho(\vec{r})$. This is determined from the near-Hartre-Fock wavefunction of the target at its equilibrium geometry,

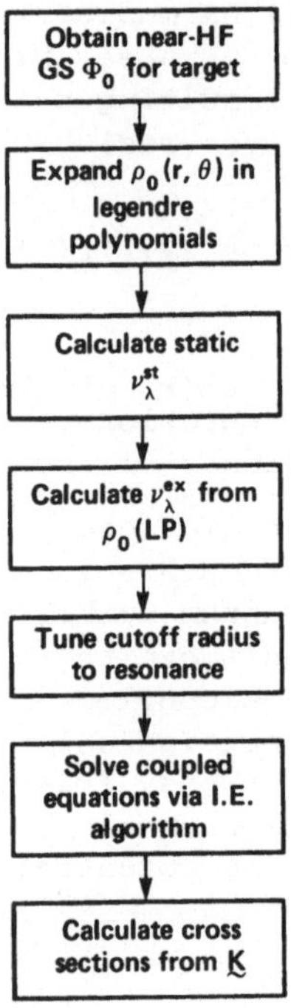

Fig. 1. Outline of the coupled-channels procedure for electron-molecule collisions. I.P. in step 4 is the target ionization potential, which is required for the calculation of the HFEGE potential.

$$\Phi_0(\tau;R) = \mathcal{A} \prod_{i=1}^{N_{occ}} \phi_i(\vec{r}_i,\sigma_i) , \tag{3.5}$$

where N_{occ} is the number of occupied spatial orbitals $\phi_i(r_i,\sigma_i)$ and $\mathcal{A}$ represents the full antisymmetrizer. Thus for the $X^1\Sigma_g^+$ ground state of N_2 we have the Slater determinant

$$\Phi_0(\tau;2.068a_0) = 1\sigma_g^2\ 1\sigma_u^2\ 2\sigma_g^2\ 2\sigma_u^2\ 3\sigma_g^2\ 1\pi_u^4 \tag{3.6}$$

These functions are readily available for a wide variety of molecules.

Given the molecular orbitals $\Phi_i(\vec{r})$, the charge density can be calculated from

$$\rho(\vec{r}) = \sum_{i=1}^{N_{occ}} N_i |\phi_i(\vec{r})|^2 , \qquad (3.7)$$

where N_i is the occupation number of the ith spatial orbital (e.g., $N_{1\sigma_g} = 2$, $N_{1\pi_u} = 4$). For computational convenience, the density (3.7) is expanded in Legendre polynomials, taking advantage of the axial symmetry of the molecules of interest here, viz.,

$$\rho(r,\theta) = \sum_{\lambda=0}^{\lambda_{max}^{el}} a_\lambda(r) P_\lambda(\cos\theta) \qquad (3.8)$$

The expansion coefficients are evaluated by Gauss-Legendre quadrature.[27]

The most time-consuming part of this step is the calculation of the molecular charge density ρ. Hence each a_λ is calculated on a "minimal" mesh of r-values which is selected to allow these coefficients to be interpolated as needed using a cubic spline.[28] For N_2 (CO_2), we require only even-λ terms in (3.8) and use $\lambda_{max}^{el} = 14$ (28).

The third step is calculation of the static contribution, V_{st} of Eq. (2.7). With the expansion of (2.9) and (2.10), each electronic term can be determined from the corresponding a_λ by numerical quadrature as

$$v_\lambda^{el}(r) = \frac{4\pi}{2\lambda+1} \int_0^\infty a_\lambda(r') \frac{r_<^\lambda}{r_>^{\lambda+1}} r'^2 dr' , \qquad (3.9)$$

where $r_<$ ($r_>$) is the minimum (maximum) of r and r'. In general, the mesh of r-values for which we require $v_\lambda^{st}(r)$, the "integration mesh," is more dense than that required to determine the coefficients a_λ in (3.8). One way to deal with this problem is to use the aforementioned cubic-spline interpolation in evaluating (3.9). Typically, a fairly dense mesh of points $\{a_\lambda(r_i)\}$ is required in any region of r where the slope of a_λ is changing rapidly (e.g., near the nuclei).

The other contribution to v_λ^{st} is due to to the electron-nuclei interaction and is trivially calculated. For example, for a homonuclear diatomic molecule, we have

$$v_\lambda^{nuc}(r) = -2Z\, \xi_<^\lambda / \xi_>^{\lambda+1} \qquad (3.10)$$

where $\xi_< = \min(\frac{1}{2}R, r)$, etc., and Z is the nuclear charge.[38] Fortunately, v_λ^{nuc} increasingly dominates v_λ^{st} as λ increases. So although we require expansion coefficients of the static potential up to λ_{max}^{st} = 28 in low-energy e-N_2 scattering calculations, we only need to calculate v_λ^{el} for 8 values of λ, i.e., up to λ^{el} = 14.

Once the coefficients v_λ^{st} have been determined, the potential matrix elements in (3.2) are easily calculated as

$$V_{\ell\ell'}(r) = \sum_{\lambda=0}^{\lambda_{max}^{st}} f_\lambda(\ell\ell'\Lambda) v_\lambda^{st}(r) \; , \qquad (3.11)$$

where $f_\lambda(\ell\ell'\Lambda)$ is a numerical coefficient which is calculated from Clebsch-Gordan coefficients.[2]

Turning now to the calculation of the exchange contribution to the interaction potential, we recall[4,12] that if exchange is treated rigorously, an additional term of the form $\int_0^\infty \underset{\sim}{K}(r,r')\underset{\sim}{f}(r')dr'$ appears on the right-hand-side of the coupled scattering equations (3.1), with $\underset{\sim}{K}(r,r')$ the exchange kernel. In the present application, this term is replaced by a local, energy-dependent, short-range attractive potential. One such potential which has been applied to a variety of problems is the Hara-Free-Electron-Gas-Exchange (HFEGE) potential.[45] It is based on two assumptions. First, the target electrons can be approximated by a free-electron-gas. Thus they are treated as non-interacting fermions, satisfying the Pauli Exclusion Principle. But the mutual forces between electrons are ignored. Second, the distortion of the scattering function is neglected in so far as exchange is concerned, and $\underset{\sim}{f}(r')$ in the exchange integral $\int_0^\infty \underset{\sim}{K}(r,r')\underset{\sim}{f}(r')dr'$ is replaced by its first Born Approximation, a diagonal matrix of Ricatti-Bessel functions. Two further modifications in the exchange potential are made. First, the quantum mechanical $\vec{r}$-dependent molecular charge density, $\rho(\vec{r})$ of (3.8), is used in the free-electron-gas formula. Second, distortion of the scattering function in the exchange potential is accommodated in a very crude way by replacing the asymptotic momentum of the electron, k, by an $\vec{r}$-dependent local momentum.

Applying these assumptions to the exchange terms leads to the aforementioned HFEGE potential. It, too, is expanded in Legendre polynomials. This process leads to expansion coefficients of the exchange potential for values of λ from 0 to λ_{max}^{el}, the maximum order coefficient for the electronic contribution to the static potential [cf., (2.9) and (3.8)].

Although the exchange potential is energy-dependent, the only time-consuming part of its evaluation for a particular system is the calculation of $\rho(\vec{r})$. This step has already been completed as part of the calculation of the coefficients $a_\lambda(r)$ of (3.8) and is not repeated. The expansion coefficients of the exchange potential are evaluated using the quadrature routines which were implemented earlier in the calculation of the a_λ's.

The final constituent of the interaction potential is that due to induced polarization. Using the semi-empirical adiabatic form of (2.11), we need only determine the cutoff radius r_c in (2.12). For systems with a well-established feature, such as a narrow shape resonance, we "tune" r_c to reproduce the feature at the proper energy.[29] Thus, in e-N_2 collisions there is a d-wave shape resonance in the Π_g symmetry at roughly 2.3 eV. This resonance is present in the static-exchange approximation,[4] although at a higher energy (3.75 eV). We carry out scattering calculations in the full static-exchange-polarization (SEP) potential for the resonant symmetry (Π_g for e-N_2) for several values of r_c, thereby determining the value which positions the resonance energy at the experimentally determined location.

Once this "resonance tuning" has been carried out, r_c is not changed. For example, in e-CO_2 collisions we find[2] that $r_c \simeq 2.6\ a_0$ positions a resonance peak at 3.8 eV in the Π_u symmetry. This value of r_c is used in scattering calculations over roughly three decades of energy, including very low energies where Σ_g is overwhelmingly the dominant symmetry.

If the target is highly aspherical (e.g., CO_2), the resulting cross sections can be very sensitive to the value selected for r_c. This effect is illustrated in Figure 2 by SEP integrated cross sections for e-CO_2 scattering in the vicinity of the 3.8 eV p-f shape resonance. In such cases, special care is called for in determining r_c. The most efficient procedure we have found is to calculate the eigenphase sum in the resonant symmetry as a function of r_c for fixed energy equal to the known position of the resonance. (See ref. 4 for examples of this procedure.)

It is important to note that such a tuning procedure must be performed only once for a given system. An evident disadvantage of this way of introducing polarization effects is that it is difficult to implement if there is no known well-defined feature of the cross section for tuning or if the feature is very broad (e.g., a very short-lived shape resonance). From the standpoint of theory, the principal advantage of the procedure is that it is semi-empirical.

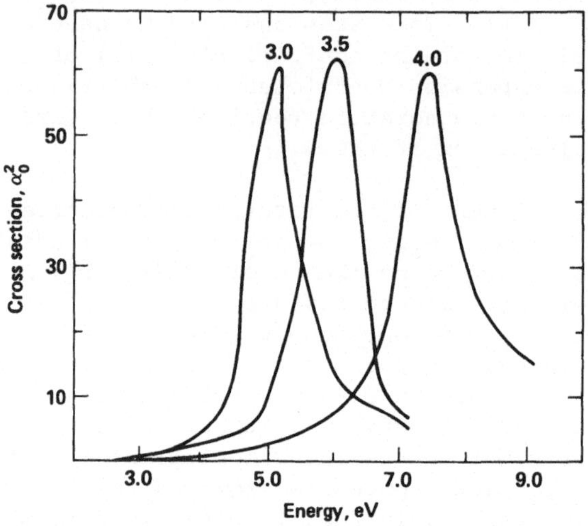

Fig. 2. Integrated cross sections (in a_0^2) for e-CO_2 collisions in the SEP approximation for three values of the cutoff radius, r_c in (2.12).

The SEP interaction potential is thus fully determined in the present application by the above considerations. The full SEP matrix $\underset{\sim}{V}(r)$ in the coupled equations (3.1) is easily determined from the coefficients $f_\lambda(\ell\ell'\Lambda)$ of (3.11). The full expansion coefficient of V(r) is given by

$$v_\lambda(r) = v_\lambda^{st}(r) + v_\lambda^{exch}(r)\mathcal{O}_\lambda(\lambda_{max}^{el}) + v_\lambda^{pol}(r)\mathcal{O}_\lambda(2), \quad (3.12)$$

where each term on the right-hand side arises from the Legendre-polynomial expansion of the relevant component of $V(\vec{r})$ and where

$$\mathcal{O}_\lambda(n) = \begin{cases} 0 & \text{if } \lambda > n \\ 1 & \text{if } \lambda \leq n \end{cases} . \quad (3.13)$$

The potential energy need only be calculated once per system <u>except</u> for the exchange contribution, which is energy-dependent in the HFEGE model.

The next step is to solve the set of coupled homogeneous second-order ordinary differential equations (3.1). This will be the subject of the next section. We then extract the K-matrix from the solution matrix in the asymptotic region as in (3.4). The number of these coupled equations N (i.e., the dimension of the solution matrix) and the value of r at which we can fit $\underset{\sim}{f}(r)$ to its asymptotic form depend on the system, the scattering energy, and the degree of accuracy desired. Typical values of N for electron collisions with H_2, N_2, and CO_2 are 4, 14, and 32, respectively. Typical values of the final integration point, r_{max}, for these molecules in the Σ_g (electron-molecule) symmetry[39] are $10.0a_0$, $85.0a_0$, and $130.0a_0$.

Having extracted the K-matrix, we can easily determine the real and imaginary parts of $\underset{\sim}{T}$ as

$$\underset{\sim}{T}_{imag} = -2\, \underset{\sim}{K}\, (\underset{\sim}{1} + \underset{\sim}{K}^2)^{-1} \tag{3.14a}$$

$$\underset{\sim}{T}_{real} = -\underset{\sim}{K}\, \underset{\sim}{T}_{imag} \tag{3.14b}$$

and the total (elastic + rotational) integrated cross section in the body frame as

$$\sigma^{tot} = \frac{\pi}{k^2} \sum_{\ell\ell'\Lambda} |T_{\ell\ell'}|^2 . \tag{3.15}$$

In practice, the upper limits on the summations over ℓ and ℓ' in (3.15) are determined by the number of channels included for each symmetry (Λ). However, not all channels required to converge the solution of the coupled equations (3.1) contribute to the cross sections. (This is principally due to the presence of the centrifugal potential terms[30] $\underset{\sim}{L}/r^2$ in (3.1) for $\ell \neq 0$.) For example, for e-N_2 collisions at energies below 1.0 Ryd. we include "asymptotic channels" with Λ = 0, 1, and 2 and values of ℓ and $\ell' \leq 6$ in the summation of (3.15). We can also calculate differential and momentum-transfer cross sections from the body-frame T-matrix (3.14).

Cross sections for rotational excitation are calculated in the present formulation by implementing the adiabatic-nuclei approximation for rotation. This entails transforming the T-matrix from the body-frame to the lab-frame by means of a unitary transformation, viz.,

$$T^J_{j\ell,j'\ell'} = (-1)^{\ell+\ell'} \sum_{\Lambda} C(J\ell j;-\Lambda,\Lambda)C(J\ell' j';-\Lambda,\Lambda)T^{\Lambda}_{\ell\ell'} , \tag{3.16}$$

and then calculating the desired cross section from the formula

$$\sigma(j\to j') = \frac{\pi}{k^2(2j+1)} \sum_J (2J+1) \left|T^J_{j\ell,j'\ell'}\right|^2 . \qquad (3.17)$$

Some care must be exercised in setting up these calculations. In (3.17) J is the quantum number for the total (orbital and rotational) angular momentum of the electron-molecule system, and we must include a sufficient number of "J-terms" to converge $\sigma(j\to j')$ for the excitation of interest. For each value of J, the laboratory-frame T-matrix elements $T^J_{j\ell,j'\ell'}$ are obtained as in (3.16) by a sum over Λ, the quantum number corresponding to the projection of the electron's orbital angular momentum along the internuclear axis. One must include enough values of Λ in this summation to converge each of the required lab-frame T-matrix elements. Typically, this means that we shall need body-frame T-matrix elements for values of $|\Lambda|>2$. However since $\ell_{min} = |\Lambda|$, it is likely that all components of the body-frame T-matrix for such large values of $|\Lambda|$ can be determined in the Born Approximation.[2]

To illustrate this situation, we shall consider a calculation of $\sigma(0\to2)$ for e-N_2 collisions at 0.01 Ryd. In a separate laboratory-frame rotational-close coupling calculation of this cross section with J=0,1,...,4 we find $\sigma(0\to2) = 0.919\ a_0^2$. In the corresponding body-frame fixed nuclei calculation, we obtain $\sigma(0\to2) = 0.919a_0^2$ if we include all values of $|\Lambda| \leq 4$ for these values of J in (3.16). However, if we include only $|\Lambda| = 0$, 1, and 2, we find that the body-frame calculation produces $\sigma(0\to2) = 0.856a_0^2$. The apparent discrepancy is simply a consequence of the fact that the cross section calculated from lab-frame T-matrices in the latter "reduced" calculation is not adequately converged in J. Indeed, in a second lab-frame calculation including only J = 0, 1, and 2 we find $\sigma(0\to2) = 0.855a_0^2$. We note that these calculations also provide a clear demonstration of the validity of the adiabatic-nuclei theory for this energy.

C. Solution of the Radial Equations

The derivation of the basic integral equations algorithm for the solution of equations of the form (3.1) has been discussed in a number of references[2-4] since its presentation by Sams and Kouri[5] in 1969. In Sec. IV, we shall describe the modifications to this procedure which must be considered for electron-molecule scattering problems and special situations which can arise. In order to provide background for this discussion, we here briefly recapitulate the high points of the derivation.

The coupled body-frame equations (3.1) are converted to a matrix integral equation using standard Green's function techniques, viz.,

$$\underset{\sim}{f}(r) = \underset{\sim}{G}^{1}(r) - \int_0^\infty \underset{\sim}{G}^{12}(r_<,r_>)\underset{\sim}{U}(r')\underset{\sim}{f}(r')dr' , \qquad (3.18)$$

where $\underset{\sim}{U} = 2\underset{\sim}{V}$ and

$$\underset{\sim}{G}^{12}(r_<,r_>) \equiv \underset{\sim}{G}^{1}(r_<)\underset{\sim}{G}^{2}(r_>) , \qquad (3.19)$$

with $r_< = \min(r,r')$, etc. The Green's function matrices $\underset{\sim}{G}^1$ and $\underset{\sim}{G}^2$ in (3.19) are diagonal matrices chosen to correspond to the boundary conditions (3.3) and (3.4) on the solution matrix; their $\ell\ell'$ elements are

$$[\underset{\sim}{G}^{1}(r)]_{\ell\ell'} = j_\ell(kr)\delta_{\ell\ell'} \underset{r\to\infty}{\sim} \sin(kr - \ell \tfrac{\pi}{2})\delta_{\ell\ell'} \qquad (3.20a)$$

$$[G^{2}(r)]_{\ell\ell'} = \frac{1}{k}\hat{n}_\ell(kr)\delta_{\ell\ell'} \underset{r\to\infty}{\sim} \frac{1}{k}\cos(kr - \ell \tfrac{\pi}{2})\delta_{\ell\ell'} . \qquad (3.20b)$$

The integral equation (3.18) can be rewritten as

$$\underset{\sim}{f}(r) = \underset{\sim}{G}^{1}(r)[\underset{\sim}{1} - \int_0^\infty \underset{\sim}{G}^{2}(r')\underset{\sim}{U}(r')\underset{\sim}{f}(r')dr']$$

$$+ \int_0^r [\underset{\sim}{G}^{12}(r,r') - \underset{\sim}{G}^{21}(r,r')]\underset{\sim}{U}(r')\underset{\sim}{f}(r')dr . \qquad (3.21)$$

At this point it is convenient to define a "homogeneous solution" $\underset{\sim}{u}(r)$ which is related to the "physical solution" $\underset{\sim}{f}$ of (3.21) by a constant transformation as

$$\underset{\sim}{f}(r) = \underset{\sim}{u}(r)[\underset{\sim}{I}^{2}(\infty)]^{-1} \qquad (3.22)$$

where we have introduced the matrix $\underset{\sim}{I}^2$ [see (3.24)]. The homogeneous solution satisfies the simple equation

$$\underset{\sim}{u}(r) = \underset{\sim}{G}^{1}(r)\underset{\sim}{I}^{2}(r) - \underset{\sim}{G}^{2}(r)\underset{\sim}{I}^{1}(r) , \qquad (3.23)$$

where the auxiliary I-matrices are defined by

$$\underset{\sim}{I}^{i}(r) \equiv \underset{\sim}{1}\,\delta_{i2} + \int_0^r \underset{\sim}{G}^{i}(r')\underset{\sim}{U}(r)\underset{\sim}{u}(r')dr' . \qquad (3.24)$$

The K-matrices can be determined directly from the asymptotic values of the auxiliary I-matrices (3.24) as

$$\underset{\sim}{K} = -\underset{\sim}{k}^{-\frac{1}{2}}\,\underset{\sim}{I}^{1}(\infty)[\underset{\sim}{I}^{2}(\infty)]^{-1}\underset{\sim}{k}^{-\frac{1}{2}} , \qquad (3.25)$$

where $\underset{\sim}{k}^{-\frac{1}{2}} = k^{-\frac{1}{2}}\underset{\sim}{1}$. Of course, in practice we use $\underset{\sim}{I}^{i}(r_{max})$ in determining $\underset{\sim}{K}$ [and $\underset{\sim}{f}$ from (3.22) if need be], where r_{max} is the final point of the integration mesh.

The advantages of this procedure only become evident when we place these basic equations on a quadrature mesh (the "integration mesh"). We have used a trapezoidal-rule quadrature scheme[27] in most applications to date. If we introduce such a mesh of points, $\{r_j;\ j = 1, 2, \ldots, j_{max}\}$, where $r_{j_{max}} = r_{max}$ is in the asymptotic region, then we can evaluate $\underset{\sim}{u}$ at r_j non-iteratively in terms of the values of $\underset{\sim}{I}^i$ at the preceding j-1 points, viz.,

$$\underset{\sim}{u}(r_j) = \underset{\sim}{G}^1(r_j)\underset{\sim}{J}^2(r_{j-1}) - \underset{\sim}{G}^2(r_j)\underset{\sim}{J}^1(r_{j-1}) , \tag{3.26}$$

where*

$$\underset{\sim}{J}^i(r_{j-1}) = \underset{\sim}{1}\delta_{i2} + \sum_{k=1}^{j-1} G^i(r_k)\underset{\sim}{U}(r_k)\underset{\sim}{u}(r_k)w_k . \tag{3.27}$$

The reason that the "homogeneous" equation (3.23) takes on the non-iterative form (3.26) when a quadrature mesh is introduced is that the Green's functions cancel at the upper limit of integration. The advantage, of course, is that at each stage the computations leading to $\underset{\sim}{K}$ in (3.25) are all carried out using known quantities. With the initial conditions

$$\underset{\sim}{u}(0) = \underset{\sim}{0};\ \underset{\sim}{I}^1(0) = \underset{\sim}{0};\ \underset{\sim}{I}^2(0) = \underset{\sim}{1} , \tag{3.28}$$

the integral equations procedure is easy to program, as illustrated in Figure 3.

---

*) Note that $\underset{\sim}{J}^i(r_{j-1}) = \underset{\sim}{I}^i(r_{j-1})$ only if $w_j = w_{j-1}$. For example, for the trapezoidal quadrature with step-size h, the weights for the evaluation of $\underset{\sim}{I}^i(r_j)$ from (3.24) are $w_0 = w_j = \frac{1}{2}h$ and $w_k = h$ for $k \neq 0,j$. Therefore for this case, we have

$$\underset{\sim}{I}^i(r_{j-1}) = \underset{\sim}{J}^i(r_{j-1}) - \tfrac{1}{2}h\ \underset{\sim}{G}^i(r_{j-1})\underset{\sim}{U}(r_{j-1})\underset{\sim}{u}(r_{j-1}) .$$

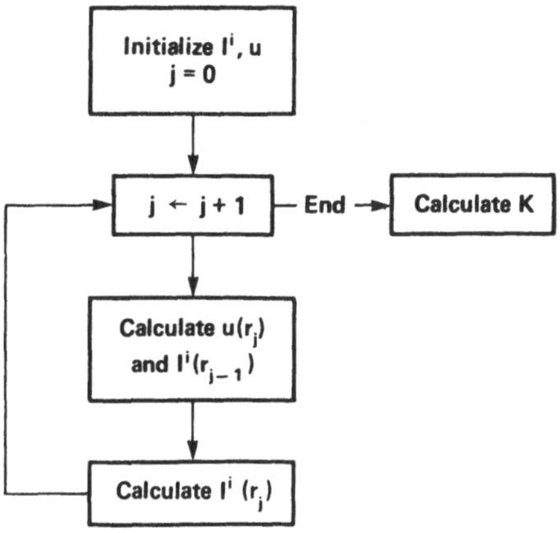

Fig. 3. Flow chart of the integral equations algorithm.

This algorithm as presented thus far requires no matrix inversions to propagate the wavefunction, is reasonably fast and efficient, and is more stable numerically than, say, the Numerov algorithm. One advantage of this formulation of the numerical problem is the relative computational ease with which the homogeneous solution can be manipulated. For example, it happens that for most electron-molecule problems it is necessarv to impose linear independence of the solutions comprising the columns of $\underset{\sim}{u}$ at a few points of the integration mesh. This is easily implemented within the integral equations procedure (see Sec. IV.B). It is also reasonably straightforward to generalize the present algorithm to take account of inhomogeneous terms in the scattering equations; such terms arise in certain treatments of exchange, for example.

IV. SPECIAL TRICKS AND PROBLEMS

The simple integral-equations algorithm summarized in Figure 3 can be applied directly to comparatively simple problems in electron-molecule scattering, e.g., elastic e-H_2 collisions. However, most systems of interest to us are characterized by a more aspherical interaction potential than that of e-H_2. This feature leads to the need for including a large number of channels in the coupled-channels calculation, which, in turn, introduces a variety of numerical problems. These difficulties will plague any eigenfunction-expansion calculation, but we shall discuss them in the context of the integral equations algorithm. In this section we shall also briefly describe a couple of "tricks" that have the potential for improving the efficiency of the scattering calculation.

A. Matters of Convergence

Special care must be used in scattering calculations on systems such as e-N_2 or e-CO_2 to ensure convergence. This is particularly important, for example, in the study of model potentials like the SEP potential that was described in Sec. II.C. or in making quantitative comparisons of theoretical and experimental cross sections. However, by itself the word "convergence" (as applied to a theoretical study) carries almost no information. To talk meaningfully about convergence, one must specify three things:

(1) the converged quantity (e.g., elastic cross section, eigenphase sum);

(2) the convergence criterion (e.g., to within 1%);

(3) the convergence parameter (e.g., number of partial waves).

One of the principal advantages of eigenfunction-expansion approaches to problems in collision theory is the ease with which one can check convergence for a variety of parameters in a straightforward, efficient, and systematic way.

A serious concern in electron-molecule studies is that convergence in certain parameters can be quite gradual. We shall use convergence in the number of channels N (i.e., number of partial waves) for illustrative purposes in this discussion. The standard way to check convergence of the cross section to M% in N is to see if the results of two calculations, one with N-1 channels and one with N channels, agree to within M%, i.e.,

$$\left| \frac{\sigma(N) - \sigma(N-1)}{\sigma(N)} \right| < \frac{M}{100} . \tag{4.1}$$

This local convergence criterion does not guarantee that σ is converged.

To illustrate this point, we consider e-N_2 scattering in the static-field approximation at an energy of 0.1 Ryd. In Table 1 we show cross sections in the Σ_g symmetry as a function of ℓ_{max}, the maximum-order partial wave included in the single-center expansion of the scattering function. We see local convergence of $\sigma(\Sigma_g)$ in ℓ_{max} to <10% by $\ell_{max} = 12$. However, the result for $\ell_{max} = 12$ differs by >24% from the converged result (for $\ell_{max} = 26$). The reason for this behavior becomes clear if we plot $\sigma(\Sigma_g)$ vs. ℓ_{max} as in Figure 4.

ℓ_{max} =	12	14	16	18	20	22	24	26	Bracket spans
	3.2%								12–26
	1.3			0.32			0.08		12–14; 18–20; 24–26
$\delta_{sum}(\Sigma_g)$	2.5321	2.5640	2.5841	2.5968	2.6051	2.6106	2.6140	2.6160	
$\sigma(\Sigma_g)$	39.2040	35.4704	33.2882	31.8938	30.9845	30.3939	30.0247	29.8169	
	9.5			2.9			0.7		12–14; 18–20; 24–26
				6.5					18–26
	24%								12–26

Table 1.

Cross sections (in a_0^2) and eigenphase sums (in radians) for e-N_2 scattering in the static field approximation at 0.1 Ryd. ℓ_{max} is the highest-order partial wave retained in the single-center expansion of the scattering function. Percentage differences are shown in this table in order to illustrate various aspects of convergence. In these calculations, $\lambda_{max}^{el} = 14$ and $\lambda_{max}^{nuc} = 24$.

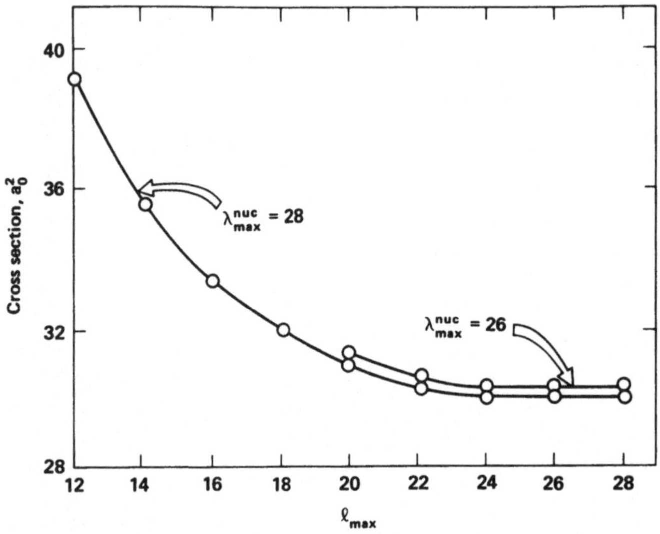

Fig. 4. Σ_g cross sections (in a_0^2) as a function of ℓ_{max} for e-N_2 scattering in the static-field approximation at 0.1 Ryd.

It is therefore necessary to impose a more stringent global convergence criterion on the converged quantity, i.e., we demand that

$$\left|\frac{\sigma(N) - \sigma(N-K)}{\sigma(N)}\right| < \frac{M}{100} , \tag{4.2}$$

where typically $K \geq 6$ affords a reliable check. Use of a global criterion avoids the neglect of a large number of small percentage errors, which can lead to highly inaccurate results.

In the example of Table 1, the converged quantity was chosen to be cross section $\sigma(\Sigma_g)$. An oft-used alternative is the eigenphase sum $\delta_{sum}(\Sigma_g)$. Unfortunately, the behavior of δ_{sum} for electron-molecule collisions can be dangerously misleading. Thus in the e-N_2 calculation of Table 1, we see that examination of δ_{sum} suggests a much higher degree of convergence in the calculation at low ℓ_{max} than is actually present in the cross sections. On the other hand, in their study of low-energy electron collisions with polar molecules, Collins and Norcross[16] found cases where more channels were required to converge δ_{sum} to a given criterion than to converge the cross section! As a rule of thumb, in a given calculation it is best to choose σ as the converged quantity unless only δ_{sum} is desired.

Related to the convergence parameter ℓ_{max} is the number of expansion coefficients in the Legendre-polynomial expansion of the static potential energy, λ^{st}_{max} in (2.9). Fortunately, it is not necessary to converge $V_{st}(r,\theta)$. It is only necessary to include a sufficient number of terms $v^{st}_{\lambda}(r)P_{\lambda}(\cos\theta)$ to converge the cross sections. As the number of channels N and consequently ℓ_{max} is increased, more expansion coefficients are required to converge each scattering calculation for a given ℓ_{max}. [This follows from the fact that for given values of ℓ and ℓ', the matrix element (3.11) includes contributions from $v_{\lambda}(r)$ for λ in the range $|\ell - \ell'| \leq \lambda \leq \ell + \ell'$.]

This fact might seem to make a thorough convergence study prohibitively time-consuming. However, it is fairly easy to determine the region of the parameters in which convergence to some specified criterion is at hand. We merely set $\lambda^{st}_{max} = \ell_{max}$ and carry out successive scattering calculations, incrementing these parameters by 1 until we arrive in the desired region.[40] Once there, we check for global convergence in ℓ_{max} and λ_{max} separately. This process is illustrated in Table 2 for e-N_2 scattering in the static-field approximation. We impose convergence to <1% in $\sigma(\Sigma_g)$; the final recommended values of the convergence parameters are $\lambda^{st}_{max} = 28$ and $\ell_{max} = 26$.

An additional feature which recommends this strategem arises from the fact that at any given stage in a convergence study, the cross section may not be in a monotonically convergent region in the parameters of interest. The present procedure guards against missing the proper values of λ^{st}_{max} and ℓ_{max} altogether, as can happen with the more arduous approach of converging each ℓ_{max} calculation separately in λ^{st}_{max}.

The optimum values of the various convergence parameters in a coupled-channels calculation depend not only on the system being studied but in the electron-molecule symmetry under consideration, and on the scattering energy, and on the presence or absence of special features in the cross section (such as a shape resonance). To illustrate this observation, we show the number of channels for four symmetries of the e-CO_2 system in Table 3.

λ_{max}^{nuc} \ ℓ_{max}	20	22	24	26	28	30
20	46.8405	---	---	---	---	---
22	---	44.6493	---	---	---	---
24	---	---	43.0830	42.7479	42.6021	42.5427
26	---	---	42.1592	41.9286	41.6784	41.5659
28	---	---	---	41.2304	41.0569	40.8660

Table 2.

Cross sections in the Σ_g symmetry for e-N_2 scattering in the static-field approximation at an energy of 0.01 Ryd. Cross sections in a_0^2 are shown for various values of ℓ_{max} and λ_{max}^{nuc} as defined in Sec. II of the text.

	Σ_g	Σ_u	Π_u	Π_g
N	32	31	15 (23)	15
ℓ_{max}	62	61	29 (45)	30
ℓ_{min}	0	1	1	2
$\lvert\Lambda\rvert$	0	0	1	1

Table 3

Convergence parameters for e-CO_2 scattering in various symmetries. Numbers in parenthesis for the Π_u symmetry apply for the energies near the 3.8 eV resonance in this symmetry. ℓ_{min} is the lowest-order partial wave in the single-center expansion of the scattering function.

We shall bypass a discussion of the determination of the integration mesh for quadrature of the integral equations except to mention that it is yet another important thing to check. The use of variable step sizes over the whole region from r=0 to $r=r_{max}$ is strongly recommended, since only in the region of the nuclear Coulomb singularity is a highly dense mesh required.

Eventually, of course, one must stop the propagation of the wavefunction by (3.26) and determine K à la Eq. (3.25). The value of r_{max} for a given problem must be determined by trial-and-error. However, as a general rule for low-energy collisions, the cross sections will not be stable in r_{max} until the potential energy is of the order of 10^{-4} hartree or less. In Figure 5 the partial cross sections $\sigma_{\ell\ell'}(\Sigma_u)$ for e-CO_2 scattering at 0.38 eV in the static-exchange approximation (using the HFEGE potential) is shown as a function of r_{max}. Note that for large values of r, the SE potential reduces to the quadrupole term $(-q/r^3)P_2(\cos\theta)$, where[2] $q \simeq -3.8\ e\ a_0^2$.

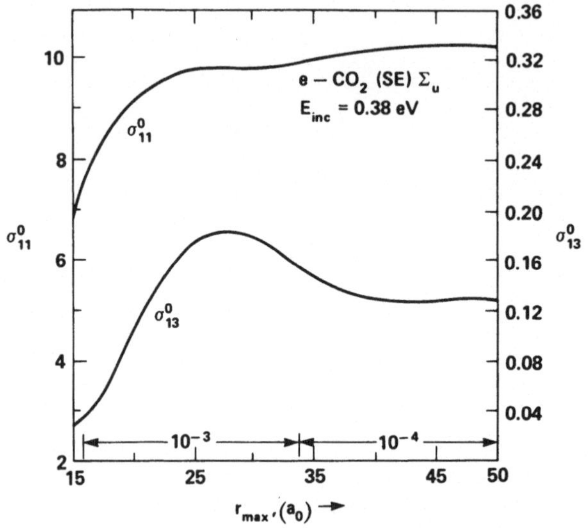

Fig. 5. Static-exchange partial cross sections $\sigma^{\Lambda}_{\ell\ell'}$, in a_0^2 for e-CO_2 scattering in the Σ_u symmetry. The incident energy is 0.38 eV. The value of r at which the K-matrix is extracted as in Eq. (3.25) is r_{max}. At the bottom of the figure is shown the order-of-magnitude of the SE potential (in hartrees).

B. Creeping Linear Dependence

In any wavefunction propagation scheme like the algorithm of Fig. 3, it is possible for numerical linear dependence to creep into the solution matrix as r increases. The source of this difficulty is apparent from the form of the two-dimensional solution matrix being propagated. The columns of $\underset{\sim}{u}$ correspond to the N linear-independent solution vectors of the coupled equations that are regular at the origin. The rows correspond to the various partial wave components of each solution vector. For large values of the row index (i.e., large ℓ) there will be locally closed, exponentially growing components in $\underset{\sim}{u}$. These arise because of the centrifugal barrier term $\underset{\sim}{L}/r^2$ in (3.1). Due to the finite word-length of the computer, these high-ℓ components can "contaminate" the important low-ℓ components [which contribute to the cross section, as in (3.15)]. This contamination leads to a loss of significance and consequent linear dependence of the columns of $\underset{\sim}{u}$. If linear dependence is infecting $\underset{\sim}{u}$, it will show up in an examination of the elements of the K matrix, which will not be symmetric. An example is given in Table 4.

ℓ ℓ'	Without Reorthogonalization	With Reorthogonalization
0 0	24.1295	24.1295
0 2	1.5632(-2)	1.5632(-2)
2 0	1.5607(-2)	1.5632(-2)
0 4	1.5637(-7)	1.5636(-7)
4 0	2.4948(-4)	1.5636(-7)
0 6	1.3548(-12)	1.3548(-12)
6 0	52.1046	1.3548(-12)

Table 4

Partial cross sections $\sigma^{\Lambda}_{\ell\ell'}$ for e-N_2 scattering in the Σ_g symmetry at 0.1 Rydberg. The static-field approximation is made, with ℓ_{max} = 24. If the solution matrix is properly linearly-independent, the S matrix should be symmetric, and we should have $\sigma^{\Lambda}_{\ell\ell'} = \sigma^{\Lambda}_{\ell'\ell}$.

To solve this problem, White and Hayes[31] introduced a re-orthogonalization procedure (also known as "stabilization") which ensures linear independence of the columns of $\underset{\sim}{u}$. The idea of this procedure is simply to apply a unitary transformation to the solution matrix which transforms it to a new basis in which the columns of the transformed matrix are guaranteed to be linearly independent.

To implement reorthogonalization at a particular stabilization radius r_s, we first factor the matrix $\underset{\sim}{u}$ into the product of a right-upper-triangular and a left-lower-triangular matrix, viz.,

$$\underset{\sim}{u}(r_s) = \underset{\sim}{u}^{(upper)}(r_s)\, \underset{\sim}{u}^{(lower)}(r_s) \quad (4.3)$$

We next replace $\underset{\sim}{u}$ by $u^{(\text{upper})}$ in the fundamental integral-equations equation (3.23), giving

$$\underset{\sim}{u}^{(\text{upper})}(r_s) = \underset{\sim}{G}^1(r_s)\bar{\underset{\sim}{I}}^2(r_s) - \underset{\sim}{G}^2(r_s)\bar{\underset{\sim}{I}}^1(r_s) \ , \quad (4.4)$$

where the transformed I-matrices are defined by

$$\bar{\underset{\sim}{I}}^i(r_s) = \underset{\sim}{I}^i(r_s)\left[\underset{\sim}{u}^{(\text{lower})}(r_s)\right]^{-1} \ . \quad (4.5)$$

The new matrix, $\underset{\sim}{u}^{(\text{upper})}$, the columns of which are linearly independent, is propagated as before for $r>r_s$. Since the K-matrix of (3.25) is unchanged by a unitary transformation of $\underset{\sim}{u}$, it can be calculated directly from the transformed matrix in the asymptotic region.

For problems characterized by highly asymmetric potentials (such as e-CO_2 or e-LiF collisions), reorthogonalization may be necessary at several different stabilization radii in the integration mesh. Even in e-N_2 scattering, we find it necessary to reorthogonalize $\underset{\sim}{u}$ as we increase the number of channels to convergence. This necessity also arises in exact static-exchange calculations such as those described in Sec. V. The required frequency of reorthogonalization for a particular calculation must be determined by the time-honored method of trial-and-error. However, for most reasonable systems even repeated stabilizations do not markedly increase the computer time for a given calculation.

C. A Truncation Procedure

A few other problems have arisen in the course of implementing the coupled-channels method which are specific to particular applications. These are discussed in the relevant journal articles and an unpublished report.[2-4,16] We have also come across some procedures which greatly enhance the efficiency of coupled-channel calculations. For example, as the solution matrix $\underset{\sim}{u}$ is propagated according to (3.26), it can be drastically truncated just beyond the position of the nuclei. This leaves a small matrix (at most a 4 x 4) which must be propagated into the asymptotic region. The truncated matrix contains all the scattering information needed to determine $\underset{\sim}{K}$, since it is prepared prior to truncation to consist of the low-ℓ components of each solution vector. This fact again reflects the observation that very large numbers of channels are needed only very near the coordinate origin. The centrifugal barriers effectively "confine" most of the high-ℓ components to this inner region, and they do not contribute to the cross section.

The truncation procedure requires that we work with the physical solution $\underset{\sim}{f}$ rather than the homogeneous solution $\underset{\sim}{u}$, which is being propagated, since in the latter matrix the partial-wave components in $\underset{\sim}{f}$ are shuffled by application of the constant transformation $[\underset{\sim}{I}^2(\infty)]^{-1}$ as in (3.22). Thus, at the truncation radius r_t, we first transform $\underset{\sim}{\mu}$ to $\bar{\underset{\sim}{\mu}}$ as

$$\bar{\underset{\sim}{\mu}}(r_t) = \underset{\sim}{\mu}(r_t)[\underset{\sim}{I}^2(r_t)]^{-1} , \tag{4.6}$$

where we note that the use of $\underset{\sim}{I}^2(r_t)$ rather than $\underset{\sim}{I}^2(r_{max})$ makes $\underset{\sim}{\mu}(r_t)$ only an approximation to $\underset{\sim}{f}(r_t)$. We then pick a truncation order ℓ_t and delete from all matrices in the algorithm (3.26) rows and columns with $\ell > \ell_t$. The resulting reduced solution matrix is propagated into the asymptotic region where $\underset{\sim}{K}$ is calculated directly from (3.25).

Unlike reorthogonalization, this truncation procedure is an approximation,[41] and care must be exercised in the choice of the truncation parameters ℓ_t and r_t. However, this tactic can reduce 32 x 32 matrices to 5 x 5 matrices, and may yield a substantial savings of computer time. A typical application of the truncation procedure is shown in Table 5.

	N_2	CO_2
N(before)	14	32
N(after)	4	5
r_t (a_0)	3.0	4.0
ℓ_t	6	8
$r_{max}(a_0)$	85.0	130.0

Table 5

Number of channels N before and after truncation of all partial-waves with orders $\ell > \ell_t$ at radius r_t for e-N_2 and e-CO_2 collisions in the Σ_g symmetry.

V. AN ITERATIVE STATIC-EXCHANGE PROCEDURE

In our most recent work with the coupled-channels method, we have turned to the question of the treatment of electron exchange in electron-molecule scattering in a rigorous formulation which proceeds from fully antisymmetrizing the electron-molecule system wavefunction with respect to pairwise electron interchange.[32]

In recent years, considerable attention has been focused on developing numerical approaches to the solution of the resulting coupled integrodifferential equations. Among the methods that have been reported[33] are

(1) a procedure introduced by Seaton (and discussed in the accompanying article by Moores) in which these equations are transformed into a set of simultaneous linear algebraic equations, which are subsequently solved by matrix inversion;

(2) a procedure presented by Smith, Henry and Burke[46] in which the exchange terms are represented by additional coupled differential equations;

(3) the Numerov algorithm;[34]

(4) the non-iterative integral equations method of Smith and Henry[35], in which auxiliary functions for each exchange term are determined followed by solution of a set of simultaneous equations.

Difficulties arise in the application of these methods to electron-molecule collisions because of the extremely large number of exchange terms which characterize these problems (e.g., more than 700 for certain e-N_2 symmetries). Furthermore, a large and locally dense integration mesh may be required due to the nuclear singularities that are important in low-energy electron-molecule scattering.

We recently reported[32] preliminary results of an <u>iterative</u> procedure for solving the static-exchange equations for electron-molecule scattering which is based on the integral equations algorithm under discussion here. We begin with the full coupled integrodifferential static-exchange equations for the radial scattering function,

$$\left(\frac{d^2}{dr^2}\underset{\sim}{1} - \frac{1}{r^2}\underset{\sim}{L} + \underset{\sim}{k}^2\underset{\sim}{1}\right)\underset{\sim}{f}(r) = 2\underset{\sim}{V}(r)\underset{\sim}{f}(r) + 2\underset{\sim}{W}(r) \quad , \tag{5.1}$$

where the non-local exchange term is given by

$$\underset{\sim}{W}(r) = \int_0^\infty \underset{\sim}{K}(r,r')\underset{\sim}{f}(r')dr' \quad . \tag{5.2}$$

This term could be treated as an inhomogeneity in an iterative procedure such as that of Seaton and John[36] for e-H scattering. In contrast, our procedure can be implemented directly within the homogeneous theory of Sec. III.C. The starting point is the observation that provided $\underset{\sim}{f}$ is non-singular, (5.1) can be written as

$$\left(\frac{d^2}{dr^2}\underset{\sim}{1} - \frac{1}{r^2}\underset{\sim}{L} + k^2\underset{\sim}{1}\right)\underset{\sim}{f}(r) = 2\left[\underset{\sim}{V}(r) + \underset{\sim}{W}(r)\,\underset{\sim}{f}^{-1}(r)\right]\underset{\sim}{f}(r) \quad . \tag{5.3}$$

To solve (5.3) we begin with the zeroth-iteration, the static equation (3.1), i.e.,

$$\left[\frac{d^2}{dr^2}\underset{\sim}{1} - \frac{1}{r^2}\underset{\sim}{L} - 2\underset{\sim}{V}(r) + k^2\underset{\sim}{1}\right]\underset{\sim}{f}_0(r) = 0 \tag{5.4}$$

Then the mth iteration for the physical solution is calculated from the $(m-1)^{st}$ iterate as

$$\left[\frac{d^2}{dr^2}\underset{\sim}{1} - \frac{1}{r^2}\underset{\sim}{L} - \underset{\sim}{V}(r) + k^2\underset{\sim}{1}\right]\underset{\sim}{f}_m(r) = \left[\underset{\sim}{W}_{m-1}\,\underset{\sim}{f}_{m-1}^{-1}(r)\right]\underset{\sim}{f}_m(r) \quad , \tag{5.5}$$

where the $(m-1)^{st}$ approximation to (5.2) is given by

$$\underset{\sim}{W}_{m-1}(r) = \int_0^\infty \underset{\sim}{K}(r,r')\underset{\sim}{f}_{m-1}(r')dr' \quad . \tag{5.6}$$

We note that the exchange kernel $\underset{\sim}{K}$ in (5.6) depends on the bound target molecular orbitals, not on the scattering function and thus does not change with m. As $m\to\infty$, $\underset{\sim}{f}_{m-1}^{-1}\,\underset{\sim}{f}_m \to 1$, and (5.5) returns to (5.3).

We have studied the application of this iterative static-exchange method to e-H_2, e-N_2, e-LiH, e-H_2^+, and e-CH^+ collisions.[32 42] The nature and rapidity of convergence of the integrated cross section with iteration number m is illustrated in Table 6 for e-H_2 collisions at three energies. In all cases, convergence to $\lesssim 1\%$ is obtained after 3 iterations. (In these calculations, each iteration required roughly 0.1 sec. of CDC-7600 computer time.)

Although our study of this method is not completed at the present writing, it appears to have certain advantages. This scheme avoids the extensive reprogramming and increased complexity which

Symmetry	Σ_u	Π_u	Σ_g
Energy (Ryd.)	0.09	0.5	0.04
0	0.2847	0.0952	255.6629
1	1.8793	2.4332	69.7894
2	1.8119	2.3951	53.5915
3	1.8113	2.3951	53.8184
4	1.8113	2.3951	53.8324
5	1.8113	2.3951	53.8335
6	1.8113	2.3951	53.8335
7	---	---	53.8335
8	---	---	53.8335

Table 6

Integrated cross sections in a_0^2 vs. iteration number m for e-H_2 collisions in various symmetries obtained with the iterative static-exchange procedure described in Sec. V of the text.

ensues if inhomogeneities are introduced into the coupled differential questions (3.1). Moreover, it provides a local function of r, Wf^{-1}, which can be compared directly to model potentials in order to study their physical characteristics. Finally, since the execution time for this method increases roughly as the cube of the number of channels rather than the cube of number of channels plus number of exchange terms, this procedure is more efficient for electron-molecule problems than many alternative methods.

VI. CONCLUSION

The coupled-channel method described in Secs. I-IV of this article has been successfully applied to date to a variety of problems in electron-molecule scattering involving non-polar and polar targets.[2-4,16] The most aspherical non-polar target studied thus far is CO_2. The presence of anomalies in the low-energy momentum-transfer data from swarm experiments on e-CO_2 collisions provided the stimulus for our initial developmental work on these numerical aspects of the theory. Although the results of these e-CO_2 coupled-channel calculations agree only qualitatively with experiment in the vicinity of the 3.8 eV resonance, because of neglect of the vibrational motion of the target, the low-energy momentum-transfer cross sections agree quite well with the swarm results.

The method has also been recently used in the first phase of a long-term study of model exchange potentials for low-energy electron-molecule collisions.[4] In the initial stage of this work, we focused on e-H_2 and e-N_2 collisions. For example, we have found that one can generate a simple local FEG exchange potential for e-H_2 collisions which reproduces $\sigma(\Sigma_g)$ for scattering energies from 0.01 to 1.0 eV.

Collins and Norcross[16] have reported an extensive study of electron-polar molecule scattering in which the integral equations method was used to solve the scattering equations for a host of model interaction potentials, from simple point-dipole approximations to SEP potentials using the HFEGE model potential. They have carried out calculations for electron collisions with LiF, LiCl, NaF, NaCl, KI, and CsF for a range of energies, exploring the utility of the various model potentials and the validity of the adiabatic approximation. A more recent report[37] contains coupled-channel calculations for e-CsOH and e-KOH scattering.

In regard to the coupled-channel method, we note that there are special features of its implementation for electron collisions with strongly polar molecules such as LiF. For example, in converging static-field calculations for e-LiF collisions, Collins and Norcross encountered a "false plateau" as they increased λ_{max}^{st}. This effect made the cross sections appear to be converged in λ_{max}^{st} for a small range of values. Further matters pertaining to convergence and other numerical features of these problems are discussed by these authors in a recent publication.

In this short review, the coupled-channels integral-equations method has been sketched, and several special features of its current incarnation have been discussed. The method is still being refined and developed. It is our hope that the present numerical approach to the scattering problem will provide a reasonably efficient and reliable tool for the study of the wide range of physical and chemical questions which pervade the field of low-energy electron-molecule scattering.

In closing, I would like to acknowledge several years of

delightful and scientifically stimulating collaboration with Drs. Neal F. Lane and Lee A. Collins on the work which is described in this article. Their contributions to the development and evolution of these procedures have been major and invaluable. In addition, I should mention the continued support of the Los Alamos Scientific Laboratory and, especially, of the members of group T-12. It is doubtful that any of this work could have been performed without the ready and plentiful access to the LASL computers which has been generously provided over the last few years.

REFERENCES

1. A complete set of references for the integral-equations method and various numerical aspects of the coupled-channel approach can be found in Refs. 2-4. Hence no attempt at completeness will be made here.
2. M. A. Morrison, N. F. Lane and L. A. Collins, Phys. Rev. A 15, 286 (1977).
3. M. A. Morrison, "Theory of Low-energy Electron-Molecule Collision Physics in the Coupled-Channel Method and Application to e-CO_2 Scattering"; LA-6328-T; Los Alamos Scientific Laboratory (available from National Technical Information Service, U. S. Dept. of Commerce, 5285 Port Royal Road, Springfield, VA 22161).
4. M. A. Morrison and L. A. Collins, Phys. Rev. a 17, 918 (1978)
5. W. N. Sams and D. J. Kouri, J. Chem. Phys. 51, 4809 (1969). See also R. G. Newton, Scattering Theory of Waves and Particles, New York: McGraw-Hill, 1966, Chap. 12 and Sec. 17.1.
6. Cf., E. F. Hayes and D. J. Kouri, Chem. Phys. Lett. 11, 233 (1971); E. F. Hayes, C. A. Wells and D. J. Kouri, Phys. Rev. A 4, 1017 (1971); J. T. Adams, R. L. Smith and E. F. Hayes, J. Chem. Phys. 61, 2193 (1974).
7. N. F. Mott and H. S. W. Massey, The Theory of Atomic Collisions, Third Edition (Oxford U. P., 1965).
8. Cf., S. Hara, J. Phys. Soc. Japan 27, 1009 (1969).
9. M. A. Crees and D. L. Moores, J. Phys. B 8, L195 (1975).
10. Cf., N. F. Lane and S. Geltman, Phys. Rev. 160, 53 (1967).
11. S. Hara, J. Phys. Soc. Japan 22, 710 (1967); A. Temkin and K. V. Vasavada, Phys. Rev. 160, 109 (1967).
12. Cf., P. G. Burke and A. L. SinFailam, J. Phys. B 3, 641 (1970).
13. D. M. Chase, Phys. Rev. 104, 838 (1956); E. S. Chang and A. Temkin, Phys. Rev. Lett 23, 399 (1969).
14. F. Faisal and A. Temkin, Phys. Rev. Lett. 28, 203 (1971); A. Temkin and E. C. Sullivan, Phys. Rev. Lett. 33, 1057 (1974).
15. E. S. Chang and U. Faro, Phys. Rev. A 6, 173 (1972).
16. L. A. Collins and D. W. Norcross, Phys. Rev. A 18, 467 (1978).
17. A. M. Arthurs and A. Dalgarno, Proc. Roy Soc. (London) A 256, 540 (1960).
18. N. F. Lane and R. J. W. Henry, Phys. Rev. 173, 183 (1968).
19. B. I. Schneider, Chem. Phys. Lett. 51, 578 (1977).

20. D. G. Truhlar and F. A. Van-Catledge, J. Chem. Phys. 59, 3207 (1973).
21. A. Klonover and U. Kaldor, J. Phys. B 11, 1623 (1978).
22. J. Callaway, R. W. LaBahn, R. T. Pu and W. M. Duxler, Phys. Rev. 168, 12 (1968).
23. M. A. Morrison and L. A. Collins, Bull. Amer. Phys. Soc. 22, 1331 (1977); M. A. Morrison and P. J. Hay (to be published).
24. M. E. Riley and D. G. Truhlar, J. Chem. Phys. 63, 2182 (1975); B. N. Brandsden, M. R. C. McDowell, C. J. Nobel and T. Scott, J. Phys. B 9, 1301 (1976).
25. P. G. Burke and N. Chandra, J. Phys. B 5, 1696 (1972).
26. W. G. Richards, T. E. H. Walker, L. Farnell and P. R. Scott, Bibliography of ab initio Molecular Wave Functions. Supplement for 1970-1973 (Clarenden Press, Oxford, 1974) and references therein.
27. F. S. Acton, Numerical Methods That (Usually) Work (Harper and Row, New York, 1970), Chapter 4.
28. T. E. Greville, in Mathematical Methods for Digital Computers, edited by A. Ralston and H. S. Wilf (Wiley, New York, 1967), Vol. II, p. 156.
29. For a recent review of such features in electron-molecule systems see G. J. Schulz, Rev. Mod. Phys. 45, 47 (1973).
20. U. Fano, Comm. At. Mol. Phys. 1, 140 (1970).
31. R. A. White and E. F. Hayes, J. Chem. Phys. 57, 2985 (1972).
32. L. A. Collins, W. D. Robb and M. A. Morrison, "Low-Energy Electron Scattering by H_2 and N_2: An Iterative Static-Exchange Calculation," J. Phys. B 11, L777 (1978).
33. See the review by P. G. Burke and M. J. Seaton, in Methods in Computational Physics, 10 edited by B. Alder, S. Fernbach and M. Rotenberg (Academic Press, New York, 1971), p. 2 for references and descriptions. See also R. Marriot, Proc. Phys. Soc. 72, 121 (1958) and K. Omidvar, Phys. Rev. A 133, 970 (1964).
34. R. J. W. Henry and N. F. Lane, Phys. Rev. 183, 221 (1969).
35. E. R. Smith and R. J. W. Henry, Phys. Rev. A 7, 1585 (1973).
36. M. J. Seaton, Proc. Roy. Soc. A 218, 400 (1953); T. L. John, Proc. Phys. Soc. (London) 76, 532 (1960).
37. L. A. Collins, D. W. Norcross and G. Bruno Schmid, "Electron Collisions with Highly Polar Molecules: Integrated and Momentum-Transfer Cross Sections and Conductivity Integrals for KOH and CsOH" (J. Phys. B. accepted for publication).
38. For electron collisions with heteronuclear molecules, we have $v_\lambda^{nuc}(r) = -2 \sum_\alpha z_\alpha \xi_<^\lambda(\alpha)/\xi_>^{\lambda+1}(\alpha)$, where the sum over α runs over all nuclei and where $\xi_<(\alpha) = \min(R_\alpha, r)$, R_α being the location of the αth nucleus with respect to the origin.
39. The value of r_{max} depends on the system symmetry. Thus for e-H_2 in Σ_u symmetry, we have $r_{max} = 50.0\ a_o$.
40. In the iterative-static-exchange procedure, which is described in Sec. V, we find that $\lambda_{max}^{st} = 2\ell_{max}$ is more cost-effective.

41. It should be noted that in general this procedure is not applicable to electron collisions with heteronuclear molecules, where the strong dipole coupling necessitates retention of many partial waves (typically up to $\ell = 16$).

42. L. A. Collins, W. D. Robb, and M. A. Morrison, Bull. Amer. Phys. Soc. 23, 1082 (1978).

43. See also N. F. Lane and R. J. W. Henry, Phys. Rev. 173, 183 (1968) for ab initio calculations on e-H_2 collisions and A. Dalgarno and N. Lynn, Proc. Phys. Soc. London A 70 (1957) for e-He scattering.

44. See M. E. Riley and D. G. Truhlar, J. Chem. Phys. 63, 2182 (1975); B. N. Bransden, M. R. C. McDowell, C. J. Nobel, and T. Scott, J. Phys. B 9, 1301 (1976).

45. S. Hara, J. Phys. Soc. Jpn. 22, 710 (1967).

46. K. Smith, R. J. W. Henry, and P. G. Burke, Phys. Rev. 147, 21 (1966).

DISCUSSION

Truhlar: I think it might be helpful if you said a few more words about the way you treat exchange.

Morrison: If we were actually solving the static-exchange equations, we would get a wave-function that satisfied orthogonality. Of course we are using a numerical procedure and consequently we do have to wrestle with the question of whether the resulting scattering function is orthogonal to the bound molecular orbitals. In our H_2 work, we found that by simply building in orthogonality, things worked pretty well, but this is rather difficult to implement in N_2. At the moment, that is the stumbling block in the development of the method. Derek Robb, Lee Collins, and I have calculations for N_2 that do converge, but I don't think any of us is entirely happy with the way we are doing it. What I'm saying is that the numbers look correct, but there are features of the convergence I'm not really very pleased with and there are some questions that I think need to be addressed.

Bardsley: Do you impose orthogonality to bound-state orbitals as a matter of course?

Morrison: We do in the H_2 calculations for Σ_g. What I was trying to say is that there we don't seem to have any difficulty. It's when we move to the N_2, which has more bound orbitals, that we have not yet decided on the best procedure.

Poe: I wonder it it's possible in this workshop to have some kind of concensus as to when we call something converged?

Morrison: That's what I'm trying to stress. If you say what you're converging, what your convergence parameter is and what your criterion is, then you've said what you've done precisely. What we tried to do is to treat as much of the numerical aspects of the scattering problem as accurately as we possibly could with our resources. The idea is to study the physical questions that arise and remove as much uncertainty as possible from the numerical side of the calculation. That's why we impose what some people might consider to be ludicrously stringent convergence criteria. In certain cases - for example, if you're trying to see whether a resonance is there or not - you probably don't care to the level that we care, how nearly-converged you are. But you can always specify that if you just say what your criterion was.

Lane: May I just interject a comment along the same lines? You also see frequently in the literature the statement that we have this new approach and that our results agree with Groups A and B. If you go to the literature, Groups A and B often don't agree. It's a question of degree, and there is some confusion in the literature already in our field which all of us are a little bit guilty of.

ROUNDTABLE ON NUMERICAL METHODS

Chairman: Aaron Temkin

Temkin: I'm going to take the perogative, being the chairman of this session, to take my five minutes asking Mike Morrison a question, and it seems to me, a very relevant question. In the one-electron exchange model, we know that there's such a thing as the Levinson and Swann theorem which says that scattering orbitals of the same symmetry as doubly occupied orbitals must be orthogonal to the latter. That's the essence of the Levinson-Swann theorem. Therefore, I ask you the question, when you put in the free-electron gas exchange potential and look at, for example, the σ_u scattering orbitals in the $e^- - N_2$ scattering -- there are two occupied σ_u orbitals in N_2 -- do you get three nodes at zero energy in the free-electron gas approximation?

Truhlar: While he's thinking, let me comment. We did some tests on atoms and in every case that I remember -- the results were published -- we did get the right number of multiples of π in the phase shift which means we did get the right number of nodes.

Morrison: In CO_2 we do and in N_2, as I recall, we do, but I'd have to look at it very carefully.

Temkin: O.K., that's all I wanted to ask. Let me now call on Barry Schneider.

Schneider: I was curious about whether anybody here would like to comment on these iterative methods of solving large sets of linear equations. There are at least two people in the audience who are experts -- Bob and Ernie -- they've done these kinds of things for eigenvalue problems.

Nesbet: I'd like to make a comment, if it's relevant, to Dave Moores' work. In the engineering world, people have lots of experience with this kind of problem of fairly large scale. Well, there is a powerful technique that people should be aware of in this context and that is the use of sparse matrix techniques. They use it in such a way that you put in some criterion for elements which are rather small and throw them away completely. If you do that, then you can get quite a lot of mileage out of sparseness even though you're throwing away a lot of relatively small numbers. But of course that gives results that are not quite right. Then you use an iterative method to refine the result and you can get second order methods that work very well once you are close to the solution. So a method that is very much worth considering when you get to the large inhomogeneous linear systems is first to do the usual triangular decomposition or Gaussian elimination with pivoting by using a crude sparse matrix technique where you really throw away a lot of elements and then go back and use an iterative refinement which at that point might converge very rapidly. However, if you start an iterative technique when you don't already have a pretty good solution, it can very easily diverge.

Schneider: But aren't you guaranteed, at least under certain conditions, that iterative methods will always converge?

Nesbet: Only if you have pretty well determined matrices and only if you're lucky in the sense that round-off errors don't accumulate enough to throw that out. You have to be very careful.

Davidson: You can always get eigenvalues but you can't always get the solution of linear equations.

Truhlar: How sparse is the matrix once you put exchange in?

Moores: It's still fairly sparse.

Truhlar: I mean what percentage of the elements would be below some reasonable threshold?

Moores: They're often either zero or of the same order of magnitude. You don't get lots of very small elements. It's hard to say really, maybe half the elements are zero. But without exchange, more are zero than are non-zero.

Truhlar: Well, without exchange, you just have a few bands right?

Moores: Yes, that's right.

Rumble: But if you do as suggested, you can get a starting solution in no time at all. I've done cases where you have 2000 by 2000 very sparse matrices, read in 5000 non-zero elements and it just takes no time at all. It's a very fast method for getting a starting solution for an iterative method which is really important. I don't think it can be over-emphasized that if you don't know a good starting solution for an iterative method, you might never get there.

Robb: The advantage of the method that we've been using is that, as Mike showed you, you multiply the inhomogeneous term, W, by f^{-1} and that essentially gives you a local exchange potential for iteration. Then once you get it converged at one energy, when you move to another energy, you start with that previous solution. That usually gives us convergence in a few steps.

Nesbet: But how do you deal with the singularities in f^{-1}?

Robb: Well, there are a few tricks. It's particularly good to use the integral equation method because there you're doing integrals of the form gVf where V is our effective local potential that may have singularities -- it's not just the static potential.

Nesbet: Do you take a principal value integral or something like that?

Robb: No, just a simple computational extrapolation that we've been using recently. V contains $f^{-1}W$ and what you're relying on is the holes caused by the node in f_{N-1} being cancelled in the integral by the node in the new f_N. We know where that node is, so when we get near it we just extrapolate right through that point. Everything else is going smoothly.

Temkin: Let me call now on Brian Buckeley to make a remark.

Buckley: Really, I'm a convert to the analytic basis set method, so I don't really have any vested interest in what I'm going to say. About the replacement of the exchange integrals by these second-order equations that are used by Burke, SinFailam and myself--Dave Moores mentioned that for a 5-channel case you need to include up to, say, 100 exchange terms. Now this isn't quite the case in practice. For a start-off, you don't always have to include all the exchange terms in N_2, for example. I just included the $3\sigma_g$, $2\sigma_u$ and $1\pi_u$ terms. The others contribute negligibly to exchange. So that brings you down from 100 to about 40 terms. The second thing is that the partial wave expansion is very quickly convergent. You need only include 2 or 3 terms, in contrast to the static terms which can converge very slowly. But I think the most important contribution comes from Raseev himself when he pointed out that the exchange terms, which can be written as a sum of partial waves, in my case, ℓ', ℓ'' and ℓ''', can be replaced with new terms where the sum ℓ'' and ℓ''' can be done beforehand. So what you end up with is down by a factor of 3 or 4 at least in the number of exchange terms. So you generate a new set of second-order equations and you end up with something like 10 or 15 exchange equations. So it turns out that exchange is simple and it's the partial wave expansion of the static part that's going to be the problem for us.

Temkin: Are you saying that you forget about the exchange terms in the equations which pertain to $1\sigma_g$, $1\sigma_u$ and $2\sigma_g$?

Buckley: Yes.

Temkin: But do you then get the right nodal structure?

Buckley: Well, that's beside the point.

Schneider: This is getting very confusing.

Rescigno: Orthogonality to a doubly occupied orbital will change the wavefunction but it will have absolutely no effect on the S-matrix.

Schneider: That's right.

Rescigno: You can add any component you want of the doubly occupied orbitals and it will not change the scattering matrix.

Temkin: That is quite true. But the point is that if you add orthogonality plus additional potentials, then it most certainly will change things. And when he includes the exchange terms coming from $1\sigma_g^2$, say, explicitly, it's not at all clear to me that he's going to get the same answers as when he does not include that orthogonality.

Buckley: Well, we've gone through and checked and the terms we throw out contribute negligibly.

Schneider: But the effect must be energy dependent because if the electron penetrates inside, it will start seeing those orbitals.

Buckley: O.K., sure, I'm talking of energies up to about 15 eV.

Vo Ky Lan: Just one more point. This trick of Raseev is very useful. If you look at $e^- + CH^+$, dissociative recombination, he needs only 7 exchange terms, as opposed to about 80 originally, and you get exactly the same phase shifts as Burke and SinFailam would get. It's very useful.

Moores: Does this mean you have the same number of exchange terms as the number of λ's retained?

Buckley: Yes, just about.

Robb: What Brian and everyone else here are saying is quite true for these systems. But life isn't always kind to you. You go to something like LiF, even with all these tricks, it still leaves you with about 225 exchange terms. It's a much more tricky problem because you don't have just even terms coming into the expansion.

Buckley: I still maintain that the exchange isn't the problem if you're going to a single center calculation. It's the direct part of the expansion.

Robb: In the iterative technique, though, we're always dealing with matrices that are just the number of channels by number channels.

Buckley: Well, exchange should not more than double the number of channels.

Nesbet: How many iterations do you have to go through?

Robb: About twelve. In LiF, it gets more expensive. The first one usually takes up to 15 iterations to converge to better than 4 figures.

Davidson: I have one question. Don't you go through the density matrix in order to calculate the exchange operator? The density matrix as expressed in terms of the original basis set?

Bottcher: I think I know what Dr. Davidson is getting at. The density matrix is not a convenient animal to store; it just forms too many parameters, so it's not explicitly formed.

Temkin: Let me cut off the discussion since there are at least two more people who would like to make remarks. I would like Vo Ky Lan to talk a little bit about his new asymptotic program and this is relevant to what Dave Moores had to say. [See accompanying paper by M. Le Dourneuf and Vo Ky Lan.]

CONTRIBUTION OF THE VARIABLE PHASE METHOD TO THE FRAME TRANSFORMATION THEORY OF ROTATIONAL EXCITATION OF MOLECULES BY ELECTRON IMPACT

M. Le Dourneuf and Vo Ky Lan

Observatoire de Paris
92190 Meudon
France

I. INTRODUCTION

The extension of electron atom scattering techniques to molecular targets must face two major complications:

(1) the strong anisotropy of electrostatic interactions

(2) the dynamical coupling with the nuclear motion.

The *exact* description by standard collisional techniques - a close coupling (CC) type one-center expansion with *a large number of closely spaced states* of the molecule, and *a large number of partial waves* to describe the motion of the colliding electron in the laboratory frame (LF) anisotropic molecular field- becomes a prohibitive task as soon as one wants to go beyond the rotational excitation of the lighter homonuclear diatomics.[1]

A key simplification, which has opened the way to most of the present results, is the wide validity of the adiabatic approximation. Indeed, because of the large nuclear inertia, one can often assume that the nuclei have no time to move during the effective time of collisional interaction (fixed-nuclei approximation) or, more generally, that the colliding electron gets sufficiently accelerated by the molecular field to follow adiabatically the nuclear motion (adiabatic nuclei approximation). The dynamical coupling with the nuclear motion can be neglected and the single remaining complication - the anisotropy of the molecular field - is more easily faced in the moving molecular frame (MF), where only an extension of molecular bound state techniques is required. A large fraction of the present workshop has indeed focussed on these developments and the comparative efficiency of their numerous variants.

In spite of its wide applicability, the adiabatic MF formulation of electron-molecule collisions has limitations. The most famous is the case of polar molecules for which the dynamical decoupling of the colliding electron from the molecular permanent dipole field, which averages out to zero in rotation, is essential to predict the observed finite forward and total cross sections. However, if the exact LF formulation of the scattering equations, retaining the rotational kinetic energy of the nuclei in the Hamiltonian, is *essential* to describe properly the dynamical decoupling of the colliding electron from the molecular field *at large distances*, the corresponding rotational CC leads to a large number of (ℓj) coupled channels, which cannot take advantage of the long-term-development-but-efficient MF formulations.

An important step towards a simple and efficient general theory was the suggestion by Fano, then Chang and Fano[2] that the advantages of the adiabatic MF and close coupling LF formulations, adequate to describe qualitatively different aspects of the electron-molecule interaction in different spatial regions, should be combined using a frame transformation (FT). However, the interest of this suggestion for accurate quantitative studies remained to be demonstrated, since it relies upon the practical justification of a sharp splitting of the interaction space into distinct regions. A preliminary step in this direction was made by Chandra[3] in his study of CO rotational excitation. However, his use of traditional techniques, combining inward and outward integration of the coupled scattering equations, with a final matching to obtain the asymptotic K-matrix, prevented any direct determination of the FT point r_{rot}, whose location was deduced from time-consuming numerical experimentation (stabilization of the results after several trial runs). This made the whole procedure unsatisfactory and practically impossible to generalize when more than one FT is necessary (vibrational excitation ...)

A better technical approach to the problem, suggested by Chang and Fano themselves,[2] is the Variable Phase Method (VPM). By its definition,[4] it gives at each electron- molecule distance r, the rate of variation of the local K(r) matrix associated with the various parts of the interaction, and is therefore expected to help in the physical choice of r_{rot}. Furthermore, it has recently been implemented as an efficient computational technique to solve large numbers of LF coupled equations occuring in electron-atom scattering problems, remaining particularly stable in the presence of closed, or nearly degenerate channels.[5]

The remainder of this paper deals with the extension of the VPM to implement a general FT theory for the rotational excitation of rigid diatomic molecules by electron impact. The scattering equations are summarized in Section II. Starting from the two equivalent exact LF and MF formulations, we introduce the adiabatic approximation, justified at short distances ($r \leqslant r_{rot}$) where it partly decouples the MF equations, while the complete LF equations may have to be solved at large distances ($r \geqslant r_{rot}$). We describe in Section III the practical implimentation of the VPM in the local potential regions (MF for $r_{exch} \leqslant r \leqslant r_{rot}$, LF for $r_{rot} \leqslant r \leqslant r$) and the general form of the FT (ensuring the matching of the logarithmic derivative matrices at r_{rot}). In Section IV the fundamental problem of the determination of the FT point in the general case is approached, after two instructive preliminary discussions analyzing how, in the two limiting cases (high and low energies), the VPM + FT procedure identifies itself with well established approximations (the full adiabatic approximation and the generalized effective range theory, respectively). In Section V we summarize the main contributions of our VPM + FT procedure, compared with previous work, and we give the references for published checks of our numerical package and for its first applications.

II. THE SCATTERING EQUATIONS

For simplicity in the present discussion (aiming at the assessment of the VPM + FT procedure), we start from the explicit equations for a rigid diatomic in its $^1\Sigma^+$ electronic ground state, considering only its rotational degrees of freedom.

The Schrödinger equation describing the scattering of an electron by an N electron diatomic can be written

$$(H-E)\ \Phi\ (\vec{r}_1 \cdots\cdots \vec{r}_N\ \ \vec{r},\ \vec{R}) = 0 \tag{1}$$

with the total hamiltonian in a.u.

$$H\ \ = H_{inc} + H_{mol} + V \tag{2a}$$

$$H_{inc} = -\frac{1}{2}\ \Delta_{\vec{r}} \tag{2b}$$

$$H_{mol} = H_{e\ell}\ (\vec{r}_1' \cdots\cdots \vec{r}_N'\ ,\ \underline{\vec{R}'}) + H_{rot}\ (\hat{R}) \tag{2c}$$

$$V\ \ = \sum_{i=1}^{N} \frac{1}{|\vec{r}-\vec{r}_i|} - \frac{Z_A}{|\vec{r}-\vec{R}_A|} - \frac{Z_B}{|\vec{r}-\vec{R}_B|} \tag{2d}$$

All the coordinates ($\vec{r}$ for the incident electron; $\vec{r}_i$ for the molecular electrons; $\vec{R}_A$, $\vec{R}_B$ for the two nuclei) are referred to their center of mass. The unprimed coordinates refer to the fixed LF, while the primed ones in (2c) have their angular parts referred to the rotating MF (with the internuclear axis $\vec{R}'$, taken as z' axis). Owing to the axial symmetry of the interactions between the components of the total system, the Hamiltonian H commutes with the total orbital momentum J^2, its fixed projection J_z, the inversion I of all coordinates and the reflexion $\sigma_{x'z'}$ w.r.t. the x'0 z' plane containing the internuclear axis $\vec{R}'$. We will therefore concentrate on the determination of eigenstates $\Phi_E^{JM\eta}$ common to H, J^2, J_z, I, $\sigma_{x'z'}$ with eigenvalues E, J(J+1), M, I, η related by[2]:

$$I = (-1)^J \eta \tag{3}$$

Since we have assumed that the molecular electrons remain in their $^1\Sigma^+$ electronic ground state with the internuclear distance fixed at its equilibrium value, $\vec{R}'$

$$[H_{e\ell}(\vec{r}_i', \underline{\vec{R}'}) - E_o]\, \phi_{e\ell}^{^1\Sigma^+}(\vec{r}_i', \underline{\vec{R}'}) = 0 \tag{4a}$$

and we concentrate on the outer regions where the exchange has become negligible,

$$r \geqslant r_{exch} \Rightarrow \Phi_E^{JM\eta}(\vec{r}_i', \vec{r}, \underline{\vec{R}}) \underset{r>>r_i}{\sim} \phi_{e\ell}^{^1\Sigma^+}(\vec{r}_i', \underline{\vec{R}'}) \times \Psi_E^{JM\eta}(\vec{r}, \hat{R}) \tag{4b}$$

the problem reduces to the solution of

$$[-\frac{1}{2}\Delta_{\vec{r}} + \sum_{\mu} V_{\mu}(r)\, P_{\mu}(\hat{r}\cdot\hat{R}) + H_{rot}(\hat{R}) + E_o - E] \times \Psi_E^{JM\eta}(\vec{r}, \hat{R}) = 0 \tag{5a}$$

where the second term corresponds to the multipole expansion of the LF direct molecular potential

$$< \phi_{e\ell}^{^1\Sigma^+}(\vec{r}_i', \underline{\vec{R}'})|V|\phi_{e\ell}^{^1\Sigma^+}(\vec{r}_i', \underline{\vec{R}'}) > = \sum_{\mu} V_{\mu}(r)\, P_{\mu}(\hat{r}\cdot\hat{R}) \tag{5b}$$

A. LF Formulation

Following Arthurs and Dalgarno,[1] $\Psi_E^{JM\eta}$ can be expanded in the coupled angular basis of the fixed frame LF

$$\Psi_E^{JM\eta}(\hat{r}, \hat{R}; r) = \sum_{\ell j} Y_{\ell j}^{JM\eta}(\hat{r}, \hat{R})\, \frac{F_{\ell j}^{J\eta}(r)}{r} \tag{6}$$

These angular eigenstates

$$Y_{\ell j}^{JM\eta}(\hat{r}, \hat{R}) = \sum_{m} Y_{\ell m}(\hat{r})\, Y_{jM-m}(\hat{R}) < \ell m\, jM-m|\ell j\, JM > \tag{7a}$$

are also eigenstates for the inversion I

$$I\, Y_{\ell j}^{JM\eta} = (-1)^{\ell+j}\, Y_{\ell j}^{JM\eta} \tag{7b}$$

for the reflexion $\sigma_{x'z'}$,following (3)

$$\sigma_{x'z'}\, Y_{\ell j}^{JM\eta} = \eta\, Y_{\ell j}^{JM\eta} \quad \text{with } \eta = (-1)^{\ell+j-J} \tag{7c}$$

and for the rotational Hamiltonian since

$$H_{rot}(\hat{R})\, Y_{jM-m}(\hat{R}) = Bj(j+1)\, Y_{jM-m}(\hat{R}) \tag{7d}$$

(B = rotational constant of the rigid molecule)

The unknown radial functions $F_{\ell j}^{J\eta}(r)$ must then satisfy

$$\left[-\frac{1}{2}\frac{d^2}{dr^2} + \frac{\ell(\ell+1)}{2r^2} - \frac{1}{2}k_j^2\right] F_{\ell j}^{J\eta}(r) + \sum_{\ell'j'} V_{\ell j\;\ell'j'}^{J\eta}(r) \times F_{\ell'j'}^{J\eta}(r) = 0 \tag{8a}$$

with $$\frac{1}{2}k_j^2 = E - E_o - Bj(j+1) \tag{8b}$$

$$V_{\ell j\;\ell'j'}^{J\eta}(r) = [2\ell+1)(2\ell'+1)(2j+1)(2j'+1)]^{\frac{1}{2}} \sum_{\mu}(-1)^{-J-\mu} \times \begin{pmatrix} \ell & \ell' & \mu \\ 0 & 0 & 0 \end{pmatrix} \begin{pmatrix} j & j' & \mu \\ 0 & 0 & 0 \end{pmatrix} \begin{Bmatrix} j' & \ell' & J \\ \ell & j & \mu \end{Bmatrix} V_{\mu}(r) \tag{8c}$$

The LF equations, which decouple in the infinite r limit, are the most satisfactory ones at large r. However, their practical drawback (large number of ℓj couplings) can partly be avoided as r decreases. Owing to the smallness of the rotational constant

$$B = 2.77\ 10^{-4} \text{ a.u. for } H_2 \text{ (largest value)} \tag{9}$$

the small non-degeneracy of the rotational channels can be neglected:

$$\frac{1}{2}k_j^2 \simeq \frac{1}{2}k^2 \tag{10a}$$

with

$$\frac{1}{2}k^2 = E - E_o \tag{10b}$$

as soon as the rapidly increasing molecular multipoles dominate. Whenever this "adiabatic" approximation becomes justified, the equations are expected to decouple partly in an angular basis which reflects the molecular field symmetries.

B. MF Formulation

The molecular field axial symmetry is fully taken into account by choosing the alternative angular basis

$$X^{JM\eta}_{\ell\lambda}(\hat{r}', \hat{R}) = [Y_{\ell\lambda}(\vec{r}')\, D^{J}_{\lambda M}(\hat{R}) + \eta\, Y_{\ell-\lambda}(\hat{r}')\, D^{J}_{-\lambda M}(\hat{R})] \times \left[\frac{2J+1}{8\Pi(1+\delta_{\lambda o})}\right]^{\frac{1}{2}} \tag{11a}$$

of eigenstates common to the absolute projection λ of the colliding electron orbital momentum (equal in our simple case to the total electronic orbital momentum) on the moving internuclear axis $\vec{R}'$ and to the $\sigma_{x'z'}$ reflexion operator with eigenvalue $\eta = \pm 1$.

The corresponding MF expansion:

$$\Psi^{JM\eta}_{E}(\hat{r}', \hat{R}; \hat{r}) = \sum_{\ell\lambda} X^{JM\eta}_{\ell\lambda}(\hat{r}', \hat{R})\, \frac{G^{J\eta}_{\ell\lambda}(r)}{r} \tag{11b}$$

which is related to the LF expansion (6) by a purely geometrical orthogonal transformation

$$\Omega^{\ell J\eta}_{\lambda j} = < X^{JM\eta}_{\ell\lambda} | Y^{JM\eta}_{\ell j} > = \left(\frac{2j+1}{2J+1}\right)^{\frac{1}{2}} \times < \ell\lambda jo | \ell jJ\lambda > \frac{1+\eta(-1)^{J-\ell-j}}{[2(1+\delta_{\lambda o})]^{\frac{1}{2}}} \tag{12a}$$

$$F^{J\eta}_{\ell j}(r) = \sum_{\lambda} G^{j\eta}_{\ell\lambda}(r)\, \Omega^{\ell J\eta}_{\lambda j} \tag{12b}$$

$$G^{J\eta}_{\ell\lambda}(r) = \sum_{j} F^{J\eta}_{\ell j}(r)\, \tilde{\Omega}^{\ell J\eta}_{j\lambda} \quad (\tilde{\Omega} = \text{transpose of the } \Omega \text{ matrix}) \tag{12c}$$

can be obtained by solving directly the exact MF equations

$$[-\frac{1}{2}\frac{d^2}{dr^2} + \frac{\ell(\ell+1)}{2r^2} - \frac{1}{2}k^2]\, G^{J\eta}_{\ell\lambda}(r) + \sum_{\lambda'} U^{\ell J\eta}_{\lambda\lambda'}\, G^{J\eta}_{\ell\lambda'}(r) + \sum_{\ell'} V^{\lambda}_{\ell\ell'}(r)\, G^{J\eta}_{\ell'\lambda}(r) = 0 \tag{13a}$$

with electrostatic couplings diagonal in λ and independent of $J\eta$, as expected,

$$V^{\lambda}_{\ell\ell'}(r) = (-1)^{\lambda} \left[(2\ell+1)(2\ell'+1)\right]^{\frac{1}{2}} \sum_{\mu} \begin{pmatrix} \ell & \ell' & \mu \\ o & o & o \end{pmatrix} \begin{pmatrix} \ell & \ell' & \mu \\ \lambda & -\lambda & o \end{pmatrix} V_{\mu}(r) \tag{13b}$$

and rotational couplings, diagonal in ℓ , but dependent on $J\eta$ and infinite range in r

$$U^{\ell J\eta}_{\lambda\lambda'} = B \sum_{j=o}^{\ell+J} \Omega^{\ell J\eta}_{\lambda j}\; j(j+1)\; \tilde{\Omega}^{\ell J\eta}_{j\lambda'} \quad . \tag{13c}$$

As expected, the MF equations are inadequate at large r (constant rotational couplings given by 13c). But as soon as these can be neglected ($r \leqslant r_{rot}$ to be justified precisely in IVc), the simplified adiabatic MF equations

$$\left[-\frac{1}{2}\frac{d^2}{dr^2} + \frac{\ell(\ell+1)}{2r^2} - \frac{1}{2}k^2\right] \bar{G}^{\lambda}_{\ell}(r) + \sum_{\ell'} V^{\lambda}_{\ell\ell'}(r)\, \bar{G}^{\lambda}_{\ell'}(r) = 0 \tag{14}$$

become decoupled in λ (smaller sets of coupled equations) and independent of J, η (smaller total number of equations to integrate). The only remaining problem (strong ℓ- mixing due to the strong anisotropy of the molecular field, exchange and other many-electron effects) is usually limited to a smaller zone ($r \leqslant r_{exch}$) where finite range specific molecular techniques (such as the R-matrix...), apply efficiently.

III. THE VPM + FT SOLUTION OF THE LOCAL ELECTRON-MOLECULE SCATTERING EQUATIONS

After a brief summary of the VPM, its applications to the adiabatic MF and to the exact LF equations are successively described, with special attention given to the boundary condition determination and to the FT formulation, and the specific advantages of the VPM pointed out in the course of the discussion.

A. The Principle of the VPM

Given a set of NCHAN second order differential equations describing the scattering by local potentials in a.u.,

$$\left[\frac{d^2}{dr^2} - \frac{\ell_i(\ell_i+1)}{r^2} + \frac{2z}{r} + k_i^2\right] F_i(r) = \sum_{i'} 2\, V_{ii'}(r)\, F_{i'}(r) \tag{15}$$

the effect of the couplings is brought into focus by expanding the solutions on a set of two independent solutions of the asymptotically decoupled equations

$$\left[\frac{d^2}{dr^2} - \frac{\ell_i(\ell_i+1)}{r^2} + \frac{2z}{r} + k_i^2\right] J_{\ell_i k_i}, N_{\ell_i k_i} = 0 \tag{16a}$$

with

$$W[J_{\ell_i k_i}, N_{\ell_i k_i}; r] = J_{\ell_i k_i}(r)\, N'_{\ell_i k_i}(r) - J'_{\ell_i k_i}(r) \times N_{\ell_i k_i}(r) \equiv +1 \quad . \tag{16b}$$

In our present case of a neutral target (residual charge $z = 0$) and all open channels ($k_i^2 > 0$), the set of renormalized (16b) Ricatti-Bessel functions

$$J_{\ell_i k_i}(r) = k_i^{-\frac{1}{2}}\,[k_i r\, j_{\ell_i}(k_i r)] \underset{r\to\infty}{\sim} k_i^{-\frac{1}{2}} \sin\left(k_i r - \frac{\ell_i \Pi}{2}\right) \tag{17a}$$

$$N_{\ell_i k_i}(r) = k_i^{-\frac{1}{2}}\,[k_i r\, n_{\ell_i}(k_i r)] \underset{r\to\infty}{\sim} k_i^{-\frac{1}{2}} \cos\left(k_i r - \frac{\ell_i \Pi}{2}\right) \tag{17b}$$

is the most suitable (regularity of J at the origin, convenient asymptotic forms). By imposing the condition of the regularity at the origin, the physical set of solutions of (15), of rank NCHAN, can be generated by

$$F_{ij}(r) = J_{\ell_i k_i}(r)\, A_{ij}(r) - N_{\ell_i k_i}(r)\, B_{ij}(r) \quad i,j = 1..\text{NCHAN} \tag{18a}$$

with the additional relationship

$$F'_{ij}(r) = J'_{\ell_i k_i}(r)\, A_{ij}(r) - N'_{\ell_i k_i}(r)\, B_{ij}(r) \tag{18b}$$

to define A and B uniquely from their boundary condition at the origin:

$$B_{ij}(0) = 0 \text{ (regularity)} \quad ; \quad A_{ij}(0) \text{ of rank NCHAN} \tag{18c}$$

The key property of the VPM is that the generalized K(r)-matrix, defined at each point by the ratio

$$\underset{\sim}{K}(r) = \underset{\sim}{B}(t)\, \underset{\sim}{A}^{-1}(r) \Longrightarrow \underset{\sim}{F}(r) = [\underset{\sim}{J}_{\ell k}(r) - \underset{\sim}{N}_{\ell k}(r)\, \underset{\sim}{K}(r)]\, \underset{\sim}{A}(r) \tag{19}$$

independently from the particular set of NCHAN independent physical solutions of (15), can be integrated outwards:

$$\underset{\sim}{K}'(r) = -[\underset{\sim}{J}_{\ell k}(r) - \underset{\sim}{K}(r)\, \underset{\sim}{N}_{\ell k}(r)]\; 2\underset{\sim}{V}\; [\underset{\sim}{J}_{\ell k}(r) - \underset{\sim}{N}_{\ell k}(r)\, \underset{\sim}{K}(r)] \tag{20}$$

from the origin (eqn. 18c, 19)

$$\underset{\sim}{K}(0) = 0 \quad \text{(consistent with the choice (17) of } \underset{\sim}{J}, \underset{\sim}{N}) \tag{21}$$

to infinity, where it becomes the usual $\underset{\sim}{K}$ - matrix, sufficient to deduce all the scattering observables.

More generally, the first-order VPM equation (20) allows the direct propagation between any two boundaries r_1, r_2 (0, finite, ∞), of the logarithmic derivative of a set of NCHAN independent solutions of any set of type-(15) equations using one of the bijective relations on the boundary

$$\underset{\sim}{F}'\underset{\sim}{F}^{-1}(r) = [\underset{\sim\ell k}{J'}(r) - \underset{\sim\ell k}{N'}(r)\,\underset{\sim}{K}(r)][\underset{\sim\ell k}{J}(r) - \underset{\sim\ell k}{N}(r)\,\underset{\sim}{K}(r)]^{-1} \tag{22a}$$

$$\underset{\sim}{K}(r) = [\underset{\sim\ell k}{J'}(r) - \underset{\sim\ell k}{J}(r)\,\underset{\sim}{F}'\underset{\sim}{F}^{-1}(r)][\underset{\sim\ell k}{N'}(r) - \underset{\sim\ell k}{N}(r)\underset{\sim}{F}'\underset{\sim}{F}^{-1}(r)]^{-1} \tag{22b}$$

From the general definition of an $\underset{\sim}{R}$ matrix

$$\underset{\sim}{F}(r) = \underset{\sim}{R}(r,\underset{\sim}{b})[r\underset{\sim}{F}'(r) - \underset{\sim}{b}\underset{\sim}{F}(r)] \Rightarrow r\underset{\sim}{F}'\underset{\sim}{F}^{-1} = \underset{\sim}{R}^{-1}(r,b) + \underset{\sim}{b} \tag{23a}$$

where $\underset{\sim}{b}$ is a diagonal matrix, whose elements are arbitrary constants which may depend on the channel quantum numbers, but are usually taken to be zero. The bijective correspondence between $\underset{\sim}{R}(r, b)$ and $\underset{\sim}{K}(r)$ is easily written explicitly

$$\begin{aligned}\underset{\sim}{R}(r,b) = {} & [r\{\underset{\sim\ell k}{J'}(r) - \underset{\sim}{K}(r)\,\underset{\sim\ell k}{N'}(r)\} - b\,\{\underset{\sim\ell k}{J}(r) - \underset{\sim}{K}(r)\underset{\sim\ell k}{N}(r)\}]^{-1} \\ & \times [\underset{\sim\ell k}{J}(r) - \underset{\sim\ell k}{N}(r)\,\underset{\sim}{K}(r)]\end{aligned} \tag{23b}$$

$$\begin{aligned}\underset{\sim}{K}(r) = {} & [\underset{\sim\ell k}{J}(r)\{1 + b\underset{\sim}{R}(r,b)\} - r\underset{\sim\ell k}{J'}(r)\underset{\sim}{R}(r,b)][\underset{\sim\ell k}{N}(r)\{1+b\underset{\sim}{R}(r,b)\} \\ & - r\underset{\sim\ell k}{N'}(r)\underset{\sim}{R}(r,b)]^{-1}.\end{aligned} \tag{23c}$$

From this brief outline, it is clear that the VPM, based upon the direct outward propagation of the generalized $\underset{\sim}{K}(r)$ matrix, should be more efficient than the standard integration techniques which deduce the asymptotic $\underset{\sim}{K}$ matrix as the final result of the matching, at r_{exch}, between the set of NCHAN physical solutions of the core region (obtained by outward integration from the origin) and the set of (NCHAN + NOPEN) solutions obtained by inward integration from physical asymptotic boundary conditions. The VPM is very similar to the multichannel log-derivative method of Johnson[6] and to the R-matrix propagator technique of Light and Walker[7] and gives at each point the global effect of the inner region potential through $\underset{\sim}{K}(r)$ (directly related to the observables), and its rate of variation associated with each type of coupling (eq.20).

It will give much physical insight into complex molecular scattering dynamics and practical help in the implementatiion of the FT.

B. The Adiabatic MF Equations

The neglect of the rotational splittings, which leads to the λ decoupling of the adiabatic MF equations (14), corresponds to the VP expansion of the adiabatic MF equations

$$\bar{\underset{\sim}{G}}^{\lambda}(r) = [\bar{\underset{\sim}{J}}_{\ell k}(r) - \bar{\underset{\sim}{N}}_{\ell k}(r)\,\bar{\underset{\sim}{K}}^{\lambda}(r)]\,\bar{\underset{\sim}{A}}^{\lambda}(r) \tag{24a}$$

and to the adiabatic MF VP equation

$$\bar{\underset{\sim}{K}}^{\lambda\prime} = [\bar{\underset{\sim}{J}}_{\ell k} - \bar{\underset{\sim}{K}}^{\lambda}\,\bar{\underset{\sim}{N}}_{\ell k}]\,2\underset{\sim}{V}^{\lambda}\,[\bar{\underset{\sim}{J}}_{\ell k} - \bar{\underset{\sim}{N}}_{\ell k}\,\bar{\underset{\sim}{K}}^{\lambda}] \tag{24b}$$

with the electrostatic coupling $V^{\lambda}_{\ell\ell'}$ given by (13b) and the bars recalling that the corresponding quantities are calculated in the adiabatic approximation (neglecting the rotational splittings, i.e., assuming all channels degenerate). This approximation allows a significant simplification of the VPM integration (24b) since the basis functions $\bar{\underset{\sim}{J}}_{\ell k}\ \bar{\underset{\sim}{N}}_{\ell k}$ are then efficiently derived by recurrence on ℓ.

The initial boundary condition $\bar{K}^{\lambda}_{\ell\ell'}(r_{exch})$, which reduces to regularity condition (21) for $r_{exch} = 0$ (no exchange or local-exchange approximations) or for large ℓ ($\ell > \ell_M$: core penetration excluded at low energy by the centrifugal barrier), is usually deduced, for the penetrating partial waves $\ell \leqslant \ell_M$, from the knowledge of the $\bar{\underset{\sim}{R}}^{\lambda}_{\ell\ell'}(r_{exch}\ , b)$ matrix (23b) of the NCHAN = $\ell_M-\lambda+1$ regular solutions, obtained either by adequate integration of the integro-differential scattering equations (see accompanying articles by Moores and Morrison) or by "R-matrix diagonalization" of the total Hamiltonian in the core region (see accompanying articles by Schneider and Buckley).

The outward integration of (24) up to the stabilization of the $\underset{\sim}{K}$ matrix (physical infinity) corresponds to an efficient implementation of the fixed-nuclei approximation, and of the adiabatic nuclei approximation by use of the MF - LF angular transformation (12). On the contrary, whenever the FT is needed (low energy, polar molecules), the integration is stopped at r_{rot} and the boundary information stored on tape in the convenient form of the symmetric $\bar{\underset{\sim}{R}}^{\lambda}(r_{rot}\ , b)$ matrix obtained from (23b).

Another advantage of the VPM is that, in total contrast to standard integration methods, it allows a direct study of the convergence in ℓ (scattering expansion 11b) and μ (potential expansion 5b) for $r \leqslant r_{rot}$, since $\bar{K}^{\lambda}_{\ell\ell'}(r_{rot})$ gives the global effect of the potential for $r \leqslant r_{rot}$. A corollary to this is that the calculation of the inner region is *automatically* extended to higher partial waves by putting

$$\overline{K}^{\lambda}_{\ell\ell'}(r_{rot}) = 0 \qquad \text{for } \lambda,\ell > \ell_M \tag{25}$$

C. The Exact LF Equations

The VP expansion of the exact LF radial functions

$$\underset{\sim}{F}^{J\eta}(r) = [\underset{\sim}{J}_{\ell k_j}(r) - \underset{\sim}{N}_{\ell k_j}(r)\,\underset{\sim}{K}^{J\eta}(r)]\,\underset{\sim}{A}^{J\eta}(r) \tag{26a}$$

leads to the LF VP equation

$$\underset{\sim}{K}^{J\eta\,\prime} = -[\underset{\sim}{J}_{\ell k_j} - \underset{\sim}{N}_{\ell k_j}\underset{\sim}{K}^{J\eta}]\,2\underset{\sim}{V}^{J\eta}\,[\underset{\sim}{J}_{\ell k_j} - \underset{\sim}{K}^{J\eta}\underset{\sim}{N}_{\ell k_j}] \tag{26b}$$

with the couplings $\underset{\sim}{V}^{J\eta}$ given by (8c)

The initialization of the outward integration is straightforward. In a pure LF calculation ($r_{rot} = r_{exch}$), $K^{J\eta}_{\ell j\,\ell' j'}(r_{rot}) = 0$ for $r_{rot} = 0$ (regularity) or for non-penetrating partial waves; for the penetrating waves, $K^{J\eta}_{\ell j\,\ell' j'}(r_{rot})$ is deduced from the log-derivative matrix (22b) or the $\underset{\sim}{R}$ matrix (23c) of the regular inner solutions. In the FT case, the inner solution $\underset{\sim}{R}^{J\eta}$- matrices are deduced from the MF stored $\overline{R}^{\lambda}_{\ell\ell'}(r_{rot},b)$ matrices by the angular transformation

$$R^{J\eta}_{\ell j\,\ell' j'}(r_{rot},b) = \sum_{\lambda=o}^{\lambda m} \tilde{\Omega}^{\ell J\eta}_{j\lambda}\,\overline{R}^{\lambda}_{\ell\ell'}(r_{rot},b)\,\Omega^{\ell' J\eta}_{\lambda j'} \tag{27a}$$

$$= 2\,[(2j+1)(2J'P1)]^{\frac{1}{2}} \sum_{\lambda=o}^{\lambda m} \begin{pmatrix} \ell & j & J \\ 0 & \lambda & -\lambda \end{pmatrix}$$

$$\times \frac{\overline{R}^{\lambda}_{\ell\ell'}(r_{rot},b)}{1+\delta_{\lambda 0}} \begin{pmatrix} j' & \ell' & J \\ 0 & \lambda & -\lambda \end{pmatrix} \tag{27b}$$

with

$$\lambda_m = \min(\ell,\ \ell',\ J,\ \lambda_M)\ . \tag{27c}$$

Although the integration of the LF VP equations is, by far, more time consuming than the adiabatic MF ones (large number of channels, basis functions depending upon the rotational channel), the VPM approach is now well known to be among the most efficient and stable ones when the number of coupled channels becomes large and has the additional advantage of focusing on the calculation of the "observable" $\underset{\sim}{K}$-matrix, which makes it very easy to stop the time consuming integration as soon as the rate of variation of $\underset{\sim}{K}$ becomes negligible.

As for the MF problem in the inner region, the convergence in ℓ and j of the outer region LF expansion can be performed locally and the straightforward extension of the inner region boundary conditions for non penetrating partial waves (eqn. 25, 23b, 27) may be particularly useful to study the slow ℓ convergence in the case of polar molecules.

IV. FURTHER DISCUSSION OF THE VPM + FT PROCEDURE

Before discussing the practical determination of r_{rot} in the general case, let us examine how the VPM + FT procedure reaches the well known results of the full adiabatic approximation in the high energy limit and those of the effective range theory near threshold.

A. High Energy Limit

From the above physical analysis, one expects the pure adiabatic approximation ($r_{rot} \to \infty$) to be adequate at sufficiently large energies, well above all the rotational thresholds:

$$Bj(j+1) << \frac{1}{2} k_j^2 \Rightarrow \frac{1}{2} k_j^2 \simeq \frac{1}{2} k^2 \qquad \forall j = 1 \cdots\cdot j_M \ . \tag{28}$$

In its usual presentation, it corresponds to the integration of the adiabatic equations up to physical infinity ($r_{rot} = r_\infty$, effective range of the interaction at the given energy $\frac{1}{2}k^2$), followed by the FT on the $\underset{\sim}{K}$-matrix

$$\bar{K}^{J\eta}_{\ell j\ \ell' j'}(r_{rot}) = \sum_{\lambda=o}^{\lambda m} \tilde{\Omega}^{\ell J\eta}_{j\lambda}\ \bar{K}^{\lambda}_{\ell\ell'}(r_{rot})\ \Omega^{\ell' J\eta}_{\lambda j'} \tag{29a}$$

and the calculation of the rotational cross-sections, with all the channels kept degenerate for consistency

$$\frac{1}{2} k_j^2 \equiv \frac{1}{2} k^2 \qquad \forall j = 1 \cdots\cdot j_M \tag{29b}$$

The equivalence with our approach, which performs the FT on the $\underset{\sim}{R}$-matrix to insure the continuity of the logarithmic derivative at the FT matching point, relies on the assumption that all the channel energies are degenerate (29b), so that

$$J_{\ell k_j}(r_{rot} = r_\infty) \equiv \bar{J}_{\ell k}(r_{rot} = r_\infty) \tag{30a}$$

$$N_{\ell k_j}(r_{rot} = r_\infty) \equiv \bar{N}_{\ell k}(r_{rot} = r_\infty) \ . \tag{30b}$$

More precisely, the practical range of validity of the pure adiabatic approximation, in its usual form, corresponds to the range of k_j for which (29a) is practically equivalent to (27a), or following (23c), the two sets of J,N functions are practically equal at r_∞. Assuming that they have then reached their asymptotic forms (17), the condition

$$J_{\ell k_j}(r_\infty) \sim k_j^{-\frac{1}{2}} \sin (k_j r - \frac{\ell\Pi}{2}) \xrightarrow{?} \bar{J}_{\ell k}(r_\infty) = k^{-\frac{1}{2}} \sin (kr - \frac{\ell\Pi}{2}) \qquad (31a)$$

$$N_{\ell k_j}(r_\infty) \sim - k_j^{-\frac{1}{2}} \cos (k_j r - \frac{\ell\Pi}{2}) \xrightarrow{?} \bar{N}_{\ell k}(r_\infty) = - k^{-\frac{1}{2}} \cos (kr - \frac{\ell\Pi}{2}) \qquad (31b)$$

seems at first sight rather difficult to achieve (rapid variation of the phase of the trigonometric functions with k for large r). In fact, the achievement of (31) rests upon the simultaneous decrease of r_∞ and $k - k_j$ ($\simeq \frac{Bj(j+1)}{k}$) when the energy increases, so that, at sufficiently large energies:

$$(k - k_j)r_\infty \simeq \frac{Bj(j+1)r_\infty}{k} \longrightarrow 0 \text{ when } k \longrightarrow \infty . \qquad (32)$$

The *quantitative* range of validity of the adiabatic approximation is then easily deduced from (32) by the VPM + FT package, the only unknown, the effective range of the interaction r_∞ appearing clearly in the step by step outwards integration of $\underset{\sim}{K}(r)$. Furthermore, the precise quantitative range of validity of the adiabatic approximation can be determined *automatically* by the comparison of a few trial integrations of the exact LF and adiabatic MF equations, performed with the same accuracy using the same numerical technique.

B. Low Energy Limit

We will now show that, in the opposite low energy limit, the kinematical effects, neglected in the pure adiabatic approximation by applying the FT on $\underset{\sim}{K}$ rather than $\underset{\sim}{R}$, become essential.

In the molecular core region where the potential is strong ($r \leqslant r_c$), the small kinetic energy terms $\frac{1}{2}k_j^2$ can be neglected. Thus, the MF equations decouple (14) and become completely energy independent for $r \leqslant r_c = r_{rot}$. It follows that $\bar{\underset{\sim}{R}}^\lambda(r_{rot}, b)$ is energy independent, as well as $\underset{\sim}{R}^{J\eta}(r_{rot}, b)$ using (23c) and the exact channel basis $J_{\ell k_j}$, $N_{\ell k_j}$, it will reflect the well known quasi-factorizable energy dependence of the Ricatti-Bessel functions

(17) for small values of the argument $k_j r_{rot}$. Defining renormalized functions $\mathcal{J}_\ell$, $\mathcal{N}_\ell$, energy independent near the origin

$$J_{\ell k_j}(r_{rot}) = k_j^{\ell+\frac{1}{2}} \mathcal{J}_\ell(r_{rot}, k_j) \text{ with } \mathcal{J}_\ell(r,k_j) = \frac{r^{\ell+1}}{(2\ell+1)!!}\left[1 + O(k_j^2 r^2)\right] \tag{33a}$$

$$N_{\ell k_j}(r_{rot}) = k_j^{-(\ell+\frac{1}{2})} \mathcal{N}_\ell(r_{rot},k_j) \text{ with } \mathcal{N}_\ell(r,k_j) = \frac{(2\ell-1)!!}{r^\ell}\left[1 + O\,(k_j^2 r^2)\right] \tag{33b}$$

one finds the threshold energy dependence of $\underset{\sim}{K}^{J\eta}$

$$\underset{\sim}{K}^{J\eta}(r_{rot}) = \underset{\sim}{k}_j^{\ell+\frac{1}{2}} \; \underset{\sim}{\mathcal{K}}^{J\eta}(r_{rot}) \; \underset{\sim}{k}_j^{\ell+\frac{1}{2}} \tag{34a}$$

where $\underset{\sim}{\mathcal{K}}^{J\eta}(r_{rot})$ is the reduced energy independent $\underset{\sim}{K}$-matrix:

$$\underset{\sim}{\mathcal{K}}^{J\eta}(r_{rot}) = \left[\mathcal{J}_\ell - r_{rot}\, \mathcal{J}'_\ell\, \underset{\sim}{R}^{J\eta}(r_{rot},b)\right] \times \left[\mathcal{N}_\ell - r_{rot}\, \mathcal{N}'_\ell\, \underset{\sim}{R}^{J\eta}(r_{rot},b)\right]^{-1}. \tag{34b}$$

If the potential vanishes identically for $r > r_{rot}$, $\underset{\sim}{\mathcal{K}}^{J\eta}$ remains constant in the outer region and the energy dependence of the asymptotic $\underset{\sim}{K}^{J\ell}(\infty)$ matrix is unchanged (effective range result). If the potential has a long range tail, the preceeding result may be slightly modified by the residual variation of $\underset{\sim}{\mathcal{K}}^{J\eta}(r > r_{rot})$, obtained by integrating the exact LF equations (26b) for $r > r_{rot}$

C. <u>General Case</u>

The requirement (30) that at the FT point r_{rot}, the exact channel basis coincides with the degenerate adiabatic basis, allowing one to perform the FT on $\underset{\sim}{R}$ or $\underset{\sim}{K}$

$$\left.\begin{array}{l} J_{\ell k_j}(r_{rot}) \simeq \bar{J}_{\ell k}(r_{rot}) \\ N_{\ell k_j}(r_{rot}) \simeq \bar{N}_{\ell k}(r_{rot}) \end{array}\right\} \Longrightarrow \text{FT on } \underset{\sim}{R} \simeq \text{FT on } \underset{\sim}{K} \tag{35}$$

which was proposed by Chang and Fano in their original description of the FT,[2] is shown to be unnecessarily stringent in our second example. The application of the FT on $\underset{\sim}{R}$ ensures the exact matching of the log-derivative matrices, as required by the principles of quantum mechanics, and leads to a finite $r_{rot} = r_c$ (range of the strong-core potential) while the condition (35) would thus impose $r_{rot} = 0$, at least for neutral targets. For ionic targets, the J, N basis of Coulomb functions satisfying the normalization condition (16b) are known to be practically energy independent in the core region. Therefore the condition (35) is not too restrictive. This explains the success of Fano[2] and Jungen-Atabek[8] in their efficient application of the original FT to the ℓ decoupling in Rydberg levels of H_2.

In the context of our more general FT which allows our VPM + FT procedure to identify with powerful analytical techniques (effective range theory) at low energies and to generalize them, the physical condition that should guide the determination of the FT point r_{rot} is that for $r > r_{rot}$, the rate of variation of $\underset{\sim}{K}$ due to the non-degeneracy of the rotational channels should be negligible in comparison with the rate of variation of $\underset{\sim}{K}$ due to the electrostatic couplings, and also on the total rate of variation of $\underset{\sim}{K}$.

This prescription is easily converted into practice in the VPM + FT procedure. Indeed, a rough localization of r_{rot} is first obtained by determining the point where the largest splitting between two adjacent rotational levels becomes comparable to the tail of the electrostatic potential, dominated by the longest range permanent dipole field in the case of a polar molecule

$$2Bj_M \sim \frac{D}{r^2} \quad \Rightarrow \quad r_{rot}^{(o)} \sim \sqrt{\frac{D}{2Bj_M}} \tag{36a}$$

A more precise value of r_{rot} is then obtained by comparing, for a given total angular momentum J, the total rate of variation of $\underset{\sim}{K}$

$$\underset{\sim}{K}^{J\eta\,\prime} = -[\underset{\sim}{J}_{\ell k_j} - \underset{\sim}{K}^{J\eta}\,\underset{\sim}{N}_{\ell k_j}]\; 2\underset{\sim}{V}^{J\eta}\,[\underset{\sim}{J}_{\ell k_j} - \underset{\sim}{N}_{\ell k_j}\,\underset{\sim}{K}^{J\eta}] \tag{36b}$$

with the approximate rate of variation of $\underset{\sim}{K}$, neglecting the rotational splittings

$$\overline{\underset{\sim}{K}^{J\eta\,\prime}} = -\,[\bar{\underset{\sim}{J}}_{\ell k} - \underset{\sim}{K}^{J\eta}\,\bar{\underset{\sim}{N}}_{\ell k}]\; 2\underset{\sim}{V}^{J\eta}\,[\bar{\underset{\sim}{J}}_{\ell k} - \bar{\underset{\sim}{N}}_{\ell k}\,\underset{\sim}{K}^{J\eta}] \quad . \tag{36c}$$

This comparison, easily implemented in the VPM + FT procedure, allows us to refine the initial static estimation (36a) by taking into account the kinematical effects (through the J, N multiplicative factors in 36b, c). The intuitively expected influence of the collisional interaction is therefore properly included.

V. CONCLUSION

We have pointed out three important contributions of the new VPM to the implementation of the FT theory of rotational excitation of diatomic molecules by electron impact.

The most fundamental one concerns the general formulation of the FT which should be applied to the log-derivative matrix, whereas previous formulations[2,8] (applying the FT on the variable phase $\underset{\sim}{K}$ and amplitude $\underset{\sim}{A}$ matrices) led to a significant loss of generality (for neutral targets, at low energy in particular).

The second important formal contribution of the VPM + FT procedure is its polyvalence, partly illustrated in IV by showing how it identifies itself with well known, but completely independent approaches (the adiabatic - nuclei approximation at high energy, the effective range theory near threshold) and builds up a continuous range of approximations relating them.

The last, but not least *contributions* of the new VPM + FT procedure are *computational*. Its *flexibility*, which allows one to solve a given problem, using a continuous range of approximations between the exact rotational CC treatment and the pure adiabatic-nuclei approximation, based on the same numerical techniques, is essential in the comparison of numerical results, in order to draw unambiguous physical conclusions. Its *efficiency*, already well established in the electron-atom case,[5] has been further demonstrated here. Its *superiority* w.r.t. standard numerical techniques is obvious and, in waiting for practical comparison, its pure numerical efficiency is expected to be similar to that of mathematically related techniques.[7,9] Already, the lack of efficiency associated with the use of a Bessel J, N basis, suspected by Shimamura,[9] seems to us rather irrelevant, since it would be straightforward to implement the VPM with a standard sine, cosine basis (by including the centrifugal barrier in the asymptotically vanishing coupling potentials, gathered on the right hand side of eqn. (15)). However, this remark - and its straightforward remede - may not be relevant : from the pure computational point of view, we have noticed that, at least in the adiabatic MF equations, all the $J_{\ell k}$, $N_{\ell k}$ functions, degenerate in k , can be efficiently generated by a mere recurrence on ℓ. Moreover, the use of the Bessel functions, which diagonalize a larger fraction of the initial equations, leads to a weaker variation of the corresponding variable phase $\underset{\sim}{K}(r)$ and eventually amplitude $\underset{\sim}{A}(r)$ matrices, therefore to a faster integration. Again, this slower variation of $\underset{\sim}{K}(r)$ towards its asymptotic value (the physical $\underset{\sim}{K}$ matrix) is essential to preserve the help of the VPM in the delicate convergence studies (in μ, ℓ, j) occuring in electron-molecule problems and the determination of the FT point r_{rot}.

Preliminary numerical checks of the VPM + FT package by comparison with Chandra's FT results on e^- - CO[3] have recently been discussed by Le Dourneuf and Vo Ky Lan.[10] Systematic illustration in the power of the VPM + FT implementation described here, as well as further extension of the adiabatic model, are currently prepared for publication.[11] Current applications to a systematic study of model problems, as well as to polar molecules (HCL, HF) are in progress.

The authors are grateful to the organizers of the Asilomar workshop for their hospitality and financial support. Special thanks are due to U. Fano who pointed out to us a few years ago the fundamental interest of the VPM formalism. We are also indepted to N. Chandra, E. Chang and A. Temkin for useful discussions.

REFERENCES

1. A. M. Arthurs and A. Dalgarno, Proc. Roy. Soc. A 256, 540 (1960).
2. U. Fano, Phys. Rev. A2, 353 (1970); E. S. Chang and U. Fano, Phys. Rev. A6, 173 (1972).
3. N. Chandra, Phys. Rev. A16, 80 (1977).
4. F. Calogero, "Variable Phase Approach to Potential Scattering" (New York: Academic Press, 1967).
5. M. Le Dourneuf and Vo Ky Lan, J. Phys. B10, L 35 (1977).
6. B.R. Johnson, J. Comput. Phys. 13, 445 (1973).
7. J. C. Light and R. B. Walker, J. Chem. Phys. 65, 4272 (1976). B. I. Schneider and R. B. Walker, submitted to J. Chem. Phys. (1978).
8. C. Jungen and O. Atabek, J. Chem. Phys. 66, 5584 (1977).
9. I. Shimamura, 1977, Proc. CECAM Workshop on Electron-Molecule Scattering, Meudon France (1977), p.69.
10. M. Le Dourneuf and Vo Ky Lan, Proc. CECAM Workshop on Electron-Scattering, Meudon France (1977), P. 17.
11. Vo Ky Lan and M. Le Dourneuf, to be published in J. Phys. B.

THE R-MATRIX METHOD FOR ELECTRON-MOLECULE SCATTERING: THEORY AND COMPUTATION

Barry I. Schneider

Los Alamos Scientific Laboratory
Theoretical Division
Los Alamos, New Mexico 87545

I. INTRODUCTION

The R-matrix method has proven itself to be a useful tool for the study of the collisions of electrons with diatomic molecules. In principle the technique can be extended to treat polyatomic species but no applications have been reported to date. The idea behind the method is to divide configuration space into regions or boxes, solve the Schrodinger equation in each box separately and to then match the solutions on the surface bounding adjacent regions. The motivation for this division of space is the recognition that the dominant physics governing the behavior of the incident particle may be different in each spatial region. This allows us to develop numerical methods which are best suited for each box. In the electron-molecule scattering problem when the incident electron is "near" the target electrons and nuclei it is subject to strong, non-central electrostatic forces. As a consequence we must take proper account of electron correlation and anti-symmetry in this internal region. The problem becomes one of calculating the electronic wave-function of a compound (N+1) particle system and is therefore quite similar to the molecular structure problems treated by quantum chemists. When the incident electron is "far" away from the target it is subject to weak electrostatic forces of multi-polar form and rotational and vibrational coupling terms which may now be of comparable strength to the electrostatic terms. Since the electron is far away and thus distinguishable from the target electrons it is possible to ignore anti-symmetry of target and incident particle. The neglect of anti-symmetry leads to a very important simplification; there are no longer any non-local potentials in the problem. The Schrodinger equation can be reduced to the solution of a set of coupled <u>differential</u> equations. The number of channels which must be included in the external region is of course infinite, but in practice can be truncated to a manageable set of equations. If rotational and vibrational terms are smaller than the electrostatic forces in the external region it is possible to reduce the complexity of the problem even further. In any case the solution of even a large set of coupled differential equations on a modern day computer is a manageable task.

In Section II we discuss the R-matrix method using the operator technique introduced by C. Bloch in nuclear reactions.[1] Section III is devoted to a detailed exposition of the computational procedure developed by the author to implement the formalism given in Section II. Since the workshop is intended to discuss and compare various methodologies, I will not dwell on specific applications in this article.[2-8] For completeness I have included these and some notes on work in progress in the reference section at the end of the article.

II. THEORY

The first step in any R-matrix theory is to divide space up into two or more regions. In most applications two regions are distinguished: an internal and external region. In the external region the particles of the target are assumed to be distinguishable from those of the projectile. The surface which separates the two regions is usually taken to be spherical but may be chosen in other ways. For example, in the scattering of electrons from homo-nuclear diatomic molecules the use of prolate spheroidal co-ordinates is a natural choice.[3] Such a co-ordinate system leads to some complications but these may be dealt with in actual applications. For simplicity in what follows we assume a spherical boundary.

If we consider the Schrodinger equation in the interior region, where exchange and correlation play a central role, it is important to realize that the Hamiltonian operator may not be self-adjoint when operating on some arbitrary set of square integrable functions within the space. This property arises because the boundary conditions satisfied by these functions may give rise to a non-vanishing contribution on the surface enclosing the internal region. Since non-self-adjoint operators are difficult to deal with in practice, Bloch devised an elegant technique to circumvent the problem.[1] By defining a singular boundary value operator called L_b, the Schrodinger equation in the internal region may be written as,

$$(H + L_b - E)|\Psi_E) = L_b|\Psi_E) \tag{1a}$$

where

$$L_b = \sum_c |c)\delta(r - a)(\frac{\partial}{\partial r} - b)(c| \tag{1b}$$

The effect of L_b is to cancel any non-vanishing surface terms arising from the arbitrary boundary conditions satisfied by the set of internal states. The channel functions, $|c)$, are a complete set of states in the space of all coordinates save r, where r is the radial distance between the incident electron and center of mass of the target. This radial coordinate is constrained to be less than or equal to the radius a of the internal region. For a diatomic molecule in the molecular frame the states $|c)$ are labelled by the quantum numbers i, S^2, S_z and Ω where:

$$\begin{aligned} i &= \text{internal state of target} \\ S^2 &= \text{total spin of composite system} \\ S_z &= \text{projection of total spin on z axis} \\ \Omega &= \text{projection of angular momentum on internuclear axis.} \end{aligned} \tag{2}$$

We formally solve equation (1a) as

$$|\Psi_E) = (H + L_b - E)^{-1} L_b |\Psi_E). \tag{3}$$

If we now introduce an arbitrary set of expansion functions, $|k)$, into equation (3) we obtain

$$|\Psi_E) = \sum_{k,k'} |\Psi_k)(\Psi_k|(H - L_b - E)^{-1}|\Psi_{k'})(\Psi_{k'}|L_b|\Psi_E). \tag{4}$$

In order to extract the scattering information we must project equation (4) onto the space of the channel states:

$$((C|\Psi_E)) = \sum_{k,k'} ((C|\Psi_k))(\Psi_k|(H + L_b - E)^{-1}|\Psi_k')(\Psi_k'|L_b|\Psi_E) \tag{5}$$

The double bracket notation denotes a partial integration over all but the radial scattering co-ordinate. The matrix element $(\Psi_k|L_b|\Psi_E)$ can be written as,

$$(\psi_k|L_b|\Psi_E) = \sum_C ((\psi_k|C))_a [(\frac{\partial}{\partial r} - b)((C|\Psi_E))]_a \tag{6}$$

If we define,

$$((C|\Psi_E)) = F_{CE}(r) \tag{7a}$$

$$((C|\Psi_E))_a = F_{CE}(a) \tag{7b}$$

we obtain

$$F_{CE}(r) = \sum_{k,k',C'} F_{Ck}(r)(\psi_k|(H + L_b - E)^{-1}|\psi_{k'})F^{*}_{CK'}(a)$$

$$\times [(\frac{\partial}{\partial r} - b)F_{C'E}(r)]_a \quad . \tag{8}$$

Setting r = a on both sides of equation (8) gives

$$F_{CE}(a) = \sum_{k,k',C'} F_{ck}(a)(\psi_k|(H + L_b - E)^{-1}|\psi_k')F^{*}_{c'k'}(a)$$

$$\times \left[\frac{\partial F_{C'F}}{\partial r}\Big|_a - bF_{C'E}(a)\right] . \tag{9}$$

We define the R-matrix as

$$R_{CC'} = \sum_{k,k'} F_{Ck}(a)(\psi_k|(H + L_b - E)^{-1}|\psi_k')F^{*}_{C'k'}(a) \tag{10}$$

then to obtain

$$F_{CE}(a) = \sum_{C'} R_{CC'} \left[\frac{\partial F_{C'E}}{\partial r}\Big|_a - bF_{C'E}(a)\right] . \tag{11}$$

This set of linear equations may be solved once the asymptotic form of F_{CE} and its derivative are specified. In order to make this specification we must be able to solve the Schrodinger equation in the exterior region. We shall return to this point a bit later in the discussion. The central problem in the internal region is the calculation of $R_{CC'}$, the R-matrix. There are two possibilities: we may diagonalize $(H + L_b)$ to obtain a spectral representation of $R_{CC'}$,

$$R_{CC'} = \sum_{\lambda} \frac{F_{C\lambda}(a)F^{*}_{C'\lambda}(a)}{E_\lambda - E} \tag{12a}$$

or solve the set of linear equations

$$\sum_{k'} (\psi_k|(H + L_b - E)|\psi_{k'})\, X_{k'C} = F^{*}_{Ck}(a) \tag{12b}$$

to get

$$R_{CC'} = \sum_{k} F_{Ck}(a)\, X_{kC'} \quad . \tag{12c}$$

In either case the bulk of the calculation is reduced to the computation of the matrix elements of $(H + L_b)$. The main advantage of using the spectral form is that a single diagonalization of $(H + L_b)$ gives the R-matrix trivially for all energies. The linear equations must be solved anew for each incident energy. We may divide the construction of the $(H + L_b)$ matrix into three major parts; integral evaluation, integral transformation and Hamiltonian formation. These three steps are, of course, inter-related. For example the form of the scattering wavefunction can limit the type of configurations and Hamiltonian matrix elements one needs to construct. This in turn has important implications in the integral and transformation steps. Let us explore this in more detail. The most general form which may be used to expand the scattering wavefunction can be written as

$$|\Psi_E) = \sum_{i,\alpha} C_{Ei\alpha} A(\Phi_i(1\ldots N)\phi_{i\alpha}(N+1)) + \sum_q C_{Eq}\Psi_q(1\ldots N+1) \tag{13}$$

The first term on the right hand side of the equation (13) couples the incident electron to all open channels and pseudostates needed to represent important long-range correlations. The second term accounts for short range correlations and is assumed to vanish on the R-matrix surface. We guarantee the orthonormality of these (N+1) particle states by imposing the following conditions on the basis:

$$< \Phi_i(1\ldots N)|\Phi_j(1\ldots N) > = \delta_{ij} \tag{14a}$$

$$< \phi_{i\alpha}(1)|\phi_{i\beta}(1) > \quad \delta_{\alpha\beta} \tag{14b}$$

$$\int \Phi_i^*(1\ldots N)\phi_{j\alpha}(1)d\vec{r}_1 = 0 \tag{14c}$$

$$\int \Psi_q^*(1\ldots N+1)\phi_{i\alpha}(1)\, d\vec{r}_1 = 0 \tag{14d}$$

This is accomplished in practice by dividing the one particle atomic basis into an "inner" and an "outer" space. The inner basis contains all those atomic orbitals necessary to describe the N electron target states, pseudostates and Ψ_q. This basis is typically the size of the basis sets used in ordinary bound state configuration interaction (CI) calculations plus a few pseudo-orbitals to represent long range correlations. The outer basis set, which may be quite large, contains all the additional orbitals necessary to accurately represent the scattering function. The outer basis only appears in the scattering orbital, $\phi_{j\alpha}$. This division of function space is motivated by the physical nature of the scattering problem. The incident electron is not bound and does not decay exponentially. Moreover, it has a significant fraction of its amplitude outside the core and valence region of the molecule. Consequently it becomes necessary to add a large, diffuse set of atomic orbitals to account for this additional degree of freedom. The structure of the wavefunction has two very important implications. First, no two electron integrals which have more than two outer space atomic basis functions need to be computed. Second, the most time consuming step in the calculation, the transformation from atomic to molecular orbitals, can be simplified and speeded up. In all of the applications of the R-matrix method to molecules both the inner and outer basis sets have been chosen as either Slater[6] or Gaussian orbitals.[2-5] There is little doubt that these functions are capable of representing the type of physical effects we associate with our inner orbital space. Their ability to represent the physics of the outer function space is more questionable. Calculations of high quality with such basis sets for the scattering functions have only been possible using very large sets with diffuse exponents. A better approach would be to augment the more usual Slater or Gaussian basis with a set of orthogonal polynomials. Polynomials have already been used successfully in atomic R-matrix calculations. If they are to be used in the molecular case it will be necessary to devise efficient techniques to integrate the two electron integrals arising from the mixed basis. For linear molecules the charge distribution method can provide the necessary route. For arbitrary polyatomic molecules the situation is more difficult and much work needs to be done.

The transformation from atomic to molecular integrals is typically the most time consuming step in the entire calculation. The reason for this is quite simple. It is intrinsically an N^5 process and it requires a great deal of input and output for large basis sets. In the scattering calculation we are aided by requiring only a subset of all possible two electron integrals. By designing transformation programs to handle this special case it is possible to perform the calculations with reasonable efficiency.

The final step in the computation involves the construction of the Hamiltonian matrix from the transformed integrals. Since there are a number of good, general, methods and computer codes available to accomplish this task I will not dwell on this point here. It is a relatively easy task to take the Hamiltonian matrix over an appropriate set of (N+1) electron determinants or spin eigenfunctions and put it in the form implied by equation (13) using matrix multiplication. All that is required is the expansion coefficients of the N electron target and pseudostates in the N electron determinantal basis and the spin/angular coupling coefficients.

In order to extract the scattering information from the R-matrix calculation we must solve a set of coupled differential equations of the form

$$\frac{d^2}{dr^2} F_{CE}(r) + \sum_{C'} V_{CC'}(r) F_{C'E}(r) = 0 \tag{15}$$

These equations may be solved using a variety of techniques including inward integration, the variable phase method[8] and the R-matrix[7] propagation technique. The latter two methods have proven to be best in actual applications. They are efficient and have no difficulty with strongly closed channels. In both approaches the coupled equations are re-cast in a form which propagates the R or K matrix on the surface of the box to "infinity". The scattering phases can then be easily extracted using equation (11) with the box radius a replaced by the true asymptotic value of r. In most cases this requires the evaluation of Bessel or Coulomb functions and their derivatives at the asymptotic matching point.

III. COMPUTATIONAL PROCEDURES

In undertaking an R-matrix calculation for electron-diatomic scattering the first step is to decide what level of approximation is to be employed in the computation. The simplest approximation which has proven to be reliable for elastic scattering is the static-exchange or Hartree-Fock approximation. The first step in such a calculation is to find the set of occupied molecular orbitals. These functions may be described very well using the inner basis set. For first row molecules a basis set of double zeta plus one polarization function is usually sufficient. The second step is to agument the inner basis with a large set of outer functions. These functions are chosen to provide a good description of the scattering orbital. Since we work in the molecular frame, where Ω is a good quantum number, the added functions need only describe a single symmetry in each calculation. Thus for example, if we are computing the σ_g contribution to elastic $e + N_2$ scattering the added basis set need only contain even partial waves about the center of mass of the molecule. If the outer basis set contains functions which are centered at the nuclei only the gerade combinations would be needed for σ_g type scattering. However, it is often more economical as well as being physically more reasonable to put the added set at the center of mass of the diatomic. If this approach is taken, only even or odd partial wave functions would have to be added for a homonuclear calculation. At low enough energy ($\leq$ 12 ev) the centrifugal barrier and the absence of even-odd partial wave coupling limit us in practice to needing no more than f-type basis functions. In many cases including only up to $\ell = 2$ in the outer basis gives accurate results. For strongly polar heteronuclear molecules the situation is more complicated. Many partial waves couple and it is possible that quite high ℓ values may be needed to provide an adequate description of the scattering. Once the outer basis set is chosen the full set of integrals needed to construct the Hartree-Fock potential can be calculated. As was mentioned earlier a great savings in time and storage can be effected by only computing those two electron integrals which have no more than two orbitals in the outer space. A matrix representation of the Hartree-Fock potential is then constructed using the calculated integrals. Since only Coulomb and exchange matrix elements are needed the transformation is much less expensive than that required for a CI calculation. In the method developed by the author a further transformation is made to a basis of floating Gaussians defined in a prolate spheroidal co-ordinate system. The Hamiltonian matrix is then diagonalized in the prolate basis and projected onto the channel functions. In the prolate system the channel states are the spheroidal angular functions $S_{\ell m}$. The R-matrix is obtained using the spectral representation of equation (12a) and used to extract the scattering information via equation (11).

In going beyond the static-exchange approximation there are a number of practical questions which must be addressed. Perhaps the most important of these is the need to transform a set of two electron integrals which have two orbitals of different symmetries in the outer space. Since the outer space is already quite large in the static-exchange approximation the expense of doing the multichannel calculation could become quite prohibitive. Basis sets such as orthogonal polynomials, which are better suited to the description of scattering functions, could alleviate this problem by reducing the size of the outer space. The size of the inner space is also increased by inclusion of excited state and polarization-type molecular orbitals. This increase, however, is only a moderate one and does not appear to present any real difficulty in practical cases. The excited orbitals are determined using the frozen Hartree-Fock core of the (N-1) electron system. These orbitals, called IVO (improved virtual orbitals) by quantum chemists, are an excellent starting point for CI calculations. The polarization functions are determined by solving the Hartree-Fock equations in the presence of an extra electron frozen at large distances from the N electron target. Such orbitals, which are closely related to coupled Hartree-Fock functions, given an excellent description of the multipole forces induced in the target by a slow incident electron. Having chosen the inner orbital space we now diagonalize the N electron target Hamiltonian in the basis to get our N electron target and pseudostates, Φ_i (1...N). The scattering orbitals, taken, for example, from a prior static-exchange calculation, are then orthogonalized to the inner set and used to form the configurations $A(\Phi_i(1...N)\phi_i(N+1))$. The correlation configurations $\Psi_q(1...N+1)$ are constructed using (N+1) electron anti-symmetrized products of only inner space molecular orbitals. Since they have at least one orbital which is different from $\phi_{i\alpha}$ they are strongly orthogonal to the open pseudostate terms. It is important to include in $\Psi_q(1...N+1)$ all those configurations which replace $\phi_{i\alpha}$ in the $A(\Phi_i(1...N)\phi_{i\alpha}(N+1)$ with one of the inner space orbitals. These terms remove the orthogonality restriction on the scattering orbitals and in many cases give rise to the core excited resonances seen in scattering experiments. The Hamiltonian matrix is constructed over elementary spin eigenfunctions of the (N+1) electron system and then transformed to the physical basis using the coefficients of the N electron diagonalization. By proceeding in this fashion, we do not allow the target states to over-correlate in the presence of the extra electron, and in addition, reduce the size of the Hamiltonian matrix which needs to be diagonalized. The R-matrix is constructed from the eigenfunctions using a trivial generalization of the method used in the static-exchange approximation. The extraction of the scattering information is complicated by the presence of excited electronic levels which increase the number of channels to be included in the outer region. However, the same numerical integration code may be used in the multi-channel problem as in the static-exchange problem. In our programs we have chosen the R-matrix propagation method of Light and Walker[7] because of its speed and stability. The final results of the calculation, the K-matrix elements, are easily used to compute the total or differential cross section.

In concluding, I would like to take this opportunity to thank a number of people for the contributions they have made to the progress of the molecular R-matrix method. These include P. Jeffrey Hay for invaluable guidance and help in using the molecular structure codes at Los Alamos, Robert B. Walker for providing the R-matrix propagation subroutines and Michael Morrison and Lee Collins for valuable discussions and applications of the technique to N_2 and polar molecules.

I would also like to mention that work in progress includes the application of the R-matrix method to electron-polar molecule collisions. (B. Schneider and L. Collins.) This work includes calculations on model finite dipoles as well as static and static-exchange calculations of LiF.

Work performed under the auspices of the U. S. Department of Energy by the Los Alamos Scientific Laboratory under contract number W-7405-ENG-36.

REFERENCES

1. C. Bloch, Nuc. Phys. 4, 503 (1957).
2. B. Schneider, Chem. Phys. Letts., 31, 237 (1975).
3. B. Schneider, Phys. Rev. A11, 1957 (1975).
4. B. Schneider and P. J. Hay, Phys. Rev. A13, 2049 (1976).
5. B. Schneider and M. A. Morrison, Phys. Rev., A16, 1003 (1977).
6. P. G. Burke, I. Mackey and I. Shimamura, J. Phys. B, 10, 2497 (1977).
7. J. C. Light and R. B. Walker, J. Chem. Phys., 65, 4272 (1976).
8. M. LeDourneuf and Vo Ky Lan, J. Phys. B, 10, L35 (1977).

DISCUSSION

Heller: You once told me that you were sort of getting the Buttle correction automatically by using a basis set which didn't satisfy fixed boundary conditions on the surface. Could you elaborate on that?

Schneider: There are two aspects to this problem. There's the boundary condition aspect of the problem, which Buttle recognized, but there's another aspect which is just having enough poles in the R-matrix in the region you're interested in which are accurately represented. Now I'm arguing that using static-exchange poles in the coupled-channel R-matrix is a lot better than using free-particle states the way you would in the Buttle correction.

Heller: You're arguing for a sort of distorted wave treatment of the higher poles.

Schneider: That's it exactly. Now whether it is still necessary to do the last step, and complete the sum to infinity with the Buttle correction, is something I've found to be unimportant. The boundary condition problem is well taken care of by the Bloch operator.

Lane: You argue that the higher energy poles really are independent of correlation effects and that the static-exchange approximation ought to do a good job there. Why not just use the static approximation?

Schneider: Remember it's not any more difficult to do a static-exchange calculation than a static calculation when you're putting the Hamiltonian on a basis set. The static poles may be fine.

THE T-MATRIX METHOD IN ELECTRON-MOLECULE SCATTERING

A. W. Fliflet

A. A. Noyes Laboratory of Chemical Physics
California Institute of Technology
Pasadena, California 91125

I. INTRODUCTION

The current need for the development of accurate ab-initio methods for electron-molecule scattering calculations, and the difficulty of applying numerical methods, has lead to renewed interest in discrete-basis-set methods. The application of discrete-basis-set methods is further suggested by the success of these methods in molecular bound state calculations. Although the standard albegraic variational methods of Hulthen,[1] Kohn,[2] and Rubinow[3] have been studied extensively for electron-atom collisions, serious technical difficulties have delayed application of these methods to electron-molecule collision processes. Recently, several other discrete-basis-set methods have been studied which seem better suited to electron-molecule scattering. These methods avoid or minimize the need to compute free-free and bound-free interaction matrix elements since these are difficult to evaluate in the absence of spherical symmetry. The new methods also appear to be free from the spurious resonances which occur in the standard variational methods.

In the separable potential approach the potential is projected onto a subspace of square-integrable functions to form

$$U^t = \sum_{\alpha,\beta=1}^{N} |\alpha><\alpha|U|\beta><\beta|, \tag{1}$$

and the scattering problem is solved exactly for U^t. This is the approach used in the J-matrix method of Heller and Yamani[4] and the T-matrix method for electron-molecule scattering introduced by Rescigno, McCurdy and McKoy.[5] The subject of this article is the T-matrix method in the form developed by McKoy and coworkers.

In the T-matrix method, the scattering problem is expressed by the Lippmann-Schwinger equation for the transition matrix,

$$T = U + U\, G_o\, T \tag{2}$$

In actual calculations it is more convenient to work with the K' matrix, which is obtained by using the principal-value free-particle Green's function G_o^P in Eq. (2). The on-shell components of K are related to the Cayley transform of the S-matrix, i.e.,

$$S = \frac{1 + iK}{1 - iK}, \tag{3}$$

by $K = -\frac{\pi}{2} K'$. Inserting the separable potential U^t into Eq. (2), and using G_o^P, leads to the finite matrix equation

$$< \alpha|K'|\beta> = < \alpha|U|\beta > + \sum_{\gamma,\delta=1}^{N} < \alpha|U|\gamma > < \gamma|G_o^P|\delta > <\delta|K|\beta> \tag{4}$$

which has the solution

$$K^t = (1 - U^t G_o^P)^{-1} U^t \tag{5}$$

The on-shell partial-wave K-matrix is obtained by the transformation

$$K^{t'}_{\ell\ell'mm'}(k) = -k \sum_{\alpha\beta} < k\ell m|\alpha > < \alpha|K|\beta > < \beta|k\ell'm'> . \tag{6}$$

The truncated potential U^t can be constructed for a multi-center Gaussian basis set using standard molecular-bound-state computer codes. Equation (4) and (5) involve Gaussian matrix elements of the free-particle Green's function. As shown by Ostlund[6] and by Levin et al,[7] these matrix elements can be reduced to expressions involving the complex error function for which efficient algorithms exist.[8] The Bessel-Gaussian overlap matrix elements occurring in Eq. (6) can also be reduced analytically to closed-form expressions.[9] Thus all the matrix elements needed to compute the partial-wave K-matrix for the truncated multi-center potential U^t can be obtained without numerical quadrature, even for polyatomic targets.

By solving Eq. (4), one obtains scattering information without calculating the scattering wave function for the associated Schrodinger equation (in a.u.)

$$(-\nabla^2 + U^t - k^2)\, \Psi^t_{\underset{\sim}{k}} = 0. \tag{7}$$

However, it is often desirable to have a representation of the wave function itself. An important application is the correction of $\langle k\ell m|K|k\ell'm'\rangle^t$ through first order for errors due to the difference $U - U^t$. This involves the variational formula[10]

$$K^s_{\ell\ell'mm'} = K^t_{\ell\ell'mm'} + k \langle \Psi^t_{k\ell m}|(U - U^t)|\Psi^t_{k\ell'm'}\rangle \tag{8}$$

Equation (8) for the variationally stable partial-wave K-matrix is analogous to Kohn's formula for the scattering amplitude in three dimensions.[2] A representation of the wavefunction is also of interest for the calculation of distorted-wave-approximation matrix elements which occur in electronic excitation by electron impact and other electron-molecule continuum processes. These applications are discussed in this paper. The calculation of variationally stable K-matrix elements for e^--H_2 scattering in the static-exchange approximation is the subject of Section II.[10] Section III describes an approach for electronically inelastic scattering in the distorted-wave approximation. Application is made to the excitation of the $B^1\Sigma_u^+$ state of H_2.[11]

II. ELASTIC SCATTERING

In the fixed-nuclei approximation the Schrodinger equation for an elastically scattered electron is of the form

$$[-\nabla^2 + U(R,\underline{r}) - k^2]\, \Psi_{\underline{k}}(\underline{r}) = 0 \tag{9}$$

where $U(R,\underline{r})$ is an optical potential for the effective interaction between the target and the scattered electron. The potential depends parametrically on the relative coordinates of the target nuclei, denoted by R. The vector subscript $\underline{k}$ indicates the dependence of the wavefunction on the directions as well as the magnitude of the incident momentum. The incident-direction dependence of the scattering wafefunction may be expanded in the partial-wave series

$$\Psi_{\underline{k}}(\underline{r}) = \left(\frac{2}{\pi}\right)^{\frac{1}{2}} \sum_{\ell m} i^{\ell}\, \Psi_{k\ell m}(\underline{r})\, Y^*_{\ell m}(\hat{k}). \tag{10}$$

The function $\Psi_{k\ell m}(\underline{r})$ is the scattering wavefunction when a particular incident partial wave is specified. For a linear target with internuclear axis along the z axis, $\Psi_{k\ell m}(\underline{r})$ may in turn be expanded in the partial-wave series

$$\Psi_{k\ell m}(\underline{r}) = \sum_{\ell'} g_{\ell\ell'm}(k,r)\, Y_{\ell'm}(\hat{r}), \tag{11}$$

where the radial continuum functions go asymptotically as,

$$g_{\ell\ell'm}(k,r) \to j_{\ell'}(kr)\delta_{\ell\ell'} - K_{\ell\ell'm}\, y_{\ell'}(kr), \tag{12}$$

$j_\ell(k,r)$ and $y_\ell(kr)$ are spherical Bessel functions and $K_{\ell\ell'm}$ is related to the plane-wave representation of the K-matrix:

$$K_{\underline{k}',\underline{k}} = -\frac{1}{k} \sum_{\ell\ell'mm'} i^{\ell-\ell} K_{\ell'\ell m}(k) \; Y_{\ell'm'}(\hat{k}') \; Y^*_{\ell m}(\hat{k}) . \tag{13}$$

If the exact potential is replaced by the truncated potential U^t, the scattering wavefunction satisfies Eq. (7) and, equivalently, the Lippmann-Schwinger equation for the wavefunction,

$$\Psi^t_{k\ell m} = \phi_{k\ell m} + G^P_o(k)\; U^t \Psi^t_{k\ell m} , \tag{14}$$

where $\phi_{k\ell m} \equiv j_\ell(kr)\; Y_{\ell m}(\hat{r})$. Substituting the identity

$$K^{t'} \phi_{k\ell m} = U^t \Psi^t_{k\ell m} \tag{15}$$

into Eq. (14) yields an expression for $\Psi^t_{k\ell m}$ in terms of the solution of Eq. (4):

$$\Psi^t_{k\ell m} = \phi_{k\ell m} + G^P_o(k)\; K^{t'} \phi_{k\ell m} . \tag{16}$$

Substituting the partial-wave expansion of the principal-value free-particle Green's function

$$G^P_o(k,\underline{r},r') = \sum_{\ell m} k\; j_\ell(kr_<)\; y_\ell(kr_>)\; Y_{\ell m}(\hat{r})\; Y_{\ell m}(\hat{r}') \tag{17}$$

into Eq. (16) leads to the asymptotic form

$$\Psi^t_{k\ell m}(r) \rightarrow \sum_{\ell'} [j_{\ell'}(kr)\; \delta_{\ell\ell'} - y_{\ell'}(kr)\; K^t_{\ell'\ell m}(k)]\; Y_{\ell'm}(\hat{r}) \tag{18}$$

as $r \rightarrow \infty$. Comparison of Eqs. (11), (12) and (18) shows that $\Psi^t_{k\ell m}$ satisfies the same asymptotic boundary conditions as the exact scattering function except for the replacement of $K_{\ell'\ell m}$ by $K^t_{\ell'\ell m}$.

To calculate a numerical representation of $\Psi^t_{k\ell m}$, Eq. (15) is substituted into Eq. (7) yielding the inhomogeneous equation

$$(-\nabla^2 - k^2)\; \Psi^t_{k\ell m} = -K^{t'} \phi_{k\ell m} . \tag{19}$$

Substituting the single-center expansion

$$\Psi^t_{k\ell m} = \sum_{\ell'} g^t_{\ell\ell'm}(k,r)\; Y_{\ell'm}(\hat{r}) \tag{20}$$

into Eq. (19), multiplying on the left by $Y^*_{\ell'm}(\hat{r})$, and integrating with respect to $\hat{r}$ leads to a set of <u>uncoupled</u> ordinary differential equations,

$$\left(- \frac{d^2}{dr^2} + \frac{\ell'(\ell'+1)}{r^2} - k^2 \right) r g^t_{\ell\ell' m}(k,r) = - r \langle Y_{\ell' m} | K^{t'} | k\ell m \rangle \quad (21)$$

where

$$\langle Y_{\ell' m} | K^{t'} | k\ell m \rangle = \sum_{\alpha\beta} \langle Y_{\ell' m} | \alpha \rangle \langle \alpha | K | \beta \rangle \langle \beta | j_\ell Y_{\ell m} \rangle \quad (22)$$

A prescription for obtaining a numerical solution of Eq. (21) subject to the boundary conditions

$$\lim_{r\to 0} r g^t_{\ell\ell' m}(k,r) = 0 \quad (23a)$$

and

$$g^t_{\ell\ell' m}(k,r) \to j_\ell(kr)\, \delta_{\ell\ell'} - y_{\ell'}(kr)\, K^t_{\ell'\ell m} \quad (23b)$$

as $r \to \infty$ is given by Fliflet and McKoy.[10] As discussed in references 9 and 12, the matrix elements $\langle Y_{\ell m} | \alpha \rangle$ and $\langle \alpha | j_\ell Y_{\ell m} \rangle$ can be evaluated analytically for arbitrary Gaussians.

This technique for calculating electron-molecule continuum wavefunctions involves considerably less computational effort than direct numerical integration of the Schrodinger equation for the exact potential U, particularly when exchange is included. Moreover, it avoids the severe convergence problem which occurs when the dynamical solution is obtained in terms of a single-center expansion. The price for computational simplicity relative to an exact numerical solution of the Schrodinger equation is the lack of point-by-point varaiational stability in the functions $g^t_{\ell\ell' m}(k,r)$. However, the function $g^t_{\ell\ell' m}$ and the K-matrix element $K^t_{\ell\ell' m}$ converge to the exact results as the basis set approaches completeness.

One does not expect to find pseudoresonances in the behavior of the approximate K-matrix elements of the type which occur in the standard variational methods of Hulthen, Kohn, and Rubinow.[13] In these methods pseudoresonances occur as a result of approximating the branch cut of the free-particle Green's function by a finite number of simple poles. In the present method the free-particle Green's function is treated exactly. This feature of our approach is discussed by Heller and Yamani in the context of the J-matrix method.[4] For an extensive discussion of algebraic variational methods, see Truhlar et al.[14]

The variational formula for the partial-wave K-matrix [Eq. (8)] is verified in Fliflet and McKoy.[10] The distorted-wave approximation form given in Eq. (8) follows from the exact treatment of the free-particle Hamiltonian in Eq. (7).

For a closed-shell, homonuclear diatomic molecule the static-exchange potential is of the form

$$U = -\frac{2Z}{|\underline{r}-\underline{A}|} - \frac{2Z}{|\underline{r}+\underline{A}|} + 2\sum_{\sigma=1}^{N}(2\,J_\sigma - K_\sigma), \tag{24}$$

where the Coulomb operator

$$J_\sigma(r) = \int d^3r' \,\frac{\phi_\sigma^*(r')\,\phi_\sigma(\underline{r}')}{|\underline{r}-\underline{r}'|} \tag{25}$$

and K_σ is the corresponding exchange operator. The nuclear charge is denoted by Z, the nuclear coordinates are located at $\pm\underline{A}$, and N is the number of occupied orbitals ϕ_σ.

Equation (8) involves the matrix elements $<\Psi^t_{k\ell m}|U|\Psi^t_{k\ell' m}>$ and $<\Psi_{k\ell m}|U^t|\Psi^t_{k\ell' m}>$. The latter is given by

$$<\Psi^t_{k\ell m}|U^t|\Psi^t_{k\ell' m}> = <k\ell m|(K^{t'}G_o^P + 1)U^t(1 + G_o^P K^{t'})|k\ell' m> \tag{26}$$

and involves only components of K, G_o^P, and U within the discrete-basis-set subspace. To evaluate the matrix element

$$<\Psi^t_{k\ell m}|U|\Psi^t_{k\ell' m}> = \sum_{pq} < g^t_{\ell pm}\,Y_{pm}|(U^{(s)} - U^{(ex)})|g^t_{\ell' qm}\,Y_{qm}> \tag{27}$$

we use the single-center expansion method formulated by Faisal for the static potential $U^{(s)}$,[15] and by Burke and Sinfailam for the exchange part $U^{(ex)}$.[16] Expressions for the right-hand side of Eq. (27) are given in ref. 12 and involve a multipole expansion of the static potential

$$U^{(s)}(\overline{r}) = 2\sum_\lambda V_\lambda(r)\,P_\lambda(\hat{r}), \tag{28}$$

and a single-center expansion of each occupied orbital

$$\phi_\sigma(\underline{r}) = \sum_s \phi_{sm_\sigma}(r)\,Y_{sm_\sigma}(\hat{r}) \tag{29}$$

The Cartesian Gaussian function used in these calculations have the form:

$$\mu^{\alpha\underline{A}}_{pqs}(r) = N_{pqs}\ (x - A_x)^p (y - A_y)^q (z - A_z)^s\ e^{-\alpha|\underline{r}-\underline{A}|^2} \quad (30)$$

Results for zero order and variationally corrected K-matrix elements in H_2 are shown in Table I. The agreement between corrected and uncorrected s-wave ($\ell=\ell'=0$) matrix elements is good but some of the higher partial-wave matrix elements differ by more than a factor of two at low incident momenta. The basis sets used to obtain these results are given in reference 10. The accuracy of the corrected results depends on the agreement between the uncorrected and corrected matrix elements. The best agreement and hence most accurate results are obtained for diagonal s-wave matrix elements. We estimate the s-wave results are accurate to 0.1% and that the other results given in Table I are accurate to 5%.

Table II compares our "s" and "dσ" variationally corrected eigenphase results with R-matrix static-exchange eigenphases calculated by Schneider[17] and with the static-exchange diagonal phase shifts of Tully and Berry.[18] The off-diagonal K-matrix elements for the cases considered here are small enough that our eigenphases are essentially equal to our diagonal phase shifts. Our s eigenphases agree very well with the other calcultions. The agreement of our dσ eigenphases with the other results is not as good, but the agreement improves as k increases and the dσ eigenphases become larger. The differences in the calculated results may be due to the use of slightly different target wavefunctions in the calculations. However, we expect that our SCF result for the H_2 ground state is close to the Hartree-Fock limit.

Figure 1 shows the first three components $g^t_{0\ell'0}$, $\ell = 0, 2, 4$ of an s-wave scattering function Ψ^t_{k00} at incident momentum $k = 0.4$. The components g^t_{200} and g^t_{220} of a dσ wavefunction Ψ_{k20} at $k = 0.4$ and the same basis set are also shown in Fig. 1 scaled by a factor of 5. Figure 2 shows the components $g^t_{1\ell'0}$, $\ell' = 1, 3, 5$ of a pσ wavefunction Ψ^t_{k10}, $k = 0.6$, and the components g^t_{111} and g^t_{131} of a pπ wavefunction Ψ^t_{k11}, $k = 0.6$. The broad shape resonance in the pσ channel is manifested by the $g^t_{1\ell'0}$ continuum orbitals being larger in the region of the nuclei than the $g^t_{1\ell'1}$ orbitals. The basis sets used to calculate these functions are given in Ref. 10.

Our results for $e^- - H_2$ scattering are in good agreement with other static-exchange calculations. Accurate static-exchange results are of interest as a necessary first step toward an ab-initio treatment of polarization effects. Moreover, the static-exchange approximation is itself a useful ab initio method which accounts, at least qualitatively, for many features of electron-molecule scattering. The approach to elastic scattering described here is valid when long-range potential effects are small. This is clearly not the case for scattering from ions and strongly polar molecules. The generalization of our techniques to treat these systems is in progress.

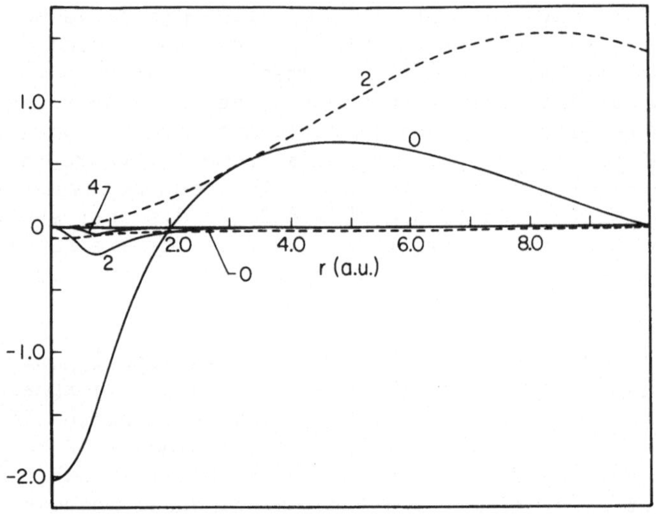

Fig. 1. Solid lines: radial components $g^t_{0\ell'0}$, ℓ' = 0,2,4 of Ψ^t_{k00} for k = 0.4 and basis set A, Table I, Ref. 10; dashed lines: radial components $g^t_{2\ell'0}$, ℓ' = 0,2 components of Ψ^t_{k20} for k = 0.4 and basis set A multiplied by 5.

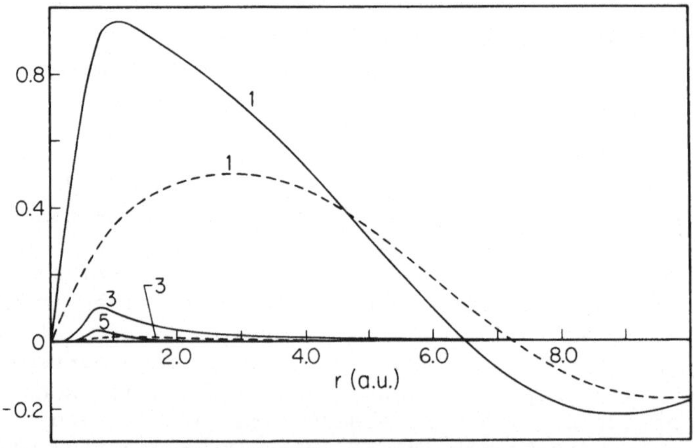

Fig. 2. Solid lines: radial components $g^t_{1\ell'0}$, ℓ' = 1,3,5 of Ψ^t_{k10} for k = 0.6 and basis set B Table 1, Ref. 10; dashed lines: radial components $g^t_{\ell\ell'1}$, ℓ' = 1,3 of Ψ_{k11} for k = 0.6 and basis set C, Ref. 10.

Table I. K-matrix results.[1]

k (a.u)	K^t_{000}	K^s_{000}	K^t_{020}	K^s_{020}	K^t_{220}	K^s_{220}
0.1	-0.2165	-0.2160	3.30(-3)	4.44(-3)	1.67(-2)	8.1(-4)
0.2	-0.4391	-0.4537	4.30(-3)	7.64(-3)	2.04(-2)	4.15(-3)
0.3	-047146	-0.7207	1.14(-2)	1.02(-2)	2.54(-2)	6.27(-3)
0.4	-1.063	-1.066	1.42(-2)	1.31(-2)	1.14(-2)	1.14(-2)
0.5	-1.522	-1.549	9.23(-3)	1.52(-2)	3.67(-2)	1.67(-2)
0.6	-2.290	-2.313	1.31(-2)	1.46(-2)	2.71(-2)	2.65(-2)
0.7	-3.765	-3.824	1.07(-2)	8.37(-3)	5.85(-2)	3.78(-2)
1.0	8.26	8.13	0.142	0.116	0.103	9.27(-2)
	K^t_{110}	K^s_{110}	K^t_{111}	K^s_{111}		
0.6	0.652	0.589	0.164	0.162		

[1] $^1\Sigma_g$ symmetry results are for basis set A, Table I, Ref. 10 pσ and pπ results are for basis sets B and C, Table I, Ref. 10, respectively.

Table II. Eigenphase results.

k	Calc.	s	dσ
0.1	FM[1]	2.929	9.0(-4)
	S[2]	2.931	
	TB[3]	2.939	
0.2	FM	2.716	4.28(-3)
	S	2.721	
	TB	2.737	
0.3	FM	2.517	6.41(-3)
	S	2.513	2.2(-3)
	TB	2.541	3 (-3)
0.4	FM	2.324	1.16(-2)
	S	2.312	5.7(-3)
	TB	2.352	7 (-3)
0.5	FM	2.144	1.68(-2)
	S	2.122	1.01(-2)
	TB	2.174	1.2(-2)
0.6	FM	1.988	2.70(-2)
	S	1.949	1.83(-2)
	TB	2.006	2.0(-2)
0.7	FM	1.827	3.78(-2)
	S	1.797	3.33(-2)
1.0	FM	1.827	3.78(-2)
	S	1.418	7.29(-2)
k	Calc.	pσ	pπ
0.6	FM	0.532	0.161
	S	0.561	0.1628
	TB	0.537	0.162

[1](Ref. 10).
[2]Schneider, Ref. 17.
[3]Tully and Berry, Ref. 18.

III. ELECTRONICALLY INELASTIC SCATTERING

In the Born-Oppenheimer and Franck-Condon approximations, and treating the target rotational levels as essentially degenerate, the differential cross section for electronic excitation by electron impact may be expressed in the form:[11]

$$\frac{d\sigma}{d\Omega}(n\leftarrow 0;\ E,R',\hat{r}') = S\sum_{v'}\frac{k_{v'}}{k_o}\,q_{v'o}\,\frac{1}{8\pi^2}\int d\hat{R}'\,|f_{k_o}(n\leftarrow 0;\vec{R}',\hat{r}')|^2 \qquad (31)$$

for impact energy $E = \frac{k_o^2}{2}$. In Eq. (31) $f_{k_o}(n\leftarrow 0;\vec{R}'\hat{r}')$ is the fixed-nuclei scattering amplitude in the laboratory frame (z'-axis in the direction of the incident electron beam), $\vec{R}'$ denotes the nuclear coordinates in the laboratory frame, and $\hat{r}'$ denotes the scattering angles. The symbol $q_{v'o}$ is the Franck-Condon factor between the v=0 ground state vibrational level and the v'level of the excited state. The momentum of the outgoing electron is

$$k_{v'} = \sqrt{k_o^2 - 2(E_{v'} - E_o)} \qquad (32)$$

where E_o is the energy of the initial target state and $E_{v'}$ is the energy of the v' level of the final target state. Eq. (31) implicitly neglects the dependence of the scattering amplitude on the vibrational level of the final target state. The factor S results from summing over final and averaging over initial spin sublevels and equals 1/2 for singlet to singlet excitation.

In the prescription for inelastic scattering proposed by Rescigno, McCurdy, and McKoy[19] the scattering is treated in a form of the distorted-wave approximation derived from the two-potential formula;[20] the initial state is the Hartree-Fock ground state; and the final target state is treated in the random phase approximation.[21] A simplifying feature of this formulation is that both the initial and final distorted-wave functions are calculated using the static-exchange potential of the ground state. As shown by Rescigno et al,[19] their prescription for inelastic scattering is equivalent to the "first order many-body formula" derived by Taylor and coworkers using the many-body Green's function technique of Martin and Schwinger.[22]

The present calculation follows the prescription of Rescigno et al.[19] except that the final state is treated in the single-channel Tamm-Dancoff approximation (TDA).[21] The single-channel TDA is equivalent to an independent-electron picture in which the excited orbital is an eigenfunction of the V^{N-1} potential[23] due to the N-1 core electrons.

In the body-fixed frame (with z-axis along the principal symmetry axis) the electronic portion of the transition matrix involves matrix elements of the form:

$$< \vec{k}_n, n|T_{el}|\vec{k}_o, 0 > = <\phi_n \Psi^{(-)}_{\vec{k}_n} |V| \phi_\alpha \Psi^{(+)}_{\vec{k}} >_a \tag{33}$$

where $\Psi^{(+)}_{\vec{k}_o}$, $\Psi^{(-)}_{\vec{k}_n}$ are initial, final Hartree-Fock (static-exchange) continuum spin-orbitals satisfying outgoing-wave, incoming-wave boundary conditions; ϕ_α is a Hartree-Fock occupied spin-orbital; and Φ_n is a spin-orbital of the V^{N-1} potential formed by removing an electron from the target orbital α. The anti-symmetrized matrix element is defined as

$$<ij|v|k\ell>_a = <ij|v|k\ell> - < ij|v|\ell k > \tag{34}$$

where

$$< ij|v|k\ell > = \int d\vec{x}_1 d\vec{x}_2 \phi_i^*(\vec{x}_i) \phi_j^*(\vec{x}_2) \frac{1}{|r_1 - r_2|} \phi_k(\vec{x}_1) \phi_\ell(\vec{x}_2). \tag{35}$$

In Eq. (35) $\vec{x}$ denotes combined space and spin coordinates; space coordinates are denoted by $\vec{r}$.

To treat the target orientation dependence of the scattering analytically, the initial and final continuum space orbitals are expanded in the partial-wave series:

$$\psi^{(+)}_{\vec{k}_o}(\vec{r}) = \sqrt{\frac{2}{\pi}} \sum_{\ell m} i^\ell \psi^{(+)}_{k_o \ell m}(\vec{r})\, Y_{\ell m}(\hat{k}_o) \tag{36a}$$

$$\psi^{(-)}_{\vec{k}}(\vec{r}) = \sqrt{\frac{2}{\pi}} \sum_{\ell m} i^\ell \psi^{(-)}_{k_n \ell m}(\vec{r})\, Y_{\ell m}(\hat{k}_n) \tag{36b}$$

This leads to a single-center expansion of the transition matrix in the body-fixed frame of the form:

$$< \vec{k}_n, n|T_{el}|\vec{k}_o, 0 > = \sum_{\ell\ell' mm'} i^{\ell'-\ell} < k_n\ \ell m, n|T_{el}|k_o \ell' m', 0 > \times Y_{\ell m}(\hat{k}_n) Y^*_{\ell' m'}(\hat{k}_o). \tag{37}$$

Introducing the fixed-nuclei dynamical coefficients

$$a_{\ell\ell' mm'}(n\leftarrow 0; k_o, R) = -\frac{\pi}{2} \sqrt{4\pi(2\ell'+1)}\ i^{\ell'-\ell} \times < k_n \ell m, n|T_{el}|k_o \ell' m', 0 >, \tag{38}$$

the laboratory frame scattering amplitude has the expansion

$$f_{k_o}(n\leftarrow 0;\vec{R},\hat{r}') = \sum_{\substack{\ell\ell' mm' \\ m''}} a_{\ell\ell' mm'}(n\leftarrow 0;k_o,\hat{R})\, D^{(\ell)}_{m''m}(\hat{R}') \times D^{(\ell)*}_{om'}(\hat{R}')\, Y_{\ell m''}(\hat{r}') \tag{39}$$

where $D^{(\ell)}_{m''m}(\hat{R}')$ is a rotational harmonic defined in Edmonds.[24] Substituting Eq. (39) into Eq. (31) and carrying out the angular integrations, we obtain

$$\frac{d\sigma}{d\Omega'}(n\leftarrow 0;E,R',\hat{r}') = S\sum_{v'} q_{v'o}\sum_{L} A_L(n\leftarrow 0;E,R')\, P_L(\theta') \tag{40}$$

where

$$A_L(n\ 0;E,R') = \sum_{\substack{\ell\ell' mm' \\ \lambda\lambda'\mu\mu'}} a_{\ell\ell' mm'}\, a^{*}_{\lambda\lambda'\mu\mu'}\, \frac{(2L+1)}{2\lambda+1}\sqrt{\frac{2\ell+1}{2\lambda+1}}\,\frac{1}{4\pi}$$
$$\times\ (L\ell 00|\lambda o)(L\ell' oo|\lambda' o)(L\ell,\mu-m,m|\lambda\mu)$$
$$\times\ (L\ell',-\mu'+m',-m'|\lambda'-\mu')\,. \tag{41}$$

The quantity $(\ell_1\ell_2\ m_1m_2|\ell_3m_3)$ is a Clebsch-Gordan coefficient, $P_L(\theta')$ is a Legendre polynomial. The symbols Ω' and $\hat{r}'$ both denote the scattering angles (θ', ϕ'). By a similar analysis the integrated cross section

$$\sigma(n\leftarrow 0;E,R') = \int d\hat{r}'\, \frac{d\sigma}{d\Omega'}(n\leftarrow 0;E,R',\hat{r}') \tag{42}$$

can be expressed in the form

$$\sigma(n\leftarrow 0;E,R') = S\sum_{v'}\frac{k_{v'}}{k_o}\, q_{v'o}\sum_{\ell\ell' mm'}|a_{\ell\ell' mm'}|^2 \Big/ (2\ell'+1)\,. \tag{43}$$

To obtain numerical representations of the continuum space orbitals $\Psi^{(-)}_{k_n\ell m}$ and $\Psi^{(+)}_{k_o\ell' m'}$, we use the discrete-basis-set method described in the preceding section. Traveling-wave boundary condition wave functions are obtained from standing-wave functions by the transformation[25]

$$\Psi^{(\pm)}_{k\ell m} = \sum_{\ell'}(1 \pm iK)^{-1}_{\ell\ell'}\, \Psi^{P}_{k\ell' m}\,. \tag{44}$$

Matrix elements of the electronic transition potential are evaluated by a straight forward generalization of the method used to calculate the matrix element $\langle \Psi^{t}_{k\ell m}|U^{t}|\Psi_{k\ell' m}\rangle$ discussed in Section II. Explicit expressions for the direct and exchange distorted-wave matrix elements are given in reference 11.

Our approach to converging the single-center expansion of the electronic transition matrix element is to use the distorted-wave approximation up to where the Born approximation is valid. Contributions to higher partial-wave matrix elements are then included in the Born approximation.

Our calculated differential cross sections for the excitation $X'\Sigma_g^+(v=0) \rightarrow B^1\Sigma_u^+(v'=2)$ of H_2 are shown in Fig. 3 along with the experimental data of Srivastava and Jensen.[26] Fig. 4 shows our integrated cross section for excitation of the $B'\Sigma_u^+$ state of H_2 summed over final vibrational levels. Fig. 4 also shows the 2-state close-coupling, Born-Ochkur, and Born results of Chung and Lin,[27] and the experimental data of Srivastava and Jensen.[26] We estimate the uncertainty in our results due to numerical round-off errors and the use of discrete-basis-set representation of the distorted-

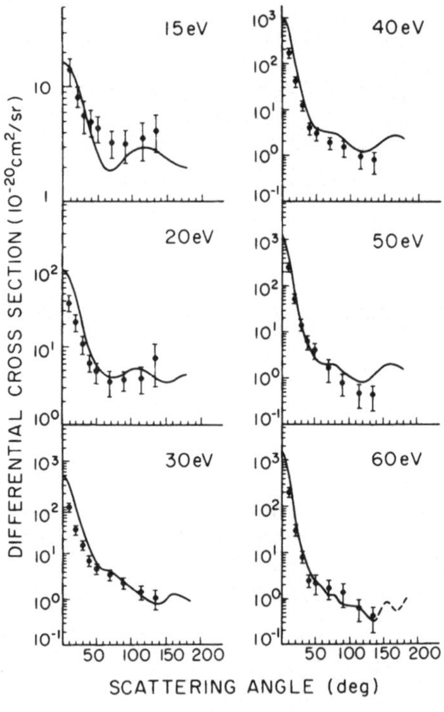

Fig. 3. Differential cross section for excitation of the $B'\Sigma_u^+(v'=2)$ state at indicated electron impact energies. The solid curves show the D results of Ref. 11. The dashed curve at 60 eV is less accurate than solid curve as discussed in Ref. 11. The solid dots with error bars are the experimental data of Ref. 26.

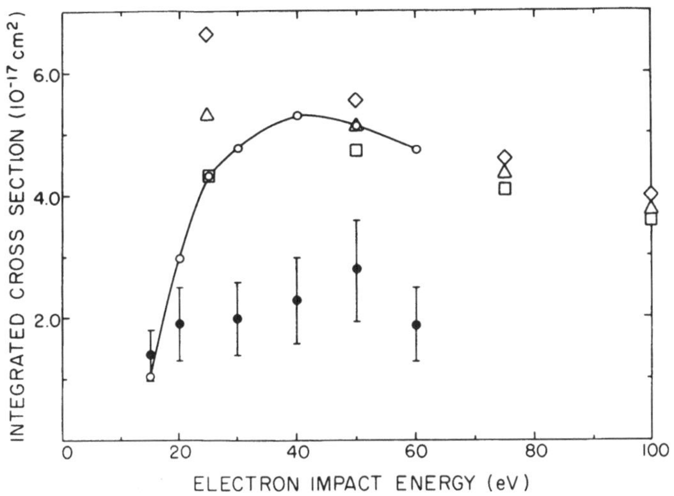

Fig. 4. Integrated cross section for excitation of the $B'\Sigma_u^+$ state by electron impact. ——: DW results of Ref. 11, calculated points are indicated by open circles; open squares: 2-state close-coupling results from Ref. 27; open triangles: Born-Ochkur results from Ref. 27; open diamonds: Born results from Ref. 27; solid dots with error bars are experimental data from Ref. 26.

wave potential at ±10% for the integrated cross section and ±20% for the differential cross section. Our differential cross section at 60 eV at large scattering angles and indicated by the dashed curve may be in error by a factor of 2. Details of this calculation and the discrete-basis-sets used are given in reference 11.

As shown in Fig. 4, at 25 eV impact energy and above our integrated $B^1\Sigma_u^+$ state cross section is in good agreement with the 2-state close-coupling calculation of Chung and Lin.[27] Our results at 60 eV appear to extrapolate towards Chung and Lin's Born and close-coupling results at higher energy. For impact energies of 30 eV and above the experimental results of Srivastava and Jensen[26] are about a factor of 2 smaller than the theoretical results, and theory and experiment do not appear to converge as the impact energy increases. Comparison of our differential cross section shown in Fig. 3 with the experimental data of Srivastava and Jensen shows that the discrepancy in the integrated cross sections corresponds to the difference in the differential cross sections at small scattering angles. For scattering angles less than 40° our results are about a factor of two larger than experiment. Small angle scattering accounts for most of the integrated cross section since the differential cross section is strongly forward peaked.

Our results show that a simple form of the distorted-wave approximation yields differential cross sections in qualitative agreement with experiment and accurate in magnitude to within a factor of two. Our approach is sufficiently straightforward that application to a wide range of systems appears feasible.

This work was supported by a grant (CHE76-05157) from the National Science Foundation.

REFERENCES

1. L. Hulthen, K. Fysiograf, Selsk. Lund. Forh. 14, 257 (1944).
2. W. Kohn, Phys. Rev. 74, 1763 (1948).
3. S. I. Rubinow, Phys. Rev. 98, 183 (1955).
4. E. J. Heller and H. A. Yamani, Phys. Rev. A 9, 1201 (1974).
5. T. N. Rescigno, C. W. McCurdy, and V. McKoy, Chem. Phys. Lett. 27, 401 (1974); Phys. Rev. A 10, 2240 (1974); 11, 825 (1975).
6. N. S. Ostlund, Chem. Phys. Lett. 34, 419 (1975).
7. D. A. Levin, A. W. Fliflet, M. Ma, and V. McKoy, J. Comp. Phys. 28, 416 (1978).
8. W. Gautschi, SIAM J. Num. Anal. 7, 187 (1970).
9. A. W. Fliflet, D. A. Levin, M. Ma, and V. McKoy, Phys. Rev. A 17, 160 (1978).
10. A. W. Fliflet and V. McKoy, Phys. Rev. A 18, 2107 (1978).
11. A. W. Fliflet and V.McKoy, "Distorted-wave approximation cross sections for excitation of the b $^3\Sigma_u^+$ and B'Σ_u^+ state of H_2 by low-energy electron impact. Submitted to Phys. Rev. A.
12. A. W. Fliflet and V. McKoy, Phys. Rev. A 18, 1048 (1978).
13. C. Schwarz, Phys. Rev. 124, 1468 (1961).
14. D. G. Truhlar, T. Abdallah, and R. L. Smith, in Advances in Chemical Physics, Vol. 25, edited by I. Prigogine and S. A. Rice (Wiley, New York, 1974), p. 211.
15. F. H. M. Faisal, J. Phys. B 3, 636 (1970).
16. P. G. Burke and A. L. Sinfailam, J. Phys. B 3, 641 (1970).
17. B. I. Schneider, Phys. Rev. A 11, 1957 (1975).
18. J. C. Tully and R. S. Berry, J. Chem. Phys. 51, 2056 (1969).
19. T. N. Rescigno, C. W. McCurdy, Jr., and V. McKoy, J. Phys. B 7, 7396 (1974).
20. J. R. Taylor, Scattering Theory (Wiley, New York, 1972) p. 418.
21. C. W. McCurdy, Jr., T. N. Rescigno, D. L. Yeager, and V. McKoy, Modern Theoretical Chemistry 3, ed. by H. F. Schaefer III. (Plenum, New York, 1977), p. 339.
22. Gy.Csanak, H. S. Taylor, and R. Yaris, Phys. Rev. A 3, 1322 (1971); Gy.Csanak, H. S. Taylor, and D. N. Tripathy, J. Phys. B 6, 2040 (1973).
23. H. P. Kelly, Phys. Rev. 136, B896 (1964).
24. A. R. Edmonds, Angular Momentum in Quantum Mechanics (Princeton University, Princeton, 1960).

25. R. G. Newton, Scattering Theory of Waves and Particles (McGraw-Hill, New York, 1966) p. 191.
26. S. K. Srivastava and S. Jensen, J. Phys. B 10, 3341 (1977).
27. S. Chung and C. C. Lin, Phys. Rev. A17, 1874 (1978).

DISCUSSION

Nesbet: It is not clear to me that there are no spurious poles or anomalies in this method.

Rescigno: I think it's pretty well known that the spurious poles that one gets (in the standard algebraic variational methods) are not due to an approximation of the interaction potential but rather result from approximation of the kinetic energy. In this method the free-particle Green's function is treated exactly and that's why you have no spurious singularities.

Nesbet: Is there any proof that the operator $1-VG_o$ is positive definite?

Rescigno: It doesn't have to be positive definite.

Schneider: It's not positive definite.

Nesbet: Then what happened to the singularities due to the inverse of the operator appearing in the equations?

Schneider: Those poles all lie in the complex plane.

Fliflet: These are poles due essentially to the K-matrix blowing up when the phase shift is $\pi/2$.

Nesbet: That's no problem.

Schneider: I think you can show that the eigenvalues of the operator $1-VG_o$ all go into the complex plane when the energy is $E + i\varepsilon$.

McCurdy: That's certainly true in the 1-term separable potential. You can show that if the pole is on the real axis it corresponds to a bound state.

Schneider: Newton has a discussion of these poles.

Nesbet: This is very much connected with the way you are representing G_o.

Bottcher: Let me say that if you took a model potential and did a Kohn variational calculation you would get pseudo-resonances.

I think Arne is right that the method is free of pseudo-resonances because it has the correct branch cut structure. The Kohn variational method replaces a branch cut with a set of poles.

Nesbet: I think you may have a problem because if you are using a finite matrix representation of the principal-value Green's functions you now have a formalism which produces pseudo-resonances

Rescigno: No. You do not.

Schneider: You're taking matrix elements of the exact free-particle Green's function, you're not inverting a basis set representation of the Hamiltonian.

Bottcher: That's my point.

Nesbet: Let me clarify what I just said with reference to that. You are taking matrix elements of the exact Green's function but you have a finite matrix of real numbers which is energy dependent. Is there a proof that that matrix does not have singularities?

Schneider: That's what Newton discusses.

Macek: Since you're solving an approximate potential exactly, you don't have any analyticity problems although the approximate potential may be poor.

ROUNDTABLE ON L^2-METHODS

Chairman: Vincent McKoy

Macek: I would like to show how it is possible to go beyond static-exchange with the separable approximation in electron-molecule scattering and do multi-channel separable approximations essentially as easily as single channels, except for the problem of the dimensions of the matrices you must manipulate. [See contribution by J. Macek and G. Gallup at the end of this discussion.]

McKoy: Which systems have you studied?

Macek: We started with the e + H_2 system and we looked at the ground state and the first excited state. We coupled these two states and we pulled the molecule apart, i.e., we did the calculation at different internuclear distances. The reason we did it at different internuclear distances is that as you increase the internuclear distance the singlet and triplet states become nearly degenerate and very strongly coupled so you can never use the distorted wave approximation. You have to treat them in one shot and hence that should be a good test of this multi-channel approximation.

Temkin: In these calculations you included just electronic states and no vibrational states?

Macek: Yes, this is just in the Born-Oppenheimer approximation. One could incorporate nuclear motion as well but with a great deal of added complexity. That would be quite difficult and I am not sure that I would want to do it this way.

Rescigno: What kind of functions did you choose for your expansion?

Macek: Cartesian Gaussian functions.

Rescigno: And how did you do the matrix elements of the Green's function?

Macek: The Green's function is translatable and we evaluated them in terms of error functions.

Hazi: Could you have a multi-configuration representation of the singlet ground state?

Gallup: Yes, and we do.

Kaldor: I will talk about how one can include the effects of polarization in a completely ab initio way without introducing any semi-empirical parameters. This will be done in the framework of the T-matrix. [See contribution by U. Kaldor and A. Klonover at the end of this discussion.]

Temkin: I asked you this question privately and I would like to ask it publicly. That is: what static polarizability is implied in the optical potential that you use? One would like to have it and I think one could get it by taking a matrix element of the terms you have in the second-order optical potential. My suspicion is that you are not getting all of the static polarizability although you are clearly getting the dynamical effects. If you are not getting all of the static polarizability then one should not expect that good agreement with experiment in your cross section, especially in the forward direction.

Heller: I find the effect of vibrational averaging interesting. Did you take it over a distribution corresponding to a vibrational wavefunction squared or did you take it over classical distributions?

Kaldor: Over the vibrational wavefunction squared.

Lane: Did it actually change the shape of your integral excitation cross section? You indicated that it did raise the results at lower energies. How do your static-exchange results agree with previous static-exchange results? Are they within about 1% or so of those?

Kaldor: Yes.

Truhlar: I don't understand your answer to Temkin's question. It seems to me that you have already calculated the static-polarizability in an equivalent basis.

Kaldor: In the calculation of the static-polarizability in the equivalent basis you have to go to third order to get good results.

Taylor: He is going to second-order in the Hartree-Fock model so it is equivalent to the uncoupled HF. If he were using for his transition densities the RPA transition densities he would have the coupled HF results and would get very good results for the polarizability.

Buckley: I want to discuss the R-matrix code which Phil Burke, I. Mackey, Vo Ky Lan and I have been developing for studying the scattering of electrons by diatomic molecule. [See contribution by B. Buckley and P. G. Burke at the end of this discussion.]

THE SEPARABLE APPROXIMATION IN MULTICHANNEL ELECTRON-MOLECULE COLLISIONS

Gordon Gallup, Department of Chemistry and
Joe Macek, Behlen Laboratory of Physics
The University of Nebraska
Lincoln, Nebraska 68588

In this talk we will formulate the separable approximation for calculating electron-molecule scattering completely generally. That is, we consider a reaction such as

$$e^- + H_2 \to e^- + H_2^* ,$$

where H_2^* is an electronically excited state of H_2. This treatment is a simple extension of the separable approximation of Rescigno, McCurdy, and McKoy who treated elastic scattering of electrons from molecules.[1]

Our derivation proceeds from an expansion of the wavefunction in terms of a set of basis states, a familiar starting point in bound state calculations. By carefully identifying singular terms peculiar to scattering calculations we obtain a Lippmann-Schwinger type integral equation for the wavefunction and the transition matrix T. The treatment is similar to the usual derivations of the Lippmann-Schwinger equation, but differs from it in one essential respect, namely we use fully antisymmetric basis states, thus the "potential" matrix V in our equation involves exchange terms. In this way we obtain the multichannel generalization of the separable approximation. Our derivation is conventional in all other regards, in particular we do not use symmetry adapted wavefunctions initially, but introduce the symmetry of the system later via a basis set transformation on the T matrix equations.

In calculations for the specific $e^- + H_2$ system we actually proceeded by expanding in symmetry adapted wavefunctions from the beginning. This turned out to lead more directly to algorithms which could use our available bound state programs. The details of this alternate procedure are given in the appendix.

We consider a molecule M described by a set of electronic eigenstates $\psi_n(\vec{r}_1 s_1, \cdots \vec{r}_N s_N)$, s_i = electron spin, in the Born-Oppenheimer approximation. The electronic energy corresponding to these eignestates is denoted by ε_n. An incident electron impinges on the molecule and causes transitions between electronic states. An appropriate complete set of states for this N + 1'th electron are the states $\phi_{\vec{k}}(\vec{r}_{N+1}, s_{N+1})$ which are plane wave states normalized per unit momentum times a spinor

$$\phi_{\vec{k}}(\vec{r}_{N+1}, s_{N+1}) = (2\pi)^{-3/2} \exp(i\vec{k}\cdot r_{N+1}) \begin{cases} \alpha \text{ spin up} \\ \beta \text{ spin down,} \end{cases}$$

where α and β are two component spinors.

These functions have the property that they represent the asymptotic situation, i.e., a free electron and a molecule in an energy eigenstate, correctly. This is essential to discuss any scattering process.

Let us now define the product function $\psi_{\vec{k}n}$ as

$$\psi_{\vec{k}n}(\vec{r}_1 s_1, \cdots \vec{r}_{N+1} s_{N+1}) = A\phi_{\vec{k}}(\vec{r}_{N+1} s_{N+1})\ \psi_n\ (\vec{r}_1 s_1 \cdots \vec{r}_N s_N)$$

where A is an operator which antisymmetrizes the wavefunction with respect to exchange of electron coordinates.

We now expand the total wavefunction for the N + 1 electron system as

$$\Psi = \sum_{\vec{k}'n'} C_{\vec{k}'n'} \Psi_{\vec{k}'n'} \tag{1}$$

We want to solve

$$(H - E)\Psi = 0 \tag{2}$$

subject to the boundary condition

$$\Psi \to \phi_{\vec{k}_o} \psi_{n_o} + \text{(outgoing wave part)} \tag{3}$$

as any electron coordinate becomes large. In Eqn. (3) $\vec{k}_o$ and n_o refer to the initial electron momentum and the initial electronic eigenstate of the N electron molecule.

Substituting Eqn. (1) into Eqn. (2) and projecting onto $\psi_{\vec{k}n}$ gives us the equation

$$\sum_{\vec{k}'n'} < \psi_{\vec{k}n} | H - E | \psi_{\vec{k}'n'} > C_{k'n'} = 0 \qquad (4)$$

Eqn. (4) looks just like the matrix version of the Schrödinger equation one might solve to find the energy levels of a molecule. There is a crucial difference, however, between the bound state problem and the scattering problem represented by (4), namely, some of the terms in Eqn. (4) have Dirac delta-function type singularities and thus have meaning only when summed (integrated) over $\vec{k}'$. One must first isolate the δ function singularities in Eqn. (4) and treat them separately. The isolation of such singularities gives us what is known as the Lippmann-Schwinger equation.

Both $<\psi_{\vec{k}n}|H-E|\psi_{\vec{k}'n'}>$ and $C_{\vec{k}'n'}$ in Eqn. (4) have δ function singularities. We must first isolate those in the matrix elements of H-E. To do this we choose a particular representation for the antisymmetrizer A in our definition of $\psi_{\vec{k}n}$. We set

$$\psi_{\vec{k}n} = A\, \phi_{\vec{k}}\, \psi_n = \frac{1}{\sqrt{N+1}} \sum_{p=1}^{N+1} \delta^P\, \phi_{\vec{k}}\,(P)\, \psi_n\, (P^{-1}) \qquad (5)$$

where

$$\delta^P = \pm 1$$

$$\phi_{\vec{k}}\,(P) = \phi_{\vec{k}}\,(\vec{r}_P, s_P)$$

and

$$\psi_n(P^{-1}) = \psi_n(\vec{r}_1 s_1 \cdots \vec{r}_{P-1} s_{P-1}, \vec{r}_{P+1} s_{P+1} \cdots \vec{r}_{N+1} s_{N+1})$$

is already antisymmetric. For definiteness we take $\delta^P = 1$, $P = 1$ and $\delta^P = -1$, $P > 1$. The choice of representation for $\psi_{\vec{k}n}$ is not unique. One could choose other fully antisymmetric representations. This choice has the advantage of explicitly exhibiting a function with the correct boundary conditions, i.e., as $r_P \to \infty$, $\psi_{\vec{k}n} \to \phi_{\vec{k}}(P)$ $\psi_n(P^{-1})$ which describe a free P'th electron and a molecule with N electrons, none of which is the P'th electron. Such a representation may not be the most convenient for calculating the matrix elements, but we can transform to a more convenient representation after deriving the equation for $C_{\vec{k}n}$. Just which representation to transform it to depends upon the symmetry of the system. In our initial derivation of the Lippmann-Schwinger equations we will not assume any particular symmetry, thus that part of our discussion is general.

Specialization to a particular system occurs when we transform to a representation convenient for calculating the matrix elements. An alternate approach is given in the appendix. We next write:

$$< \psi_{\vec{k}n} | H-E | \psi_{\vec{k}'n'} > = < \psi_{\vec{k}n} | H-E_{\vec{k}n} | \psi_{\vec{k}'n'} > + < \psi_{\vec{k}'n'} | E_{\vec{k}n} - E | \psi_{\vec{k}'n'} > \quad (6)$$

where $E_{\vec{k}n} = \frac{1}{2} k^2 + \varepsilon_n$. Substituting $\psi_{\vec{k}n}$ from Eqn. (5) into the first term of Eqn. (6) we have

$$< \psi_{\vec{k}n} | H-E_{\vec{k}n} | \psi_{\vec{k}'n'} > = \frac{1}{\sqrt{N+1}} \sum_{P=1}^{N+1} \delta^P < \phi_{\vec{k}}(P) \psi_n (P^{-1}) | H-E_{\vec{k}n} | \psi_{\vec{k}'n'} >$$

$$= \frac{1}{\sqrt{N+1}} \sum_{P=1}^{N+1} \delta^P < (H-E_{\vec{k}n}) \phi_{\vec{k}}(P) \psi_n (P^{-1}) | \psi_{\vec{k}'n'} > \quad (7)$$

$$= \frac{1}{\sqrt{N+1}} \sum_{P=1}^{N+1} \delta^P < V_P \phi_{\vec{k}}(P) \psi_n (P^{-1}) | \psi_{\vec{k}'n'} >$$

where

$$V_P = \sum_{A=1}^{N_A} \frac{Z_A e^2}{|\vec{r}_P - \vec{r}_A|} - \sum_{j=1}^{N+1}{}' \frac{e^2}{|\vec{r}_P - \vec{r}_j|} \quad (8)$$

and where $\sum_{j=1}^{N}{}'$ excludes j = P. In Eqn. (8) we also have

N_A = Number of constituent atoms in the molecule

N = Number of electrons in the molecule.

Clearly $<\psi_{\vec{k}n}|H-E_{\vec{k}n}|\psi_{\vec{k}'n'}>$ has no δ function singularities since V_P drops off faster than r^{-1}. (Note that we are explicitly assuming that the molecule is electrically neutral.)

We now consider $<\psi_{\vec{k}n}|\psi_{\vec{k}'n'}>$ which does have δ function singularities. Using $\psi_{\vec{k}n}$ from Eqn. (5) we have

$$< \psi_{\vec{k}n} | \psi_{\vec{k}'n'} > = \frac{1}{N+1} \sum_{\substack{PQ \\ P\neq Q}} \delta^P \delta^Q < \phi_{\vec{k}}(P) \psi_n (P^{-1}) | \phi_{\vec{k}'}(Q) \psi_{n'} (Q^{-1}) >$$

$$+ \frac{1}{N+1} \sum_P < \phi_{\vec{k}}(P)\psi_n(P^{-1}) | \phi_{\vec{k}'}(P)\psi_{n'}(P^{-1}) >$$

$$= < \overline{\psi_{\vec{k}n} | \psi_{\vec{k}'n'}} > + \delta_{\vec{k}\vec{k}'} \delta_{nn'} \tag{9}$$

where

$$<\overline{\psi_{\vec{k}n} | \psi_{\vec{k}'n'}} > = \frac{1}{N+1} \sum_{\substack{PQ \\ P \neq Q}} \delta^P \delta^Q < \phi_{\vec{k}}(P)\psi_n(P^{-1}) | \phi_{\vec{k}'}(Q)\psi_n(Q^{-1}) > \tag{10}$$

Since the plane wavefunctions in Eqn. (10) always overlap bound state orbitals, there are no δ function factors in $<\overline{\psi_{\vec{k}n}|\psi_{\vec{k}'n'}}>$. Thus defining

$$V_{\vec{k}n,\vec{k}'n'} = < \psi_{\vec{k}n} | H - E_{\vec{k}n} | \psi_{\vec{k}'n'} > - (E - E_{\vec{k}n}) < \overline{\psi_{\vec{k}n} | \psi_{\vec{k}'n'}} > \tag{11}$$

Eqn. (4) becomes

$$\sum_{\vec{k}'n'} V_{\vec{k}n,\vec{k}'n'} C_{\vec{k}'n'} - (E - E_{\vec{k}n}) C_{\vec{k}n} = 0 \tag{12}$$

where we have used Eqns. (9), (10), and (11) to exhibit the δ function singularities explicitly.

We now solve Eqn. (12) recalling the boundary conditions Eqn. (3) on Ψ.

We have

$$C_{\vec{k}n} = \delta(\vec{k}-\vec{k}_o)\delta_{nn_o} + \sum_{\vec{k}'n'} \frac{1}{E - E_{\vec{k}n} + i\varepsilon} V_{\vec{k}n,\vec{k}'n'} C_{\vec{k}'n'} \tag{13}$$

which is an integral equation (The Lippmann-Schwinger Equation) for C_{kn}. In Eqn. (13) it is understood that the limit $\varepsilon \to 0$ is taken after integration. Note that $C_{\vec{k}n}$ has a δ function term.

We write Eqn. (13) in operator form as

$$C = C_o + G_o VC , \tag{14}$$

where

$$C^{(o)}_{\vec{k}n} = \delta(\vec{k}-\vec{k}_o)\delta_{nn_o} , \tag{15}$$

and

$$G^{(o)}_{\vec{k}n,\vec{k}'n'} = \frac{\delta(\vec{k}-\vec{k}')\delta_{nn'}}{E-E_{kn}+i\varepsilon} \quad , \tag{16}$$

in our particular representation of basis functions.

The formal solution of Eqn. (14) is

$$C = (1 - G_oV)^{-1} C_o \tag{17}$$

Because of the C_o term in Eqn. (14) we know that C contains δ function terms, however these terms are not explicitly exhibited in Eqn. (17). To obtain an expression for C which explicitly exhibits the δ function singularities we substitute Eqn. (17) back into the right hand side of Eqn. (14) to obtain

$$C = C_o + G_o V(1 - G_oV)^{-1} C_o \tag{18}$$

where the singular term C_o representing the incoming electron-molecule state is explicitly exhibited. The second term is also singular because of the G_o factor. This term represents the outgoing scattered waves, as may be verified by substituting Eqn. (18) into Eqn. (1), and evaluating the sum over $\vec{k}'$ for large value of one of the electron coordinates, say r_p.

We now define the transition operator T as

$$T = V(1 - G_oV)^{-1} \tag{19}$$

The operator T derives its name from the circumstance that the square of its matrix element gives the rate for transitions from the initial state $\vec{k}_on_o$ to the final state $\vec{k}_fn_f$. For weak transitions, where we may expand $(1 - G_oV)^{-1}$ in a series in powers of G_oV and retain only the first term, we have

$$T \approx V \tag{20}$$

which is the well known first order perturbation result for the transition operator. In this approximation the square of the matrix element of V is proportional to the transition rate per unit time.

Since $\langle\psi_{\vec{k}n}|T|\psi_{\vec{k}_on_o}\rangle$ is the matrix element we seek, it is natural to calculate T in the basis $\psi_{\vec{k}n}$, which we are using. To find T we could evaluate the right hand side of Eqn. (19). Alternatively we note that Eqn. (19) implies that T satisfies the integral equation,

$$T = V + T\,G_oV \tag{21}$$

also called the Lippmann-Schwinger equation.

Various approximations have been used to find T from equations like Eq. (21). One promising approximation used in nuclear physics calculations and introduced into the electron-molecule scattering problem by Rescigno, McCurdy and McKoy[1] is the separable approximation where one sets

$$V_{\vec{k}n,\vec{k}'n'} = \sum_{ij} v_i^{\dagger}(\vec{k})\, V_{in,jn'} v_j(\vec{k}') \tag{22}$$

and we assume that the v's form a complete orthonormal set. The key approximation consists in truncating the sum in Eqn. (22) to include only a few terms chosen to represent $V_{\vec{k}n,\vec{k}'n'}$ reasonably well. With this approximation one readily shows that T is given by

$$T_{\vec{k}n,\vec{k}'n'} = \sum_{ij} v_i^{\dagger}(\vec{k})\, T_{in,jn'} v_k(\vec{k}') \tag{23}$$

where $T_{in,jn'}$ is a matrix element of the matrix T on the basis $v_i(\vec{k})$. The matrix T solves the truncated matrix version of Eqn. (21) with G_o replaced by the matrix with elements

$$G^{(o)}_{in,jn'} = \delta_{nn'} \int d^3k \frac{v_i(\vec{k}) v_j^{\dagger}(\vec{k})}{E - E_{\vec{k}n} + i\varepsilon} \tag{24}$$

Equations (11), (21), and (23) provide a well defined formalism amenable to numerical calculations, however the sequence of operations leads to a matrix element $V_{in,jn'}$ requiring a calculation of a matrix element in momentum space, which must then be used to calculate

$$V_{in,jn'} = \int d^3k d^3k' \; v_i(\vec{k})\; V_{\vec{k}n,\vec{k}'n'} v_j^{\dagger}(\vec{k}') \tag{25}$$

Clearly it would be preferable to work only in coordinate space. Defining

$$\chi_i(r_p s_p) = \int d^3k \; v_i(\vec{k}) \phi_{\vec{k}}(P) \tag{26}$$

where $\chi_i(\vec{r}_p s_p)$ is an orbital in coordinate space, we see that $V_{in,jn'}$ can be written in terms of matrix elements evaluated in the basis ψ_{in}, defined by

$$\psi_{in} = A\, \chi_i\, \psi_n \tag{27}$$

Upon replacing the energy $E_{\vec{k}n}$ by the operator

$$K(P) + \varepsilon_n \tag{28}$$

when operating on $\chi_i(P)\ \psi_n(P^{-1})$ in the defining equation for V, Eqn. (11), we may rewrite $V_{in,jn'}$ as

$$V_{in,jn'} = \langle\psi_{in}|H-E|\psi_{jn}\rangle + \{(E-\varepsilon_n)\langle\chi_i|\chi_j\rangle - \langle\chi_i|K|\chi_j\rangle\}\delta_{nn'} \tag{29}$$

where $K(P) = -\hbar^2/2m\ \nabla_P^2$ is the kinetic energy operator, ε_n is the energy of the n'th level of the N electron molecule, and the χ's are now one-electron spin orbitals.

Eqn. (29) is convenient for numerical calculation. Note that V is the sum of two terms. The first term $\langle\psi_{in}|H-E|\psi_{jn'}\rangle$ is a many electron matrix element similar to those one might compute in a quantum chemistry bound state calculation. The second terms, like the matrix elements of the Green's function, are peculiar to scattering problems. They represent a new feature, but since they involve only one-electron opeators, they are relatively easy to evaluate.

Eqn. (29) is not yet in the most general form, since one could subject the integral equation for T to an arbitrary nonsingular transformation U. Thus define $T' = U^\dagger TU$, then T' satisfies

$$\begin{aligned} T' &= V' + T'\ (U^{-1}\ U^{\dagger-1})\ G_o'\ (U^{-1}\ U^{\dagger-1})\ V' \\ &= V' + T'\ S^{-1}\ G_o'\ S^{-1}\ V' \end{aligned} \tag{30}$$

where $V' = U^\dagger VU$ and $G_o' = U^\dagger G_o U$. The quantity of interest $T_{kn,\vec{k}'n'}$ is then given by

$$T_{\vec{k}n,\vec{k}'n'} = \sum_{ij} v_i'^{\dagger}(\vec{k})\ T'_{in,jn'}\ v_j'(\vec{k}') \tag{31}$$

where

$$v_j'(\vec{k}') = \sum_m U^{-1}{}_{jm} v_m(\vec{k}')$$

and

$$S_{ji} = \int d^3k\ v_j^{\dagger\prime}(\vec{k})\ v_i'(\vec{k}) \tag{32}$$

Corresponding to the transformed functions $v_\ell'(k)$ there is a set of transformed coordinate function $\chi_\ell'(r,s_\ell)$ which are the Fourier transforms of the v'_ℓ's. The equation (31) for T is unchanged in form by the transformation U. The only complication introduced by using, say, a non-orthogonal basis is to introduce an overlap matrix S in Eqn. (30).

The basis transformation matrix U is arbitrary. One useful specific transformation is to transform to a basis where the total electron spin of the electron-molecule system, that is, the e^- + M system, is diagonal. Then, when H does not depend upon spin, the total spin S and it's projection M_S will be good quantum numbers. In Eqn. (31) this means that since V and G_o are diagonal in S, then so is T. Carrying out this transformation one then has

$$T^{(S)} = V^{(S)} + T^{(S)} G_o V^{(S)}$$

where

$$V^{(S)}_{in,jn'} = < \psi^{(SM_S)}_{in} | H-E | \psi^{(SM_S)}_{jn'} > + \{ (E-\varepsilon_n) < \chi_i | \chi_j > - < \chi_i | K | \chi_j > \} \delta_{nn'} \tag{33}$$

The functions χ_i in $<\chi_i|\chi_j>$ and $<\chi_i|K|\chi_j>$ refer to spatial parts only, i.e., it is understood that the spin functions have been factored out. We have included the label M_S on $\Psi^{(SM_S)}_{in}$ but the V matrix element does not, of course, depend upon M_S.

The unitary transformation which transforms Ψ_{in} to eigenstates of total spin is just the matrix whose matrix elements are the Clebsch-Gordan coefficients

$$(SM_S | sm_s S_n M_{Sn})$$

where s = 1/2, $m_s = \pm 1/2$ and S_n and M_{Sn} are the spin and spin projection of the N electron molecular energy eigenstate n.

One could also transform to a basis in which the incoming electron orbital label i is no longer distinct from the electron wavefunction label n. This, of course, is the form most convenient for adaptation of existing quantum chemistry programs to the problem of electron molecule scattering.

NOTE: The scattering cross section for the reaction

$$e^- + M(n_o) \rightarrow e^- + M(n)$$

is given by

$$\frac{d\sigma}{d\Omega} = \frac{1}{W_i} \sum_{\substack{\text{initial and} \\ \text{final spins}}} (2\Pi)^4 \frac{k}{k_o} \left| T_{\vec{k}n,\vec{k}_o n_o} \right|^2$$

W_i is the statistical weight of the spin states of the initial electron molecule system. It is given by

$$W_i = 2 \times (2S_{n_o} + 1)$$

APPENDIX A

This appendix gives a discussion of an alternative method for deriving the Lippmann-Schwinger equation for multi-channel, composite systems. This is no more general than the method given in the body, but it has the advantage of showing directly, the method of adapting a standard, non-orthogonal CI computer code to doing scattering calculations when applying the separable approximation to the Lippmann-Schwinger Equation. The principal difference between the two approaches centers upon the method used for dealing with the spin, and the order in which certain operations are performed.

In the CI method, wavefunctions for an N-electron molecule are represented as linear combinations of a basis of N-electron functions. These functions are based upon a set of configurations, and, of course, in real cases the set is finite. The configurations are built from one-electron orbitals, a_1, a_2, ..., a_M where 2M must be greater than N, for the total wavefunction to be nontrivial.

A configuration may be characterized by the vector, $\bar{\alpha}$ = $(\alpha_1, \alpha_2, \ldots, \alpha_M)$ where the actual orbital occupations are

$$a_1^{\alpha_1} \cdots a_M^{\alpha_M} .$$

Thus any one α_i is 0, 1, or 2, with higher numbers possible when orbital-degeneracy is present. The N-electron basis functions of spin S associated with the configuration $\bar{\alpha}$ will be denoted $\phi_\ell^{(S)}(\bar{\alpha})$, ℓ = 1, 2, ..., $K(\bar{\alpha})$, where $K(\bar{\alpha})$ is the number of spin couplings to the value S possible for the configuration.

In terms of this basis, the molecular CI wavefunctions are

(A1) $$\Psi_\eta^{(S)} = \sum_{\bar{\alpha},\ell} C^{(S)}_{\bar{\alpha}\ell\eta}\, \phi_\ell^{(S)}(\bar{\alpha}),$$

such that

(A2) $$H\Psi_\eta^{(S)} = E_\eta^{(S)}\Psi_\eta^{(S)}.$$

When doing the scattering calculations, we add one more "orbital" to the list, $a_{M+1}(\bar{k})$,

(A3) $$a_{M+1}(\bar{k}) = \frac{1}{(2\pi)^{3/2}} e^{i\bar{k}\cdot\bar{r}} .$$

In our present discussion, we are interested only in configurations with $\alpha_{M+1} = 1$, since other values correspond more closely either to completely bound, or to ionized systems.

With the added orbital a_{M+1}, we have a new set of $(N + j)$-electron functions,

$$\Phi_\ell^{(S')}(\bar{\alpha}'),\ \ell = 1, 2, \ldots, K(\bar{\alpha}').$$

The main problem arises from the nature of most of the algorithms that are used to generate spin eigenfunctions by projection techniques. These produce functions for which there are great difficulties in correlating the $\phi_\ell^{(S)}(\bar{\alpha})$ functions with the $\Phi_\ell^{(S')}(\bar{\alpha}')$, so far as linear combinations over ℓ are concerned.

We have solved this problem by using what is termed a <u>phantom atom</u>. We augment the basis set $a_1, \ldots, a_M$, not with a plane wave, but with a single spherical gaussian orbital, centered at a single proton which is placed some distance away (500 AU) from the molecule. It is convenient to set the scale of this gaussian atom so that the phantom atom has zero energy, and does not appear in the total energy of the system. The crucial step in the process then consists of a CI calculation using as a basis, $\Phi_\ell^{(S')}(\bar{\alpha}'')$ where $\bar{\alpha}'' = \{\alpha_1\ \alpha_2 \ldots \alpha_M 1\}$, and there is one electron on the phantom atom. The molecular states from this are

(A4) $$\Psi_\eta^{(S')} = \sum_{\bar{\alpha}''\ell} C^{(S')}_{\bar{\alpha}''\ell\,\eta}\, \Phi_\ell^{(S')}(\bar{\alpha}'').$$

It should be remembered that the functions of Eqn. (A4) collectively represent two different spin states of the molecular part of the system. If the molecule we are studying is in a singlet system, for example, then S' in Eqn. (A4) is 1/2. The eigenvalues in

$$\text{(A5)} \qquad H\Psi_\eta^{(S')} = E_\eta^{(S')}\Psi_\eta^{(S')},$$

correspond to both singlet and triplet solutions to Eqn. (A2).

We are now in a position to write our asymptotically correct molecular functions with a plane wave electron. This is very simply done by replacing the phantom atom by $a_{M+1}(\bar{k})$, and

$$\text{(A6)} \qquad \Psi_{\eta\,\bar{k}}^{(S')} = \sum_{\bar{\alpha}'\ell} C_{\bar{\alpha}'\ell\,\eta}^{(S')}\,\Phi_\ell^{(S')}(\bar{\alpha}'),$$

where $\bar{\alpha}' = (\alpha_1\,\alpha_2 \ldots \alpha_M\,1)$ is identical with the $\bar{\alpha}''$ set. It is clear that the asymptotic form of $\Psi_{\eta\,\bar{k}}^{(S')}$ is

$$\text{(A7)} \qquad \lim_{r_{N+1}\to\infty} \Psi_{\eta\,\bar{k}}^{(S')} \longrightarrow \Psi_\eta^{(S)} \frac{1}{(2\pi)^{3/2}}\, e^{i\,\bar{k}\cdot\bar{r}_{N+1}},$$

where we have included any spinfunctions in the symbol $\Psi_\eta^{(S)}$ for convenience.

As was pointed out in the body of this note, matrix elements of the sort

$$\text{(A8)} \qquad H_{\eta\bar{k}\eta'\bar{k}'} = \langle\, \Psi_{\eta\,\bar{k}}^{(S')}\,|H|\,\Psi_{\eta'\bar{k}'}^{(S')}\,\rangle$$

have some terms which contain the Dirac δ-function, $\delta(\bar{k}-\bar{k}')$, and that it is imperative to sequester these to arrive at the Lippmann-Schwinger equation. In actual practice, of course, our separable approximation involves evaluations of Hamiltonian matrix elements in a discrete basis $v_p(\bar{k})$, p = 1, 2, ..., L

$$\text{(A9)} \qquad H_{npn'p'} = \int d^3\bar{k}' \int d^3\bar{k}\; v_p(\bar{k})^*\, v_{p'}(\bar{k}')\; H_{n\bar{k}n'\bar{k}'}\,.$$

When doing a calculation, we, of course, determine $H_{npn'p'}$ using the configuration space basis which consists of the Fourier Transforms of the $v_p(\bar{k})$. These functions we denote $\chi_p(\bar{r})$, and the (N+1)-electron basis functions are now $\Phi_{\ell p}^{(S')}(\bar{\alpha}'')$, where the (M+1)th orbital is taken from the set $\chi_p(\bar{r})$, each one in turn. The $\Phi^{(S')}_{\ell p}(\bar{\alpha}'')$ are used to obtain states appropriate to the molecular states by a formula similar to Eqn. (A6), i.e.,

$$\text{(A10)} \quad \Psi^{(S')}_{np} = \sum_{\bar{\alpha}''\ell} C^{(S')}_{\bar{\alpha}''\ell\, n} \Phi^{(S')}_{\ell p}(\bar{\alpha}''),$$

and, therefore, an alternative equation for the $H_{npn'p'}$ is

$$\text{(A11)} \quad H_{npn'p'} = \langle \Psi^{(S')}_{np} | H | \Psi^{(S')}_{n'p'} \rangle .$$

In the present formalism, we obtain the potential V in the $\Psi^{(S')}_{np}$ basis as

$$\text{(A12)} \quad V_{npn'p'} = \langle \Psi^{(S')}_{np} | H-E | \Psi^{(S')}_{n'p'} \rangle + \delta_{nn'} \{ (E-E_n) \langle \chi_p | \chi_{p'} \rangle - \langle \chi_p | -\tfrac{1}{2} \nabla^2 | \chi_{p'} \rangle \}$$

which is the same as Eqn. (33) in the body of the talk.

REFERENCES

1. T. N. Rescigno, C. W. McCurdy and V. McKoy, Chem. Phys. Lett. 27, 401 (1974); Phys. Rev. A10, 2240 (1974); Phys. Rev. A 11, 825 (1975); see also the contribution by A. W. Fliflet in this volume.

NONEMPIRICAL POLARIZATION IN LOW-ENERGY ELECTRON-MOLECULE SCATTERING THEORY

Avner Klonover and Uzi Kaldor

Department of Chemistry
Tel Aviv University
Tel Aviv, Israel

The T-matrix expansion method for electron-molecule scattering originated by Rescigno, McCurdy and McKoy[1-3] is described elsewhere in this volume.[4] A method for including polarization effects in the scattering potential in an ab initio, nonparametric way is presented below. Applications to electron-H_2 elastic scattering as well as rotational and vibrational excitation are given.

The basic equation of the T-matrix expansion method[1-4] is the Lippmann-Schwinger equation in matrix form,

$$\underline{T} = \underline{U}^t + \underline{U}^t \underline{G}_o^+ \underline{T} \tag{1}$$

obtained by expanding the potential U (twice the scattering potential V in atomic units) in a set of square-integrable functions $|\alpha\rangle$,

$$U^t = \sum_{\alpha\beta} |\alpha\rangle \langle\alpha|U|\beta\rangle \langle\beta| \ . \tag{2}$$

The matrix equation (1) is solved for $\underline{T}$ and the scattering amplitude calculated by

$$f(|\vec{k}_f\rangle \leftarrow |\vec{k}_i\rangle) = -2\pi^2 \langle\vec{k}_f|T(E)|\vec{k}_i\rangle =$$

$$-2\pi^2 \sum_{\alpha\beta} \langle\vec{k}_f|\alpha\rangle \langle\alpha|T(E)|\beta\rangle \langle\beta|\vec{k}_i\rangle \ , \tag{3}$$

where $|\vec{k}_i\rangle$ and $|\vec{k}_f\rangle$ are the initial and final plane-wave states,

$$|\vec{k}\rangle = (2\pi)^{-3/2} \exp(i\vec{k}\cdot\vec{r}) \tag{4}$$

and

$$E = \frac{1}{2} k_i^{\,2} = \frac{1}{2} k_f^{\,2} . \tag{5}$$

Cross sections for transitions from the initial molecular state with vibrational and rotational quantum numbers vj to the final state v'j' are obtained in the adiabatic-nuclei approximation[5,6] by

$$\frac{d\sigma(vj \to v'j')}{d\,\Omega} = \frac{k_f}{k_i} \frac{1}{2j+1} \sum_{mm'} |\langle \chi_{v'}(R)\; Y_{j'm'}(\hat{R})|$$

$$f|\chi_v(R)\; Y_{jm}(\hat{R})\rangle|^2 , \tag{6}$$

where $\chi_v(R)$ and $Y_{jm}(\hat{R})$ are the eigenfunctions of molecular vibration and rotation respectively and the integration in (6) is over the coordinates of the molecular nuclei. Depending on whether v and j are equal to or different from v' and j', equation (6) describes elastic, rotational, vibrational or vibrational-rotational processes. The elastic cross sections are obtained by averaging over populated states,

$$\frac{d\sigma_{el}}{d\Omega} = \sum_{vj} P_{vj} \frac{d\sigma(vj \to vj)}{d\Omega} , \tag{7}$$

where P_{vj} is the fractional population of the vj level. Integral cross sections are calculated by integrating Eq. (6) or Eq. (7) over the scattering angles Ω.

McKoy and collaborators[1-4,7] calculated cross sections for H_2 and N_2 in the static-exchange approximation, using the potential

$$V_{SX}(\vec{r}) = \sum_N \frac{Z_N}{|\vec{r}-\vec{R}_N|} + \sum_i (J_i - K_i), \tag{8}$$

where Z_N is the charge of nucleus N, $\vec{R}_N$ its coordinates, and J_i and K_i are the usual Coulomb and exchange operators. The static-exchange results differ significantly from experiment, the most prominent difference being the absence of the broad peak in the e-H_2 elastic scattering near 3.5 eV. This disagreement results from neglecting the polarization of molecular charge by the scattered electron, an effect particularly important at low scattering energy.

The interaction of the scattered electron with the target molecule can be described in terms of the optical potential[8-10], the exact one-electron potential seen by the scattered electron.

The optical potential is of course very complicated, non-local and energy-dependent. Bell and Squires[11] showed that it may be expanded in a diagrammatic perturbation series consisting of all linked, proper diagrams (i.e. diagrams which cannot be unlinked by cutting a single internal particle or hole line). Pu and Chang[12] and Kelly[13] implemented the method in numerical calculations of electron scattering by He and H atoms. The complexity of numerical, single-center expansion methods for electron-molecule scattering makes it difficult to include polarization this way, but the expansion of the T-matrix in an L^2 basis set[1-4] provides an ideal framework for including polarization via diagrammatic representation of the optical potential. Diagrammatic perturbation methods in discrete basis sets proved useful in studies of molecular ground[14,15] and excited[16,17] states, as well as polarizabilities and transition moments.[14]

The optical potential may be wirtten as

$$V_{op} = V_{SX} + V'_{op} , \tag{9}$$

with V'_{op} consisting of all linked, proper diagrams of second and higher orders. Second order diagrams are shown in Fig. 1. The matrix elements of the optical potential represented by these diagrams are

$$< \alpha|V^{(2)}_{op}|\beta > = \sum_{ijp} \frac{<\beta p|ij > < ij|\alpha p>_x}{\varepsilon_k+\varepsilon_p-\varepsilon_i-\varepsilon_j} - \sum_{ipq} \frac{< pq|\alpha i > <\beta i|pq >_x}{\varepsilon_p+\varepsilon_q-\varepsilon_k-\varepsilon_i} \tag{10}$$

where the first sum comes from diagrams (a) and (b) in Fig. 1 and the second - from (c) and (d). The indices p and q in (10) go over occupied orbirals ("holes") and i, j go over unoccupied orbitals ("particles"). The electron-repulsion integrals are

$$< ij|\alpha p >_x = < ij|\alpha p > - < ij|p\alpha > \tag{11}$$

and

$$< ij|\alpha p > = \int \frac{\phi^*_i(1)\ \phi^*_j(2)\ \phi_\alpha(1)\ \phi_p(2)}{r_{12}} d\vec{r}_1 d\vec{r}_2 \tag{12}$$

ε_i is the energy of orbital ϕ_i. The usual diagram rules[18] apply in translating Fig. 1 to Eq. (10), except that the energy associated with the incoming orbital $|\beta>$ is $\varepsilon_k = k^2/2$, the energy of the scattered electron. The Gaussian basis orbitals used in the calculation are are given in Table I. The inner basis set is relatively

Table 1. Basis Sets

Type	Origins	Exponents: Inner Set*	Exponents: Outer Set
s	H atoms	1685.517(0.0011993)+249.9584(0.009656) +55.65834(0.0515089)+15.2743(0.230237) +4.8628(0.778537),1.7316,0.66805, 0.27437,0.11698,041133	1685.517,249.9584,55.65834,15.2743, 4.8628,1.7316,0.66805,0.27437, 0.11698,0.041133
p_z	H atoms	4.8(0.10315)+2.53(0.07621)+1.33(0.86207), 0.701,0.369	4.8,2.53,1.33,0.701,0.369
s	H_2 center	0.015,0.005	0.15,0.06,0.02,0.009,0.0054,0.0032, 0.0019,0.001166,0.0006
p_z	H_2 center		0.15,0.05,0.015,0.005,0.0015
p_x	H atoms	4.0,1.5,0.5,0.15,0.05	4.0,1.5.0.5
p_x	H_2 center		0.15,0.05,0.015,0.005,0.0015,0.0006
d_{xz}	H_2 center		0.15,0.05,0.015,0.005,0.0015

*Contraction Coefficients (coefficients of the Cartesian Gaussian in the contracted basis function) are given in parentheses.

small (30 functions), and its task is to describe adequately the occupied and excited H_2 orbitals (the latter figure in the polarization). The outer set spans the orbital of the scattered electron throughout the region of space where the scattering potential is significant. It is therefore considerably larger (61 functions) and includes many diffuse orbitals. In terms of Fig. 1 and Eq. (10), the orbitals p, q, i and j are spanned in the inner set, while α and β belong to the outer set. The use of two different sets leads to great computational savings, as the integrals in (10) involve three orbitals of the small inner set and only one belonging to the larger outer set. The computation of the matrix elements of the static-exchange potential V_{SX} took 8 minutes on a CDC6600, and the second-order optical potential required 5 minutes (with an increment of 2 seconds for each additional scattering energy).

The integral cross sections for elastic, rotational ($j = 1 \rightarrow 3$), vibrational ($v = 0 \rightarrow 1$) and vibrational-rotational ($v = 0 \rightarrow 1$, $j = 1 \rightarrow 3$) processes are shown in Fig. 2. All cross sections were obtained by integrating over the orientation of the molecule $\hat{R}$ as well as internuclear separation R with appropriate rotational and vibrational functions, using the adiabatic-nuclei equation (6).

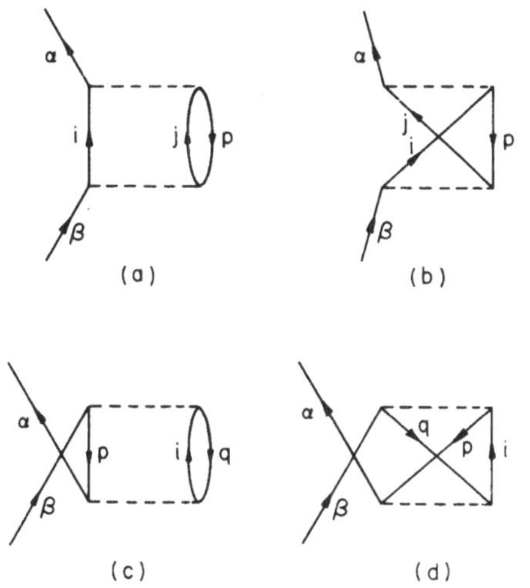

Fig. 1. Diagrams contributing to the second-order optical potential. Diagrams (b) and (d) are exchanges of (a) and (c), respectively. See Eq. (10) for explicit expressions.

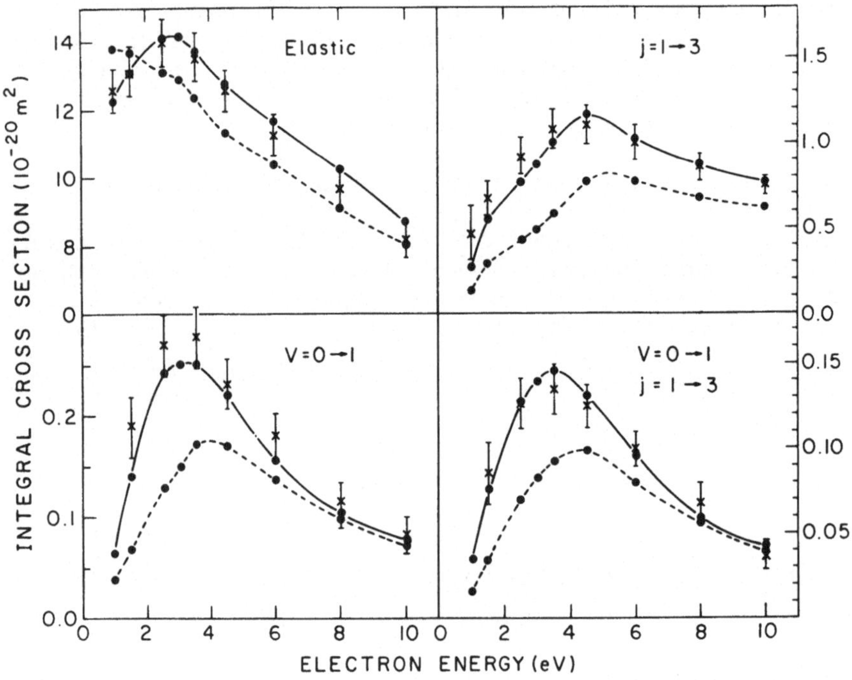

Fig. 2. Integral cross sections for elastic (upper left), rotational (upper right), vibrational (lower left) and vibrational-rotational (lower right) processes. Dashed curves are static-exchange results, solid lines include second-order polarization. Crosses show the experimental results of Linder and Schmidt.[23]

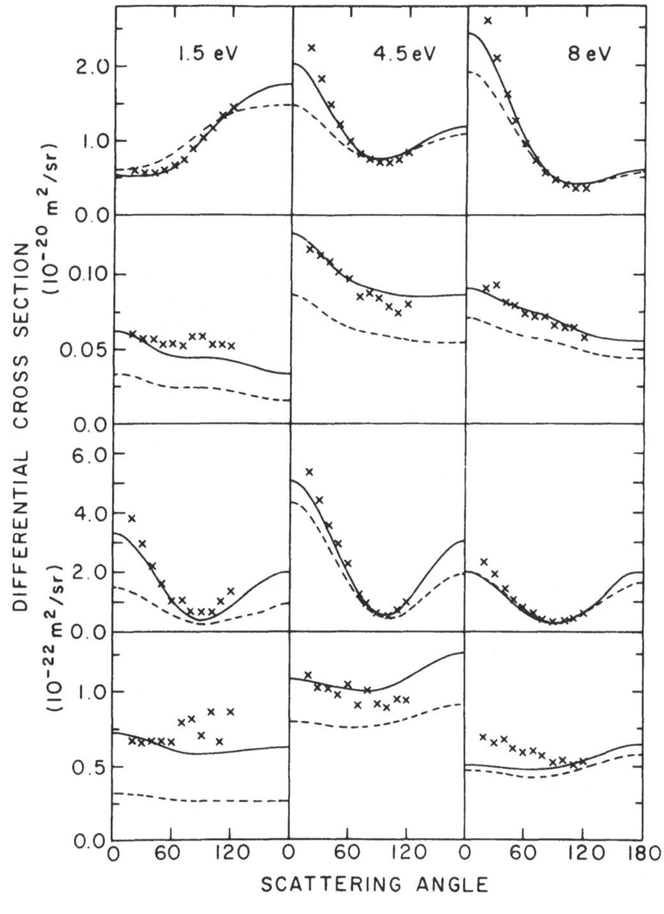

Fig. 3. Differential cross sections for elastic scattering (upper row), rotational excitation ($j = 1 \rightarrow 3$, second row), vibrational excitation ($v = 0 \rightarrow 1$, third row) and rotational-vibrational excitation ($v = 0 \rightarrow 1$, $j = 1 \rightarrow 3$, bottom row). The columns are (from left to right) for scattering energies of 1.5, 4.5 and 8 eV. Symbols as in Fig. 2.

Six integration points for each of the angles θ and ψ in the first octant and 10 points for R between 0.6 and 3.0 bohr were taken. The vibrational funxtion χ_o was obtained from the Kolos-Wolniewicz potential.[19] The elastic scattering and rotational excitation cross sections in Fig. 2 differ significantly from our previously reported results,[20] obtained from calculations at the equilibrium distance R_e only using a fixed-R formula,

$$\frac{d\sigma(j\ j')}{d\Omega} = \frac{k_f}{k_i} \frac{1}{2j+1} \sum_{mm'} |\langle Y_{j'm'}(\hat{R})|f(R_e,\hat{R})|Y_{jm}(\hat{R})\rangle|^2 \qquad (13)$$

with integration over the angles $\hat{R}$. There are also differences between previous[20] and present (Table 1) basis sets, but their effect is insignificant. Using the full adiabatic-nuclei expression (6), on the other hand, increases integral cross sections for elastic scattering by up to 5% and for rotational excitation by up to 40% over the single-R results, giving much better agreement with experiment. The subject of adiabatic-nuclei vs. fixed-R calculations for vibrationally elastic processes is discussed in detail elsewhere.[21] Finally, the vibrationally inelastic cross sections in Fig. 2 differ slightly from previously reported values[22] because the R-integration there was over a smaller range ($0.8 \leqslant R \leqslant 2.0$ bohr).

The integral cross sections shown in Fig. 2 demonstrate clearly the importance of polarization. The qualitatively incorrect behavior of the elastic cross section as a function of energy is rectified, and the peak appears at the right place. Other cross sections are increased as much as twofold. It is remarkable that inclusion of lowest-order polarization brings integral cross sections for all processes investigated over the whole energy range (1-10 eV) within the experimental[23] error limits or very close to them. This indicates that higher order corrections are small, at least for H_2. We intend to include third-order effects in the future.

Differential cross sections at three energies are shown in Fig. 3. Polarization effects are again important, and good agreement with experiment is obtained with the second-order optical potential. The forward peak in the elastic and pure vibrational cross sections is not fully reproduced; this is probably due to inclusion of lowest-order polarization diagrams only, and should be rectified when higher-order effects are properly taken into account.

We would like to thank Professor V. McKoy, Dr. A. W. Fliflet and Ms. D. A. Levin for helpful discussions and for the use of Green's function matrix elements programs. We also acknowledge the support of the U. S.-Israel Binational Science Foundation (BSF).

REFERENCES

1. T. N. Rescigno, C. W. McCurdy and V. McKoy, Chem. Phys. Lett. 27, 401 (1974).
2. T. N. Rescigno, C. W. McCurdy and V. McKoy, Phys. Rev. A 10, 2240 (1974).
3. T. N. Rescigno, C. W. McCurdy and V. McKoy, Phys. REv. A 11, 825 (1975).
4. A. W. Fliflet
5. D. M. Chase, Phys. Rev. 104, 838 (1956).
6. D. E. Golden, N. F. Lane, A. Temkin and E. Gerjuoy, Rev. Mod. Phys. 43, 642 (1971), especially Sec. IV and references therein.
7. A. W. Fliflet, D. A. Levin, M. Ma and V. McKoy, Phys. Rev. A 17, 160 (1978).
8. L. S. Rodberg and R. M. Thaler, Introduction to the Quantum Theory of Scattering (Academic Press, New York 1967), pp. 363-379.
9. B. I. Schneider, H. S. Taylor and R. Yaris, Phys. Rev. A 1, 855 (1970).
10. Gy Csanak, H. S. Taylor and R. Yaris, Adv. Atom. Molec. Phys. 7, 287 (1971).
11. J. S. Bell and E. J. Squires, Phys. Rev. Lett. 3, 96 (1959).
12. R. T. Pu and E. S. Chang, Phys. Rev. 151, 31 (1966).
13. H. P. Kelly, Adv. Chem. Phys. 14, 129 (1969).
14. U. Kaldor, J. Chem. Phys. 62, 4634 (1975).
15. R. J. Bartlett and P. M. Silver, J. Chem. Phys. 62, 3258 (1975); ibid 64 4578 (1976).
16. U. Kaldor, J. Chem. Phys. 63, 2199 (1975).
17. P. S. Stern and U. Kaldor, J. Chem. Phys. 64, 2002 (1976).
18. B. H. Brandow, Rev. Mod. Phys. 39, 771 (1967).
19. W. Kolos and L. Wolniewicz, J. Chem. Phys. 43, 2429 (1965).
20. A. Klonover and U. Kaldor, J. Phys. B 11, 1623 (1978).
21. A. Klonover and U. Kaldor, submitted to J. Phys. B.
22. A. Klonover and U. Kaldor, J. Phys. B, in press.
23. F. Linder and H. Schmidt, Z. Naturf. 26a, 1603 (1971).

R-MATRIX CALCULATIONS FOR ELECTRON SCATTERING BY DIATOMIC MOLECULES

B. D. Buckley* and P. G. Burke†*

*Science Research Council, Daresbury Laboratory, Warrington WA4 4AD, England. †Also Department of Applied Mathematics & Theoretical Physics, Queen's University, Belfast BT7 1NN, N. Ireland.

I. INTRODUCTION

Burke and Robb[1] have demonstrated rather convincingly the suitability of R-matrix techniques for describing such atomic processes as electron scattering, photoionization, polarizabilities and Van der Waals coefficients, electron affinities, and non-linear optical coefficients. In reactive and non-reactive heavy particle scattering also, the R-matrix approach has been shown applicable.[2] More recently Schneider et. al[3] have considered the scattering of low energy electrons by molecular targets using an R-matrix method, as described elsewhere in this volume. In this paper we present the results for an alternative formulation of the problem, described in detail in references 4 and 5.

Wigner and Eisenbud[6] were the first to introduce the idea of dividing configuration space into an external and internal region. In the former, the scattering particles are far removed from each other, the interactions correspondingly weak, and well represented by local potentials. In the internal region, on the other hand, the particles are in close proximity to each other and the interactions normally require non-local representation. In the R-matrix approach the Schrodinger equation in each region is solved separately, the final solutions over all space being formed by matching the logarithmic derivatives of the respective solutions in each region on the R-matrix boundary.

II. THEORY

A detailed discussion of an R-matrix theory for low energy electron scattering by diatomic molecules is given in references 4 and 5.

In the external region, the absence of exchange allows the Schrodinger equation to be reduced to a set of coupled second order homogeneous differential equations which are solved in these calculations using the variable phase method of LeDourneuf and Vo Ky Lan.[10]

In the internal region the continuum (and bound) wavefunctions are expanded in terms of Slater type orbitals (STO's) centered on the two nuclei and on the center-of-mass. Only these STO's on this latter center are permitted to be sufficiently diffuse to be non-negligible near the R-matrix boundary at r = a. In this way the multi-centered approach in the internal region becomes single-centered in the external region. Solution of the Schrodinger equation in the internal region requires the determination of the eigenvectors and eigenvalues of

$$\left[H + L_b \right]_{\text{all } r \leqslant a} \tag{1}$$

where H is the (N+1) electron Hamiltonian, and L_b the Bloch operator which subtracts the non-Hermitian terms in H.

The Bloch operator, L_b, also ensures that the eigenvectors of Eq. (1) satisfy the required logarithmic boundary condition on the R-matrix surface. This approach is referred to as the arbitrary boundary condition (ABC) method, in contrast to the fixed boundary condition (FBC) method adopted in electron-atom scattering.[7] In the FBC method the Bloch operator is not considered and the eigensolution of H in the internal region is determined numerically. The actual number of eigensolutions required in this method for a converged R-matrix is too large for any practical calculation. The approach adopted is to evaluate a truncated set of eigensolutions and approximate the omitted ones by using the Buttle correction.[8] In contrast, the use of the ABC method appears to obviate the need for a Buttle correction. This fact is well illustrated in electron-helium scattering, Table 1, and shall be returned to in the section on results.

III. COMPUTATION

A detailed description of the codes being developed for this work is given in reference 5. The bulk of the computational effort lies in the determination of the eigenvectors and eigenvalues of Eq. (1), and bears a marked resemblance to bound state configuration interaction calculations, the only difference being that integration is over a finite rather than an infinite range.

The choice of STO's in the center-of-mass has been found to be crucial in effecting a maximization of the number of eigenvectors which can be generated by minimizing the linear dependence among the

STO basis. The optimal choice has been found to lie in choosing STO's with high principle quantum numbers.

The computation is carried out in four stages. Atomic integrals over the chosen STO basis are first evaluated; a two and four index transformation determines the one- and two-electron molecular integrals; the matrix elements of Eq. (1) are calculated; and finally the matrix (1) is diagonalized, the R-matrix constructed and the integration in the external region commenced.

IV. RESULTS

A. Electron-Atom Scattering

The first calculations performed were on s-wave scattering by atomic helium and hydrogen within the static-exchange approximation. These served as a test of the theory itself as well as the codes for both a closed- and open-shell target. The results are given in Tables I and II.

The eigenphases for s-wave scattering by helium were calculated using eight continuum orbitals generated from eight STO's. They are

TABLE I

S-WAVE EIGENPHASES FOR ELECTRON-HELIUM SCATTERING IN THE STATIC-EXCHANGE APPROXIMATION

k^2	Berrington	Present
0.1	2.6891	2.6886
0.2	2.5121	2.5114
0.3	2.3827	2.3819
0.4	2.2788	2.2778
0.5	2.1914	2.1901
0.6	2.1159	2.1142
0.7	2.0495	2.0474
0.8	1.9902	1.9878
0.9	1.9368	1.9343
1.0	1.8882	1.8860
1.5	1.6978	1.6958
2.0	1.5636	1.5511
2.5	1.4625	1.4499
3.0	1.3830	1.3786
3.5	1.3182	1.2748
4.0	1.2638	1.1307

TABLE II

S-WAVE EIGENPHASES FOR ELECTRON-HYDROGEN SCATTERING IN THE STATIC-EXCHANGE APPROXIMATION

k^2	^1S		^3S	
	John	Present	John	Present
0.01	2.396	2.396	2.908	2.908
0.02	2.153	2.152	2.812	2.812
0.04	1.871	1.870	2.679	2.679
0.06	1.693	1.692	2.580	2.580
0.08	1.563	1.562	2.498	2.498
0.10	1.460	1.459	2.427	2.427
0.20	1.135	1.135	2.167	2.167
0.30	0.949	0.949	1.987	1.987
0.40	0.825	0.825	1.849	1.849
0.50	0.737	0.737	0.739	0.738
0.80	0.589	0.858	1.501	1.501
1.00	0.543	0.541	1.391	1.389

compared with a fixed boundary condition R-matrix calculation performed by Berrington in which ten eigensolutions of the internal Schrodinger equation were included as well as a Buttle correction. It is clear that only at the highest energies in Table I is there any loss of accuracy in the arbitrary boundary condition method and this can be easily rectified, if required, by the inclusion of one or two more continuum orbitals.

In the case of ^1S and ^3S electron scattering by atomic hydrogen, comparison is made with the static exchange calculations of John[12] (Table II). In the present calculation ten continuum orbitals were included and the $1s^2$ configuration allowed for explicitly in the singlet case. The agreement between the two calculations is excellent.

B. Electron-molecule Scattering

Electron scattering by molecular hydrogen in the $^2\Sigma_g^+$ and $^2\Sigma_u^+$ (molecular frame) channels has been studied within the static-exchange approximation. Although a multi-centered expansion was employed in the internal region, only the leading angular momentum term was considered asymptotically in each case, that is, s-wave and p-wave respectively. In the $^2\Sigma_g^+$ channel two nuclear-centered orbitals were generated, while in the $^2\Sigma_u^+$ channel three were used. In each case eight center-of-mass STO's were introduced to represent the outgoing scattered waves.

Comparison with the single-center calculations of Collins et. al.[13] (who used the same target wavefunction) and with the R-matrix calculation of Schneider[3] (who used different target wavefunctions) shows good agreement among all the calculations in the $^2\Sigma_g^+$ channel (Table III). The slight discrepancy between the present calculations and the others above 0.8 Rydberg can be rectified by the inclusion of a further continuum orbital in the internal wavefunction expression. At these energies the higher partial wave terms also begin to contribute more significantly to the eigenphase sum and need to be included in any further calculations.

Similar agreement has been found in the $^2\Sigma_u^+$ channel (Table IV). The disagreement between the present eigenphases, the single-center calcultions, and the results of Schneider at the lowest energies is puzzling, but has probably been caused by a failure by the latter author to integrate sufficiently far in the external region. It has been found possible to reproduce Schneider's low energy eigenphases if the integration is stopped at 10 a.u. instead of 50 a.u. as has been found necessary by Collins et. al.[13] and the present authors.

In molecular nitrogen, the $^2\Pi_g$ and $^2\Pi_u$ scattering channels have been considered in a similar fashion to the hydrogen case: that is, multi-centered internal wavefunctions are generated, but only the

TABLE III

THE $^2\Sigma_g^+$ S-WAVE EIGENPHASES FOR ELECTRON-MOLECULAR HYDROGEN SCATTERING IN THE STATIC-EXCHANGE APPROXIMATION

k^2	Schneider	Collins	Present
0.01	2.9313	2.9273	2.9280
0.04	2.7212		2.7182
0.09	2.5132	2.5117	2.5159
0.10			2.4840
0.16	2.3116	2.3226	2.3243
0.20			2.2376
0.30			2.0617
0.36	1.9494	1.9769	1.9770
0.40			1.9271
0.50			1.8146
0.60			1.7021
0.70		1.6384	1.5823
0.80			1.4684
0.90			1.3836
1.00	1.4179	1.4464	1.3439

TABLE IV

THE $^2\Sigma_u^+$ P-WAVE EIGENPHASES FOR ELECTRON-MOLECULAR HYDROGEN SCATTERING IN THE STATIC-EXCHANGE APPROXIMATION

k^2	Schneider	Collins	Present
0.01	0.0043	0.0128	0.0126
0.04	0.0333	0.0463	0.0462
0.09	0.1037	0.1180	0.1177
0.10			0.1338
0.16	0.2243		0.2388
0.20			0.3104
0.25	0.3891	0.3985	0.3973
0.30			0.4788
0.36	0.5610		0.5675
0.40			0.6193
0.49	0.7005		0.7065
0.50		0.7202	0.7136
0.60			0.7661
0.64	0.8057		0.7832
0.70			0.8116
0.80			0.8684
0.81	0.8840		0.8738
0.90			0.9072
1.00	0.9271	0.9301	0.8888

leading partial wave terms (d-wave and p-wave respectively) have been included in the external region. The Hartree-Fock wavefunction of Ransil[14] was used to represent the target.

In Table V the need for a judicial choice of basis set is illustrated. In both calculations the continuum basis was chosen to span the space between the molecule and the R-matrix boundary. The six term expansion, however, has the now apparent advantage over the four term expansion of placing a sufficient number of terms near the important region close to the R-matrix surface leading to an adequate representation of the continuum wavefunction and its derivative on the surface itself. The augmentation of the basis with further terms has a far less dramatic effect on the eigenphase.

In the $^2\Pi_u$ channel, eight center-of-mass STO's were used to make this comparison with the calculations of Schneider and Morrison.[3d] The agreement is quite satisfactory, any discrepancies being probably due to the use of different target wavefunctions (Table VI).

TABLE V

THE $^2\Pi_g$ D-WAVE EIGENPHASES FOR ELECTRON-MOLECULAR NITROGEN SCATTERING IN THE STATIC-EXCHANGE APPROXIMATION. (i) 5-CONTINUUM ORBITALS, (ii) 6-CONTINUUM ORBITALS

k^2	(i)	(ii)
0.1	0.0001	0.0235
0.2	0.1205	0.1429
0.3	0.5693	0.6697
0.4	1.9112	2.0069
0.5	2.3151	2.3684
0.6	2.4364	2.4419
0.7	2.3017	2.4646
0.8	2.0364	2.4723
0.9	1.7251	2.4496
1.0	1.4117	2.3799

TABLE VI

THE $^2\Pi_u$ P-WAVE EIGENPHASE FOR ELECTRON-MOLECULAR NITROGEN SCATTERING IN THE STATIC-EXCHANGE APPROXIMATION

		Schneider	
k^2	Present	k^2	Eig
0.01	3.1550	0.01	3.134
0.05	3.1767		
0.10	3.1517	0.09	3.091
0.15	3.0996	0.154	3.011
0.20	3.0418		
0.25	2.9850	0.25	2.893
0.30	2.9316		
0.35	2.8820		
0.40	2.3365		
0.45	2.7945		
0.50	2.7545		
0.55	2.7141		
0.60	2.6709		
0.65	2.6228		
0.70	2.5692		
0.75	2.5105		
0.80	2.4485	0.799	2.302
0.85	2.3856		
0.90	2.3250		
0.95	2.2702		
1.00	2.2249	1.0	1.335

V. CONCLUSIONS

It has been demonstrated that the R-matrix theory of electron-molecule scattering can be successfully applied to scattering within the static-exchange approximation employing a very small number of basis orbitals. This gives rise to the very reasonable hope that the theory and computer codes can be further extended to more complicated and physically realistic calculations for electron-molecule processes. This work is in progress.

We would like to thank Drs. M. LeDourneuf and Vo Ky Lan for their invaluable assistance with the variable phase method. One of us (BDB) would like to thank W. D. Robb and L. A. Collins for a very fruitful two week collaboration at the Los Alamos Scientific Laboratory this summer.

VI. REFERENCES

1. P. G. Burke and W. D. Robb, Adv. Atom. Molec. Phys. 11, 143 (1975).
2. O. Crawford, J. Chem. Phys. 55, 2563 (1971); J. C. Light and R. B. Walker, J. Chem. Phys. 65, 4272 (1976).
3a. B. I. Schneider, Chem. Phys. Letts. 31, 237 (1975);
b. B. I. Schneider, Phys. Rev. A11, 1957 (1975);
c. B. I. Schneider, and P. J. Hay, Phys. Rev. A13, 2049 (1976);
d. B. I. Schneider, and M. A. Morrison, Phys. Rev. A16, 1005 (1977).
4. P. G. Burke, I. Mackey, and I. Shimamura, J. Phys. B: Atom. Molec. Phys. 10, 2497 (1977).
5. B. D. Buckley, P. G. Burke, I. Mackey, and Vo Ky Lan, Proc. CECAM Workshop on Electron-Molecule Scattering, August-September 1977.
6. E. P. Wigner, and L. Eisenbud, Phys. Rev. 72, 29 (1947).
7. P. G. Burke, A. Hibbert, and W. D. Robb, J. Phys. B: Atom. Molec. Phys. 4, 153 (1971).
8. P. J. A. Buttle, Phys. Rev. 160, 719 (1967).
9. I. Shimamura, Invited Paper at the X Int. Conf. on Physics of Electron and Atomic Collisions, Paris, 1977. (North-Holland, 1978).
10. M. Le Dourneuf and Vo Ky Lan, J. Phys. B: Atom. Molec. Phys. 10, L35 (1977).
11. K. A. Berrington, private communication.
12. T. L. John, Proc. Phys. Soc. 76, 532 (1960).
13. L. A. Collins, M. A. Morrison and W. D. Robb, private communication.
14. B. J. Ransil, Rev. Mod. Phys. 32, 245 (1960).

DISCUSSION ON ELECTRON-MOLECULE SCATTERING

Chairman: Barry Schneider

Schneider: If you think about the various L^2-methods, you will see that we've done some fairly significant problems. On the N_2 problem, for example, a low ℓ-spoiling calculation has been done, an R-matrix calculation has been done and also a T-matrix calculation, and they all agree and those are the only three calculations in the literature that I know of that do agree.

Henry: What does agreement mean here?

Schneider: We get the resonances in the same place, and the cross sections are within 5%. And another point is that the only multichannel calculations that I know of are the one I've done on H_2 and the one that Chung's done on H_2. I'm talking about multi-electronic-channel calculations now.

Truhlar: There's Black and Lane's work in 1968 on H_2.

Rescigno: What is the reference?

Truhlar: By Ken Black and Neil Lane in the Boston ICPEAC Proceedings, 1968.

I want to make some comments about the work we are doing with semi-empirical polarization potentials. In these potentials, we take the asymptotic form of polarization attractions and a cut-off function, called C(r), whose functional form is largely arbitrary and then the cut-off parameter is adjusted empirically. For example, in N_2, Buckley and Burke included exchange and adjusted the cut-off parameter to the 2.4 eV resonance in the $^2\Pi_g$ channel. That's an empirical polarization potential. Well, we've been using the model potential approach, as opposed to L^2-methods and our results are summarized elsewhere in these proceedings [see contribution by Truhlar et al. at the end of this discussion].

Bardsley: If you take the static-exchange and the static plus polarization and subtract the static-exchange plus polarization, do you get the static results?

Truhlar: No, it doesn't work that way. I do have some results that show different polarization potentials and what happens if you make the polarization stronger. These results just sum up the previous ones--it's the sum of the elastic and the rotational excitation so that you can compare with what the experimentalists measure.

Dill: By elastic, do you mean $R = R_e$?

Truhlar: These results are for $R = R_e$.

Dill: So it may be elastic.

Truhlar: Well, I consider $R = R_e$ to be an approximation using vibrationally averaged potentials. In the past, we've sometimes used vibrationally averaged potentials. I don't think for this purpose that that's our biggest error.

Taylor: Are the electronic inelastic cross sections at 30 eV so small that the fact that you have no sinks in your calculation not important?

Truhlar: I think that's one of our most serious errors. We should be using a complex polarization potential. Looking at the experimental results, there is certainly room for a complex potential.

Herzenberg: When you do one of these high energy scattering calculations--beginning at 30 to 50 volts--with a polarization potential, how high do you have to go before the polarization potential just disappears because the target electrons don't have time to follow?

Truhlar: That seems to depend on exactly what you're looking at. But for something like elastic scattering, I think that you need the polarization potential at least up to 100 eV. In some studies on H_2 with the polarized Born approximation between 100 and 900 eV, it was a little bit hard to tell, because the experimental data doesn't always go to small enough angle where you see the differences. But the effects seemed to vanish at a few hundred eV.

Herzenberg: If you don't do a scattering calculation, but you simply calculate the polarization potential and you do it honestly and put in the velocity dependence, at what energies do the velocity dependences start to hurt you?

Taylor: In helium, we used a velocity dependent optical potential and at 50 eV we needed it. I don't know how much higher you have to go, but at 50 eV at angles below 50° you need it. If we didn't have it, we wouldn't agree with electron-helium data at 50 eV; we need the energy-dependent polarization potential.

Poe: We have an unpublished but completed calculation on electron-N_2, electron-CO and CO_2 from 50 eV to 500 eV. Now it is a much simpler method. [See contribution by B. Choi and R. Poe at the end of this discussion.] From 50 eV up, the contribution from long-range polarization starts decreasing. In fact, above 200 eV, the forward peaking is not caused by the polarization potential, although there is a tremendously large forward peak.

McKoy: When you pick your parameters for your static plus polarization potential and static-exchange plus polarization potential, how different are those parameters?

Truhlar: There's only one parameter in this whole calculation and it's a 2.3 a.u. cut-off in the polarization potential determined by Buckley and Burke in a calculation with non-local exchange.

Dill: How high did you have to go before you could forget about exchange altogether?

Truhlar: Well, exchange is basically a 1/E effect. At high energy, exchange goes like 1/E, so it goes away about at that speed.

Dill: 1/E compared to what?

Truhlar: Well, as you go to high energy, the exchange can be modeled very well by a 1/E potential, so it will be roughly half as big at 200 eV as at 100 eV. At high energy, you can use that as the leading term in the expansion. In the s-wave, you probably always need it.

McKoy: How low an energy did you go to?

Truhlar: We've done N_2 down to 5 eV, but I wasn't going to talk tonight about lower than 20 eV.

McKoy: I guess at 5 eV you see a much bigger difference between static-polarization and static-exchange polarization?

Truhlar: Yes, at 5 volts you see a bigger difference because the $^2\pi_g$ wave, for example, is still very sensitive at 5 volts. In fact, it's still very sensitive at 10 volts. And some of the other partial waves are a little bit more sensitive as well.

Temkin: I will present some recent results on the calculation of vibrational excitation cross sections for N_2 using our hybrid theory. [See contribution by A. Temkin at the end of this discussion.]

Bardsley: Are you implying that you don't doubt that the experimental results with which you are comparing are right?

Temkin: Yes, I think that the experiment is correct. It was done by Golden. It has now been repeated by Kennerly and Bonham at a much finer energy and although there may be slight differences, the normalization is absolutely the same to within a percent. It's a transmission experiment.

Taylor: Kennerly and Bonham used a time-of-flight device.

Temkin: Yes. So unless you have other information, I don't see why to believe that the experiment is still not trustworthy.

Bardsley: There are other experiments that give very different results.

Temkin: Such as?

Bardsley: Swarm data experiments.

Temkin: I have to make a choice at a certain point as to what experiment I choose to believe. I choose to believe Golden's experiment because I think it's an easy experiment and because there are two very good experiments which are in excellent agreement.

Herzenberg: Let me emphasize that our normalization (see Fig. 2 of Temkin's article) was gotten by adding to our resonant contribution the non-resonant contribution of Chandra and yourself, which is actually the same as that of Burke and Chandra. By comparing your cross sections to the experiment below the resonant energy, one sees that your non-resonant contribution was too large. This suggests that the reason our σ_T is too large is because of the exaggerated non-resonant contribution. If this is so, then our inelastic cross sections, which are purely resonant, as you mentioned, support Wong's normalization. (See Fig. 1c of Temkin's article.)

Temkin: What you say is perfectly possible: one can get an agreement with $\sigma_T = \sigma_{oo} + \sum_v \sigma_{o\to v}$ by reducing σ_{oo} and leaving the σ alone, which is not what our calculation suggests. From the calculational point of view, I would be happier if your results were more ab-initio in character, in particular the lack of ℓ-coupling even within context of the boomerang model is questionable. From the experimental point of view, it is clear that an absolute measurement of σ_{00} would be invaluable in determining which alternative is correct. The only measurement that I am now aware of is that of Srivastava, Chutjian, and Trajmar (see Fig. 11 of Temkin's article). As you can see, their results are really at higher energies than where we believe our calculation to be correct. Secondly, they are based on the e-He normalization of the McConkey and Preston whereas I think the new e-He normalization of Kennerly and Bonham is much more definitive and non-negligibly different. S. F. Wong has informed that he is measuring the $0 \to 0$ cross section absolutely. That is very good, because I (and most other people) would have great confidence in his results. The sine qua non of those results is when they are added to his $\sigma_{0\to v}$ results, they must give the experimental σ_T.

Morrison: I want to mention, in response to one of your comments, that David Golden and his group at Oklahoma are now gearing up to do those, as well as other measurements, with a time-of-flight apparatus and one hopes that within a year or so those numbers will come out.

Temkin: I think that they're tremendously important numbers.

Lane: You can also do a fixed nucleii calculation in the resonance channel in which case you can, in principle, reproduce the N_2^- curve as a function of internuclear separation, because that is the position of the resonance at each internuclear separation. In fact you did such a curve in your work with Chandra originally. Now my question is, with the current size calculation you are doing, do you know how that curve compares with the semi-empirical curve or Birtwistle and Herzenberg?

Temkin: We have not done that calculation. My suspicion is that it wouldn't change drastically.

Lane: The point I wanted to emphasize is that because of the semi-empirical nature of the polarization potential, the cut-off itself is a function of internuclear separation. And one can tune that precisely, if one wants to stay semi-empirical, by simply forcing your N_2^- curve to sit on the semi-empirical potential curve that we know reproduces the structure in a very precise fashion.

Temkin: Well, with all due respect to Arvid, I have some reservations. In other words, the approximations that go into the boomerang model are such that the $\Gamma(R)$ and the other parameters which go into his calculation are not really that model-independent.

Lane: If you're willing to do the calculation - demonstrating that, one way or the other, would be very important for those of us who are trying to better understand the answers to these questions.

Temkin: So what you're saying is that we should obtain Γ as a function of R in the fixed-nuclei sense. We will do that.

Temkin: The thing that was really surprising to us was the importance of the ℓ-coupling. I think it was somewhat of a coincidence that when Chandra and I stopped at $\ell = 3$, that the resonance came out right at the experimental value. I think that was misleading. So I think this business of an absolute normalization is obviously important and that it's a real challenge to theory and I would like very much to have a definite answer. We are the first to say that this is still not a converged calculation and Mike Morrison and Neal Lane have written an excellent paper, as you know, which gives a lot of tricks as to how we may in fact include the necessary coupling to really get something which we can call convergence. So we are trying to incorporate some of their ideas in our calculation.

Lane: That was Morrison and Collins?

Temkin: Yes.

Schneider: Let's move on to a different subject. I believe Bob Nesbet wanted to make some remarks.

Nesbet: Methods for doing calculations with fixed nucleii are apparently coming along very well for electron-molecule scattering. As you know, the adiabatic theory works quite well, if done properly, for rotational excitation. However, it's still a very serious difficulty, as we've seen in this work presented by Aaron, to include enough vibrational states in close-coupling to get any kind of meaningful coverged results, especially in the presence of a resonance. So the question is, is it possible to find some way to use fixed nuclei results from first principles calculations of the various sorts that have been discussed here and extract from them information which is physically meaningful and quantitatively correct. Now I'm going to discuss a modification of the adiabatic approximation which has the promise of actually doing this. [This method is discussed in detail in Phys. Rev. A. 19, (1979).]

Hazi: Can you say something about why there is a need for an energy modified adiabatic approximation since the adiabatic approximation has never been correctly applied to vibrational excitation in N_2 with ab initio resonance parameters?

Nesbet: My reason was to try to understand, for my own edification, just what really is behind the adiabatic approximation, how far it can be carried and so forth. I have used this modified adiabatic approximation to look at resonant vibrational excitation in N_2 and I do see the resonance substructure. In fact, if you look at the $0 \rightarrow 1$ structures, you see that the peaks are in the right place, the splittings are right and the threshold law looks very much like the experimental data and because this is an absolute calculation using only the Π_g partial waves, there's actually a definite result here which is very close to twice George Schultz's last normalization.

Dill: Can you say, in terms that you can describe to college freshmen, what those vibrational structures are due to?

Nesbet: Yes, those structures are a manifestation of a large number of resonances that are all overlapping.

Dill: Are these the vibrational states of the negative ion?

Nesbet: No, that's only in a limiting case as shown in the work of Birtwistle and Herzenberg. If you go to a limiting case where the resonances become very narrow, then these structures will be a manifestation of essentially the long-lived bound levels of the upper state. There are two curves involved here. The upper one is the negative ion state. But N_2 is an intermediate case where everything is mixed up with everything else and there are strong interference effects. What's actually happening is that something that would be a single resonance at fixed nucleii now becomes a resonance for each vibrational/rotational state; in fact -- we're leaving the rotational structure out of this but that's the implication of the theory. These are overlapping resonances, the width is at least twice as large as the spacings, so there is a lot of interference going on here. But the point is quite simply, if you can think of a resonance in terms of say the elementary introduction to scattering theory, where we speak of a pole in the scattering matrix corresponding to either a bound state or a resonance, then if the pole corresponds to a resonance then it's close to the real axis but a little below--which means that the lifetime is less than infinity and the thing decays. Now if you have an S-matrix in which you've left out the internal degrees of freedom of the system, what you really have is a pole of this S-matrix which depends upon internuclear distance. But that then becomes a matrix because it really depends on the nuclear Hamiltonian and in that matrix representation the single resonance must split into a manifold of poles of the overall S-matrix, one for each vibrational state. And that's what we are looking at here.

Dill: Each vibrational state of the compound negative ion?

Nesbet: Each vibrational state of the target molecule. But there is a transition from something which is dominated by the vibrational structure of the neutral to something which dominated by the vibrational structure of the negative ion, depending on the width of the resonances. So it's an intermediate situation. But there's this sort of one to one relationship between the number of functions that are available for internal states of the system. I hope I've answered your question.

Temkin: Let me just ask you something. When Birtwistle and Herzenberg did their calculation, they had one equation to solve which was essentially a radial equation for, if you like, a perturbed vibrational function which then goes into an overlap element. Now as I understand it, you have no radial equation to solve.

Nesbet: Right. My calculations are easier to perform than those of Birtwistle and Herzenberg because my calculation consists of laying out a set of Morse oscillator functions and evaluating matrix elements, which I do by numerical quadrature, of the resonance energy which becomes a matrix.

Bardsley: You could expand the solutions--that is the differential equation of Birtwistle and Herzenberg--in any set you like.

Nesbet: Yes, but what I'm doing is not exactly equivalent to Birtwistle and Herzenberg, particularly because I've replaced the function of R by a function of energy. That makes it different. It's a function of a different operator, a width function. I think physically it is correct to make it a function of the energy.

McKoy: Have you seen the paper by Cederbaum and Domcke?

Nesbet: Very much so.

Hazi: How do you compare with their results?

Nesbet: Their's is a procedure in which they build an operator model and then solve the operator model exactly. It's an implementation if you wish of the boomerang model but it has the correct physics of the problem in it. It is not expressed in terms of an adiabatic approximation. What I'm doing here is showing that one can start with the adiabatic approximation and go directly to results like this without building an intermediate parameterized model as such. Now the next step is to take something like Aaron's best fixed-nuclei K-matrix, then actually carry out the analysis of the corresponding S-matrix into the foreground and the background and carry out this analysis with those functions. And I would expect to get quite good results doing that. The same physics is in the Domcke and Cederbaum paper. But they actually have to build models where mine is a technique which allows us to go directly to the ab initio calculated K-matrix and use it.

Hazi: No, I beg to differ with you. The Cederbaum model is the correct adiabatic way of treating the resonance problem because it solves for the adiabatic nuclear wavefunction in the negative ion state, as suggested originally in the work of Herzenberg and others. One has to do this in the adiabatic nuclei approximation. The only input they require is exactly the same input that you require without this energy modification, that is, you need $\Gamma(R)$ and $E_{res}(R)$. You can take an ab initio calculated T-matrix, you can parametrize the resonance, get $\Gamma(R)$ and $E_{res}(R)$ and that is the correct adiabatic comparison.

Nesbet: No, if you execute the adiabatic approximation in the sense that it's understood in the literature, it doesn't work, because it doesn't yield the residues properly. You have to modify it in some way by putting in an operator representation.

Hazi: But that's exactly what the Cederbaum model does, as well as older work by O'Malley and Chen--people have worked on this problem for a long time.

Nesbet: What we're doing here is to apply ideas which go a long way back to the theory of resonances and that have been applied to the e^- - N_2 problem, particularly by Herzenberg and Mandl quite a long time ago and by Joe Chen. But we're applying it in a context where we can use ab initio calculations.

Herzenberg: I'd like to just make one point in response to the question about what one can tell a college freshman about why this thing works the way it does. The essential point is to look at the nodes and anti-nodes of the nuclear wavefunction, and to see how they shift about when you change the energy of the incident electron. You can draw a picture and show how the nodes and anti-nodes shift.

Dill: But what does the wavefunction correspond to? Is it the negative ion?

Herzenberg: It's the negative ion, yes.

POLARIZATION POTENTIALS FOR ELECTRON SCATTERING

D. G. Truhlar, D. A. Dixon, Robert A. Eades,
F. A. Van-Catledge and K. Onda†
Department of Chemistry, University of Minnesota
Minneapolis, Minnesota 55455;
†and also Jet Propulsion Laboratory
California Institute of Technology
Pasadena, California 91103

Charge polarization effects (due to polarization of the target charge distribution by the incident electron) are important for low and intermediate-energy electron scattering (these energy ranges corresponds to roughly $E \lesssim IP$ and $IP \lesssim E \lesssim 10$ IP, where E is the impact energy and IP is the target ionization potential). There are two approaches to the inclusion of such polarization effects in electron scattering. In the many-body approach, the scattering wavefunction for the whole system (incident electron plus target) is represented explicitly by basis functions or products of basis functions and numerically determined radial functions. Algebraic variational methods[2] and R matrix[3] methods are some particularly powerful variants of this approach. In this approach charge polarization effects enter by configuration mixing. Because of this and because polarization effects are of long range, basis sets are required to be large and the scattering wavefunction must be represented over a big region. To avoid the associated computational problems, most electron-molecule scattering calculations using basis functions have been restricted to the single-configuration level, i.e., the static-exchange approximation, in which polarization effects are neglected.[4] The second approach to including polarization effects is the use of effective potentials, also called optical potentials or model potentials. In this approach, electronically elastic scattering is reduced from a many-body configuration-mixing problem to single-particle scattering from an effective potential. It is too difficult to calculate the exact optical potential for electron-molecule scattering, so one must use approximations.

One model in use is to assume that the effective potential is the sum of a static potential, an exchange potential, and a polarization potential. The static potential may be calculated straightforwardly from accurate[5-8] or simple[6,9,10] target wavefunctions. The exchange potential may be taken as the nonlocal continuum-Hartree-Fock exchange potential[11-17] or as a local (but energy-dependent) approximate exchange potential.[8,17-22] These are now well tested against nonlocal exchange;[17,20,22-29] they lead to great computational simplifications yet they have been shown to be capable of good accuracy at intermediate energy and even, in some fortuitous cases or with tuning,[17] at low energy. Their success at intermediate energy makes an effective potential approach particularly appealing in that energy range. The biggest source of difficulty is the treatment of charge polarization. In this report we discuss our recent work on the polarization potential for electronically and vibrationally elastic electron-molecule scattering.

For electron scattering by linear molecules, it has been popular to use the following semiempirical functional form to represent the polarization potential $V^P(\vec{r},\vec{R})$:[8,16,17,19,20,30-34]

$$V^P(\vec{r},\vec{R}) \left[-\frac{\alpha_0(R)}{2r^4} - \frac{\alpha_2(R)}{2r^4} P_2(\hat{r}\cdot\hat{R}) \right] C(r) \tag{1}$$

where

$$C(r) = \left[1 - \exp -(r/r_c)^n \right] \tag{2}$$

$\vec{r}$ is a vector from the center of mass of molecule to the scattering electron, n is an integer (whose value is 6 for all cases discussed in this paper), and r_c is a parameter whose value is determined semiempirically. We have assumed a diatomic target so that $\vec{R}$ is the internuclear vector of the molecule and $\alpha_0(R)$ and $\alpha_2(R)$ are the isotropic and anisotropic components of the static electric dipole polarizability tensor for internuclear distance R. The main justification for this polarization potential is that the terms in brackets in equation (1) are known to provide the correct large-r form of the exact optical potential for E less than the lowest electronic excitation threshold,[35,36] but eventually at small r these terms blow up so it must overestimate the exact optical potential. But there are several difficulties with this form of the polarization potential: (i) There is no justification for $C(r) \leq 1$; in fact the bracketed part of equation (1) probably underestimates the exact optical potential at medium r where the quadrupole polarizability should be included. (ii) There is no justification for C(r) being independent of E; in fact nonadiabatic effects (the inability of the target electrons to respond adiabatically to the scattering electron at $\vec{r}$ due to the fact that $d\vec{r}/dt$ is not infinitesimal in a scattering event) decrease the target response at high E. So C(r) should be a decreasing function of E at least at some r and at high E. (iii) There is no justification for C(r) being independent of R and r·R. These assumptions are made for simplicity. (iv) There is no justification for neglecting other terms in the expansion

$$V^{P}(\vec{r}\,\vec{R}) = \sum_{\lambda=0}^{\infty} V_{\lambda}^{P}(r,R)\; P_{\lambda}(\hat{r}\cdot\hat{R}) \tag{3}$$

at small r, although these other terms do decrease more rapidly than r^{-4} at large r. Despite these difficulties, equations (1) and (2) do seem to represent the dominant physical effect of the scattering electron's interaction with the induced dipole, and these equations have been used successfully by various workers at low energy. One can argue that C(r) need not be too sensitive to E at low energy and small r because nonadiabatic effects should be a function mainly of the local kinetic energy which can be approximated (to zero order in the polarization potential) as

$$T_{loc}(\vec{r},\vec{R}) = E - V^{SE}(\vec{r},\vec{R}) \tag{4}$$

where $V^{SE}(\vec{r},\vec{R})$ is a local (possibly energy-dependent) approximation of the static-exchange potential. At small r, $V^{SE}(\vec{r},\vec{R})$ has a large negative value, so $T_{loc}(\vec{r},\vec{R})$ is not too sensitive to E at small r at low E. Intermediate energy calculations have now been performed[37,38] which partly test whether this argument can be extended to cover the intermediate-energy range by using r_c values determined empirically at low energy. Some of the results[37] are shown in Figure 1.

The quantity shown in Figure 1 is the electronically and vibrationally elastic differential cross section, i.e., the sum of the differential cross section for elastic scattering and rotational excitation. We plot this quantity so we can compare to the experimental results,[39-43] for which rotational excitation is not resolved. The calculated results in Figure 1 are converged close coupling calculations for given approximations to the effective potential. The convergence tests are detailed elsewhere.[37] For Figure 1 we used the INDOXI/1s approximation[6] for the static potential and the electronic density, the semiclassical exchange approximation,[24] and the semiempirical polarization potential of equations (1) and (2). The parameter r_c is given the value of (2.308 a_o) determined by Buckley and Burke,[16] who used a more accurate static potential and wavefunction and nonlocal exchange and semiempirically adjusted r_c to the resonance at 2.4 eV. For comparison, Figure 1 also shows the results obtained with $r_c =$, i.e., no polarization potential. (The effect of a more attractive polarization potential will be considered below.) If the experimental uncertainty is estimated as the difference between the various experimental results shown,

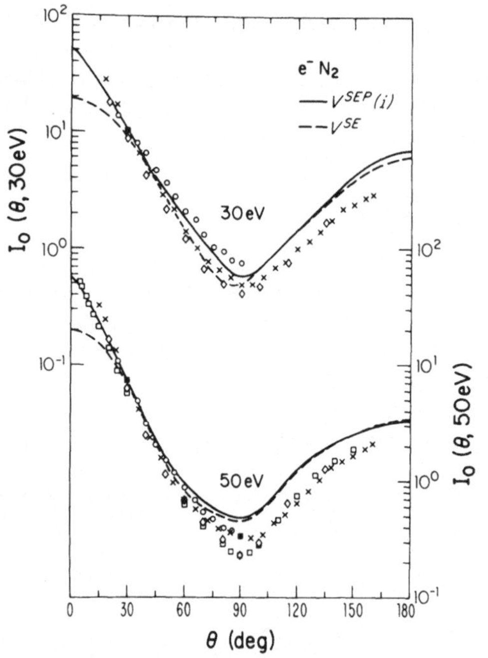

Fig. 1. The vibrationally elastic differential cross section for electron-N_2 at impact energies of 30 and 50 eV. The calculations (from reference 37) involved scattering basis set number IX and are for a fixed internuclear distance ($R=R_e$). For the calculations represented by the dashed line we used the INDOXI/1s method for the static potential and target density and the semiclassical exchange approximation. For the calculations represented by the solid line we also added the Buckley-Burke semiempirical polarization potential. This static-exchange-polarization potential will be called potential i. The various symbols represent the various experimental results. ◇ represents the experimental results of Srivastava, Chutjian and Trajmar (reference 42) which were obtained as ratios to the differential cross sections for He. They have been placed on an absolute scale for this figure by using the preliminary results of Register, Trajmar, and Srivastava (Trajmar, private communication) for He. ▢ represents the absolute measurements of DuBois and Rudd (reference 43). X and O, respectively, represent the relative measurements of Shyn, Stolarski, and Carignan (reference 40) and Finn and Doering (reference 41), normalized to the results of potential i at 30° and 30 eV. The measurements of Kuchitsu and Kambara are not included since their measurement was primarily designed for higher energies and their results at 50 eV may be less accurate than their higher-energy ones (Kuchitsu, private communication).

then both choices of r_c agree with experiment within experimental uncertainty for $\theta \geqslant 30^\circ$ eV although agreement is worse near 90° at 50 eV. However, polarization is clearly necessary at small θ; neglecting it severely underestimates the differential cross section there. We have obtained a similar level of accuracy for electron scattering by CO_2 at 20 eV[38] using a polarization potential of the form of equations (1) and (2) with the value of r_c determined by Morrison, Lane, and Collins,[8] who used the Hara free-electron-gas approximation for exchange and adjusted r_c to the resonance at 3.8 eV.

Despite the empirical success of equations (1) and (2) and their usefulness for qualitatively correct calculations of the cross sections, the difficulties with their theoretical foundation are unsatisfactory, and we require a theoretically more justified model to further our understanding of the physics, to increase our predictive capability, and to treat vibrational excitation and resolved rotational excitation more reliably.

We suggest a different model for the polarization potential which does not have difficulties (i) to (iv) detailed above. In this model[44]

$$V^P(\vec{r},\vec{R}) = V^{adP}(\vec{r},\vec{R})\ g[E,V^{SE}(\vec{r},\vec{R})] \tag{5}$$

where $V^{adP}(\vec{r},\vec{R})$ is the adiabatic polarization potential and $g[E,V^{SE}(\vec{r},\vec{R})]$ is a nonadiabaticity function. The dual role of the function C(r) in equation (1) is accomplished by two separate functions in equation (5). The adibatic polarization potential has the asymptotic form

$$V^{adP}(\vec{r},\vec{R}) \underset{r\to\infty}{\to} -\frac{\alpha_0(R)}{2r^4} - \frac{\alpha_2(R)}{2r^4} P_2(\hat{r}\cdot\hat{R}) \tag{6}$$

but it includes effects of higher multipoles at middle-range r and the breakdown of the multipole expansion at small r. When expanded in Legendre polynomials

$$V^{adP}(\vec{r},\vec{R}) = \sum_{\lambda=0}^{\infty} V_\lambda^{adP}(r,R)\ P_\lambda(\hat{r}\cdot\hat{R}) \tag{7}$$

it includes nonzero contributions from all λ except odd λ for homonuclear molecules for which odd λ terms in $V^P(\vec{r},\vec{R})$ are also zero. The nonadiabaticity function g can mimic the energy dependence of $V^P(\vec{r},\vec{R})$ in a realistic way, and the product in equation (5) should be able to mimic the dependence of $V^P(\vec{r},\vec{R})$ on R and $\hat{r}\cdot\hat{R}$ in a realistic way.

The first step in creating a model potential of the form of equation (5) is the calculation of realistic adiabatic polarization potentials for electron-molecule scattering. The adiabatic polarization potential can be calculated analytically for electron-hydrogen atom scattering[45] but otherwise requires a numerical calculation. The only results available for molecules for a long time were for H_2.[46-47] We have now made calculations emphasizing the small-r range for H_2,[48] N_2,[44,48,49] and CO:[44] and Morrison and Hay[50] have made a preliminary report of calculations for N_2 and CO_2 emphasizing $r \geqslant 5\ a_0$. These calculations are all self-consistent-field (SCF) single-configuration molecular orbital (SCF MO) calculations. The SCF MO's, n-electron wavefunction Ψ_0, and total electronic energy E_0 are calculated for the usual fixed nuclei electronic Hamiltonian $H_0(\vec{r}_1, \ldots, \vec{r}_n, \vec{R}_A, \vec{R}_B)$ for an n-electron target with nuclei of charges Z_A and Z_B at $\vec{R}_A$ and $\vec{R}_B$. Then the test-charge-added Hamiltonian is defined by

$$H(\vec{r}_1, \ldots, \vec{r}_n, \vec{R}_A, \vec{R}_B, \vec{r}) = H_0(\vec{r}_1, \ldots, \vec{r}_n, \vec{R}_A, \vec{R}_B) + \sum_{i=1}^{n} \frac{e^2}{|\vec{r}_i - \vec{r}|} - \frac{Z_A}{|\vec{R}_A - \vec{r}|} - \frac{Z_B}{|\vec{R}_B - \vec{r}|} \tag{8}$$

and new SCF MO's, a new n-electron wavefunction $\psi_{\vec{r}}$, and n-electron energy $E_{\vec{r}}$ are calculated. The static potential $V^S(\vec{r}\cdot\vec{R})$ and polarization potential are then found from

$$\begin{aligned} V^S(\vec{r},\vec{R}) &= \langle \Psi_0 | H | \Psi_0 \rangle - \langle \Psi_0 | H_0 | \Psi_0 \rangle \\ &= \langle \Psi_0 | H | \Psi_0 \rangle - E_0 \end{aligned} \tag{9}$$

$$\begin{aligned} V^P(\vec{r},\vec{R}) &= \langle \Psi_{\vec{r}} | H | \Psi_{\vec{r}} \rangle - \langle \Psi_0 | H | \Psi_0 \rangle \\ &= E_{\vec{r}} - \langle \Psi_0 | H | \Psi_0 \rangle \\ &= E_{\vec{r}} - E_0 - V^S(\vec{r},\vec{R}) \end{aligned} \tag{10}$$

The SCF MO calculations in our group have been carried out at three different levels of accuracy: the INDO and INDOXI semiempirical molecular orbital schemes employing a minimum basis set of Slater-type functions have been applied to N_2 and CO and ab initio calculations employing extended basis sets of Gaussian-type functions have been carried out for H_2 and N_2. For the calculations discussed here R equals R_e.

Our ab initio calculations for H_2 used a Gaussian basis set with 6 s functions and 4 p functions on each nucleus. The s-function exponential parameters are those of Huzinaga's 5 s set[51] plus 0.03, and the exponential parameters of the p functions are the four smallest in the s set. The two tightest s functions on each nucleus are contracted; all other functions are uncontracted. The

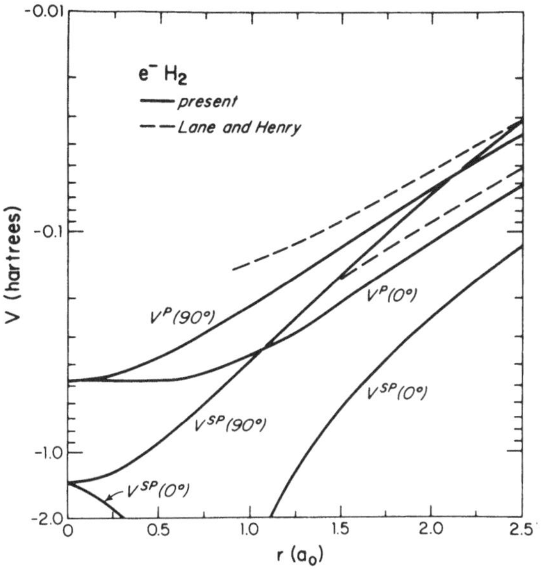

Fig. 2. Adiabatic polarization potentials and static-plus-adiabatic-polarization potentials for collinear (0°) and perpendicular-bisector (90°) approaches of electrons to H_2. V^P is the polarization potential and V^{SP} is the sum of the static and polarization potentials. The solid lines are present results and the dashed lines are the adiabatic polarization potentials of Lane and Henry.

polarization potentials for collinear and perpendicular-bisector approaches of the electron to H_2 are given in Figure 2 where they are compared to the results of Lane and Henry (Lane and Henry did not publish their adiabatic polarization potentials at small r). Considering the large differences in computational approach, the agreement in the overlapping region is good.

The *ab initio* calculations for N_2 use 9 s functions and 5 sets of p functions centered on each N with exponential parameters ξ from Huzinaga,[51] contracted to a [53] set by Dunning's rules,[52] and augmented by 4 s functions, 3 p sets, and 1 sd set. The additional s functions consist of bond-centered functions with $\xi = 1.13$ and 0.27 and nuclear-centered functions with $\xi = 0.065$. The additional p sets are all bond-centered and have exponents 0.68, 0.19, and 0.0515. The sd set consists of 6 bond-centered functions [x^2, y^2, z^2, xy xz, and yz times $\exp(-\xi r^2)$] with $\xi = 0.11$. The parameters 1.13 and 0.68 are from Vladimiroff[53] and are chosen to represent the bond region in the unperturbed target. The parameters 0.065 and 0.0515 are chosen from the smallest parts of s and p ξ's in the Huzinaga basis by extending the sequence as a geometric series. The ξ's 0.27 and 0.19 are then chosen as geometric means of the larger and smaller ones. The 0.11 value for the sd set is chosen as Werner and Meyer's polarizability optimized value for an N-centered function in NH_3.[54] We have found[48] by performing calculations at larger r that bond-centered functions are very useful for polarizability calculations.

The basis sets for the INDO/1s and INDOXI/1s calculations involved 2 s functions and 1 p set on each nucleus, with exponential parameters determined by Slater's rules.[55] This is the standard choice for INDO calculations.[56,57]

The three calculations of the adiabatic polarization potential and the static-plus-adiabatic-polarization potential for collinear and perpendicular-bisector approach of an electron to N_2 are shown in Figure 3. The INDO and INDOXI methods underestimate the spherical average of the polarization potential but show a qualitatively correct anisotropy. Further, the INDO and INDOXI methods lead to an adiabatic polarization potential that is too weak near r = 0 but the error in the static potential partly compensates for this in the INDO approximation.

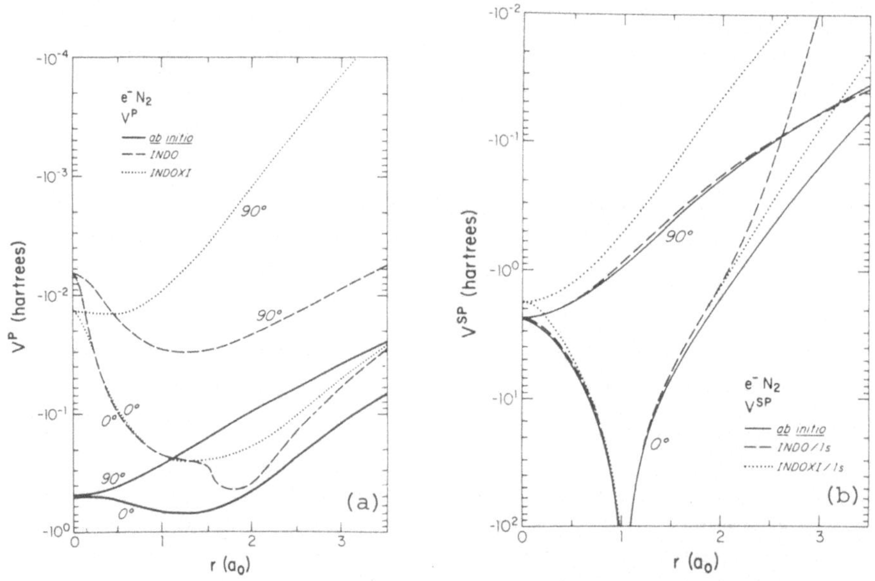

Fig. 3. Adiabatic polarization potentials (a) and static-plus-adiabatic polarization potentials (b) for collinear (0°) and perpendicular-bisector (90°) approaches of electrons to N_2. The solid line is the ab initio result, the long-dashed line is the INDO/1s result, and the dotted line is the INDOXI/1s result.

INDO calculations for other angles of approach of the electron to N_2 show that terms in equation (7) with $\lambda > 2$ are not completely negligible. For example, the INDO method yields V^{adP} $(r = 1.2a_0) = -0.094$ hartrees and V_4^{adP} $(r = 2.0\ a_0) = -0.024$ hartrees. Two additional *ab initio* calculations at the latter distance show an effect of the same order of magnitude for terms with $\lambda > 2$. Another interesting qualitative feature of the results is the size of $V_1^{adP}(r)$ for CO. In the INDO approximation the magnitude of this term exceeds 0.2 hartrees for r in the range 0.8 to 1.3 a_0. In previous effective potential calculations, terms in $V^P(r)$ with $\lambda \neq 0,2$ have almost always been ignored (see, however, reference 20).

It is interesting to write

$$V_0^{adP}(r) = -\ f_0(r)\ \lim_{r\to\infty}(2r^4)V_0^{adP}(r) \tag{11}$$

and

$$V_2^{adP}(r) = -\ f_2(r)\ \lim_{r\to\infty}(2r^4)V_2^{adP}(r) \tag{12}$$

thereby defining $f_0(r)$ and $f_2(r)$. If this is done, one finds that $f_0(r) \neq f_2(r)$ and that both $f_0(r)$ and $f_2(r)$ may exceed unity. This supports criticisms (i) and (iii) of the usual model.

To test the sensitivity of the vibrationally elastic differential cross section to the form of the polarization potential we repeated the electron-N_2 scattering calculations at 30 eV with several different static-exchange-polarization potentials.[49] The polarization and static-plus-polarization parts of three of these are shown in Figure 4 for the collinear and perpendicular-bisector geometries, where they are compared to the potentials used for the calculations in Figure 1. The corresponding vibrationally elastic differential cross sections are shown in Figure 5. Potential i is the potential used for Figure 1 and is repeated in Figures 4 and 5 for reference. Potential iii is the INDOXI/1s static-exchange potential plus the $\lambda = 0$ and $\lambda = 2$ components of the INDOXI adiabatic polarization potential, with $V_0^P(r)$ and $V_2^P(r)$ modified for r greater than 5.9 a_0 and 5.0 a_0, respectively, to have accurate asymptotic forms. Thus the static-exchange parts of potentials i and iii differ only due to the small effect of the polarization potential on the exchange potential. The scattering predicted by the two potentials is very similar except for $\theta < 30^\circ$ where potential iii underestimates the scattering. This is attributable to the fact that the INDOXI method underestimates α_0 by a factor of 5.98. Although potential iii is adjusted for $r > 5.9_0$ it is still not attractive enough at medium r and so it underestimates the forward scattering just as complete neglect of polarization does. Potential vi is the same as potential iii except that V_0^P (r) is deeper in the range 1.25 a_0 to 5.9 a_0. The INDOXI calculation of V_0^{adP} (r) exhibits a minimum at r = 1.25 a_0, where it equals $-\alpha_0/[2(2.6\ a_0)^4]$ for the accurate α_0. So we arbitrarily replaced $V_0^{adP\,(INDOXI)}(r)$ in the range 1.25 - 2.6 a_0 by $V^{adP\,(INDOXI)}(r = 1.25\ a_0)$ and in the range 2.6 - 5.9 a_0 by $-\alpha_0/(2r^4)$ with the accurate α_0. This modification increases the forward scattering sufficiently to yield good agreement with experiment, as was obtained with the semiempirical potential i. Thus we have achieved good agreement with experiment without any semiempirical parameters by using a simple model in which $V_0^P(r)$ and $V_2^P(r)$ are set equal to $V_0^{adP}(r)$ and V_2^{adP} (r) as calculated by the INDOXI method except that V_0^{adP} (r) is joined smoothly from its minimum to an accurate asymptotic form. Potential vii (for which the results are not shown in Figure 5) is the same as potential vi except the $V_\lambda^P(r)$ for $\lambda > 2$ are set equal to the INDOXI adiabatic calculated values without modification. This produces very little effect on the scattering. Thus although these higher-order anisotropies of $V^{adP}(\vec{r})$ are not always small, they seem to have only a small effect on the vibrationally elastic differential cross section.

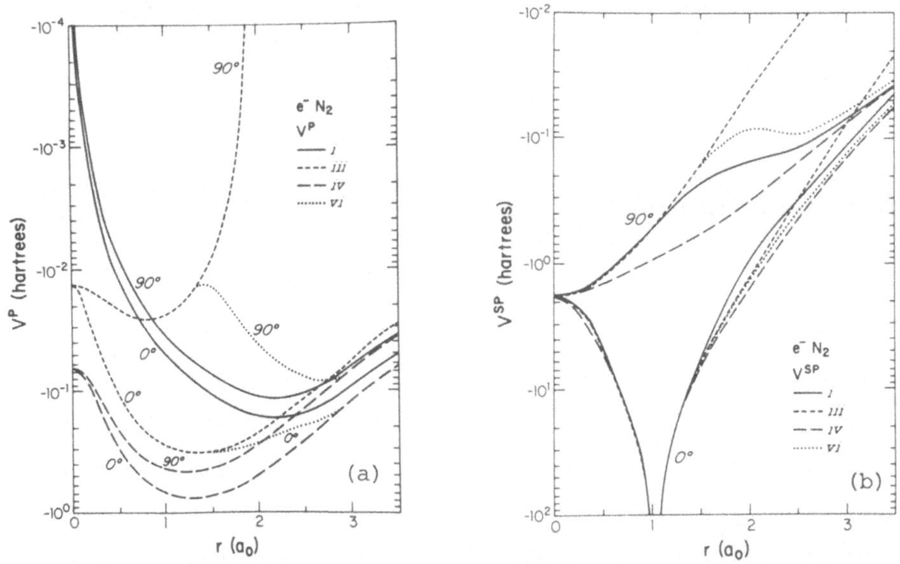

Figs. 4(a) and (b) Adiabatic polarization potentials (a) and static-plus-adiabatic polarization potentials (b) for collinear (0^0) and perpendicular-bisector (90^0) approaches of electrons to N_2. The solid line is for potential i, the short-dashed line is for potential iii, the long-dashed line is for potential iv, and the dotted line is for potential vi.

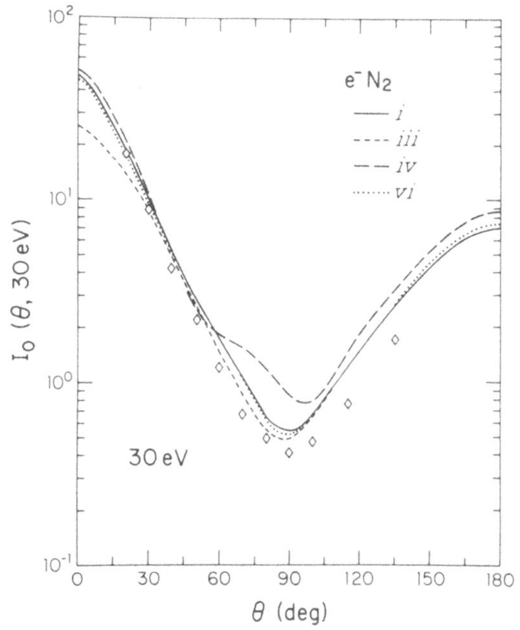

Fig. 5. Vibrationally elastic differential cross section for electron scattering by N_2. The calculations (reference 49) involved scattering basis set number XIII and are for a fixed internuclear distance ($R = R_e$). The solid, short-dashed, long-dashed, and dotted curves correspond to potentials i, iii, iv, and vi as in Figure 4. ◇ represents the same experimental results as it represents in Figure 1. At large angles the short-dashed curve (iii) merges with the solid one (i).

Finally we consider the effect of making the polarization potential stronger. Potential iv is like potential iii except that $V_0^P(r) = 4.80\ V_0^{P(INDOXI)}(r)$ for $r < 5.9\ a_0$ where the correction factor is $[-\alpha_0/(2r^4)]/V_0^{P(INDOXI)}(r)$ evaluated at $5.9\ a_0$ for the accurate α_0. For both potentials $V_0^P(r)$ is joined smoothly to the accurate large-r limit for $r > 5.9\ a_0$. Comparison of Figures 3 and 4 shows that potentials iii and iv generally bracket the *ab initio* adiabatic polarization potential. Figure 5 shows that this larger polarization potential seriously overestimates the sideways scattering. Thus the vibrationally elastic differential cross section is sensitive to making the polarization potential stronger. The error could be corrected by including a nonadiabaticity function. In fact, the spherical average of the adiabatic polarization potential (either the ab initio one or the modified ones iv or vi, all of which, when spherically averaged, tend to the same asymptotic form) shows relatively good agreement with the semiempirical polarization potential for r greater than about 3 a_0 but it becomes deeper at small r where nonadiabaticity effects may be large due to the deep static potential. Thus the adiabatic polarization potentials already provide a rough explanation of the shape of the semiempirical polarization potentials. But the semiempirical ones involve the spherical cutoff function C(r) which artificially constrains their angle dependence. The nonadiabaticity argument may also explain why the energy-independent potential of reference 1 is less accurate for sideways scattering at 50 eV than 30 eV (see Figure 1); this may be because nonadiabaticity decreases the true effective potential more at 50 eV than 30 eV.

We hope that the use of adiabatic polarization potentials and reasonable nonadiabaticity functions will lead to more realistic effective potentials. Of course the potential in equation (5) does have limitations. First, the exact optical potential is complex (has a nonzero imaginary part) at energies above the first electronic excitation threshold. It requires an additional model to estimate the imaginary part of the effective potential. Second, an accurate estimate of the nonadiabaticity function requires a dynamical calculation. For example, Kaldor and Klonover[59] have estimated the optical potential by using many-body perturbation theory in a scattering calculation. This leads to a nonlocal approximation to the optical potential. (The exact optical potential is also nonlocal, and the energy dependence in the nonadiabaticity function is an attempt to include this aspect.) In contrast to the SCF approach to the adiabatic polarization potential discussed here, the approach taken by Kaldor and Klonover requires third order terms to include the full polarizability.[59] Since their calculation of an approximate optical potential is part of a basis-set scattering calculation, it provides a suitable link between effective potential methods and the basis-set configuration-mixing approaches mentioned in the first paragraph.

This work was supported in part by grant CHE 77-27415 from the National Science Foundation, by a partial computing time subsidy from the University of Minnesota Computing Center, and by the National Aeronautics and Space Administration under contract No. NAS 7-100 to the Jet Propulsion Laboratory. David A. Dixon is an Alfred P. Sloan Foundation Research Fellow. F. A. Catledge is presently at Central Research and Development Department, Experimental Station, E. I. DuPont de Nemours & Co., Wilmington, Deleware.

REFERENCES

1. See, e.g., D. G. Truhlar, J. K. Rice, S. Trajmar, and D. C. Cartwright, Chem. Phys. Lett. 9, 299 (1971), and references therein.
2. D. G. Truhlar, J. Abdallah, Jr., and R. L. Smith, Adv. Chem. Phys. 25. 211 (1974).
3. B. I. Schneider and P. J. Hay, Phys. Rev. A 13, 2049 (1976); B. I. Schneider, Chem. Phys. Lett. 51, 578 (1977).
4. See, e.g., A. W. Fliflet, D. A. Levin, M. Ma, and V. McKoy, Phys. Rev. A 17, 160 (1978).
5. F. H. M. Faisal, J. Phys. B 3, 636 (1970); F. H. M. Faisal and A. L. V. Tench, Comput. Phys. Commun. 2, 261 (1971), erratum 5, 396 (1973).
6. D. G. Truhlar, F. A. Van-Catledge, and T. H. Dunning, Jr., J. Chem. Phys. 57, 4788 (1972).
7. F. A. Gianturco and N. Chandra, in Chemical and Biological Reactivity: The Jerusalem Symposium on Quantum Chemistry and Biochemistry, VI (Israel Academy of Sciences and Humanities, Jerusalem, 1974), p. 219; N. Chandra, Phys. Rev. a 12, 2342 (1975); F. A. Gianturco, Comput. Phys. Commun. 11, 237 (1976).
8. M. A. Morrison, L. A. Collins, and N. F. Lane, Chem. Phys. Lett. 42, 356 (1976); M. A. Morrison, N. F. Lane, and L. A. Collins, Phys. Rev. A 15, 2186 (1977).
9. D. G. Truhlar and F. A. Van-Catledge, J. Chem. Phys. 59, 3207 (1973); D. C. Truhlar and F. A. Van-Catledge, J. Chem. Phys. 65, 5536 (1976).
10. D. G. Truhlar, Chem. Phys. Lett. 15, 486 (1972).
11. H. S. W. Massey and R. O. Ridley, Proc. Roy. Soc. Lond., Ser. A 69, 659 (1956).
12. R. W. B. Ardill and W. D. Davison, Proc. Roy. Soc. Lond., Sec. A 304, 465 (1968).
13. R. J. W. Henry and N. F. Lane, Phys. Rev. 183, 221 (1969); R. J. W. Henry, Phys. Rev. A 2, 1349 (1970).
14. P. G. Burke and A.-L. Sinfailam, J. Phys. B 3, 641 (1970).
15. Y. Itikawa and O. Asihara, J. Phys. Soc. Japan 30, 1461 (1971).
16. B. D. Buckley and P. G. Burke, J. Phys. B 10, 725 (1977).
17. M. A. Morrison and L. A. Collins, Phys. Rev. 17, 918 (1978).
18. S. Hara, J. Phys. Soc. Japan 22, 710 (1967).
19. D. G. Truhlar, R. E. Poling, and M. A. Brandt, J. Chem. Phys. 64, 826 (1976).
20. D. G. Truhlar and M. A. Brandt, J. Chem. Phys. 65, 3092 (1976).
21. L. A. Collins and D. W. Norcross, Phys. Rev. Lett. 38, 1208 (1977).
22. P. Baille and J. W. Darewych, J. Phys. B 9, L1 (1977).
23. J. B. Furness and I. E. McCarthy, J. Phys. B 6, 2280 (1973).
24. M. E. Riley and D. G. Truhlar, J. Chem. Phys. 63, 2182 (1975).
25. R. Vanderpoorten, J. Phys. B 8, 926 (1975).
26. M. E. Riley and D. G. Truhlar, J. Chem. Phys. 65, 792 (1976).
27. B. H. Bransden, M. R. C. McDowell, C. J. Noble, and T. Scott, J. Phys. B 9, 1301 (1976).

28. P. Baille and J. W. Darewych, J. Chem. Phys. 67, 3399 (1977).
29. D. G. Truhlar and N. A. Mullaney, J. Chem. Phys. 68, 1574 (1978).
30. E. L. Breig and C. C. Lin, J. Chem. Phys. 43, 3839 (1965).
31. D. G. Truhlar and J. K. Rice, J. Chem. Phys. 52, 4480 (1970), erratum 55, 2005 (1971); S. Trajmar, D. G. Truhlar, and J. K. Rice, J. Chem. Phys. 52, 4502 (1970), erratum 55, 2004 (1971); S. Trajmar, D. G. Truhlar, J. K. Rice, and H. Kuppermann, J. Chem. Phys. 52, 4516 (1970); D. G. Truhlar, Phys. Rev. A 7, 2217 (1973); D. G. Truhlar and J. K. Rice, Phys. Lett A 47, 372 (1974).
32. P. G. Burke and N. Chandra, J. Phys. B 5, 1696 (1972); N. Chandra and P. G. Burke, J. Phys. B 6, 2355 (1973); N. Chandra, J. Phys. B 8, 1338 (1975).
33. M. A. Brandt, D. G. Truhlar, and F. A. Van-Catledge, J. Chem. Phys. 64, 4957 (1976); M. A. Brandt and D. G. Truhlar, Chem. Phys. 13, 461 (1976); D. G. Truhlar, M. A. Brandt, A. Chutjian, S. K. Srivastava, and S. Trajmar, J. Chem. Phys. 65, 2962 (1976).
34. N. Chandra and A. Temkin, Phys. Rev. A 13, 188 (1976); N. Chandra and A. Temkin, J. Chem. Phys. 65, 4537 (1976); A. Temkin, Phys. Rev. A 17, 1232 (1978)
35. M. H. Mittleman and K. M. Watson, Phys. Rev. 113, 198 (1959); M. H. Mittleman, Ann. Phys. (N.Y.) 14, 94 (1961).
36. C. J. Kleinman, Y. Hahn, and L. Spruch, Phys. Rev. 165, 53 (1968).
37. K. Onda and D. G. Truhlar, J. Chem. Phys. 69, 1361 (1978).
38. K. Onda and D. G. Truhlar, J. Phys. B. 12, 283 (1979).
39. H. Kambara and K. Kuchitsu, Jpn. J. Appl. Phys. 11, 609 (1972).
40. T. W. Shyn, R. S. Stolarski, and G. R. Carignan, Phys. Rev. A 6, 1002 (1972).
41. T. G. Finn and J. P. Doering, J. Chem. Phys. 63, 4399 (1975).
42. S. K. Srivastava, A. Chutjian, and S. Trajmar, J. Chem. Phys. 64, 1340 (1976).
43. R. D. DuBois and M. E. Rudd, J. Phys. B 9, 2657 (1976).
44. D. G. Truhlar and F. A. Van-Catledge, J. Chem. Phys., in press.
45. R. J. Drachman and A. Temkin, in Case Studies in Atomic Collision Physics, edited by E. W. McDaniel and M. R. C. McDowell (North-Holland Publishing Co., Amsterdam, 1972), p. 399.
46. N. F. Lane and R. J. W. Henry, Phys. Rev. 173, 183 (1968).
47. S. Hara, J. Phys. Soc. Japan 27, 1262 (1969).
48. D. G. Truhlar, D. A. Dixon, and R. A. Eades, to be published.
49. K. Onda and D. G. Truhlar, J. Chem. Phys. 70, 1681 (1979).
50. M. A. Morrison and P. J. Hay, Bull. Amer. Phys. Soc. 22, 1331 (1977).
51. S. Huzinaga, J. Chem. Phys. 42, 1293 (1965).
52. T. H. Dunning, J. Chem. Phys. 53, 2823 (1970).
53. T. Vladimiroff, J. Phys. Chem. 77, 1983 (1973).
54. H.-J Werner and W. Meyer, Mol. Phys. 31, 855 (1976).
55. J. C. Slater, Phys. Rev. 36, 57 (1930).

56. J. A. Pople, D. L. Beveridge, and P. A. Dobosh, J. Chem. Phys. 47, 2026 (1967).
57. G. Klopman and R. C. Evans, in Semiempirical Methods of Electronic Structure Calculations, Part A, edited by G. A. Segal (Plenum Press, New York, 1977), p. 29.
58. A. Klonover and U. Kaldor, J. Phys. B 11, 1623 (1978).
59. U. Kaldor, J. Chem. Phys. 62, 4634 (1975).

VIBRATIONAL EXCITATIONS OF LOW ENERGY e-CO SCATTERING

B. H. Choi and Robert T. Poe

Department of Physics
University of California
Riverside, California 92521

Hybrid theory, proposed by Chandra and Temkin[1] and reformulated by Choi and Poe,[2] was applied to the study of the vibrational transitions for low energy e-CO scattering. In this theory, the vibrational states of target molecules were coupled dynamically through the close-coupling approximation but with the direction of molecular axes fixed. The simultaneous vibrational and rotational transitions are then obtained from the adiabatic-nuclei approximation.

The number of channels of the coupled differential equation is, then, given by

$$N = V_{max} \times (\ell_{max} - m+1)$$

where V_{max}, ℓ_{max} are the number of vibrational states and the maximum partial wave, respectively, included in the calculation. The m=0, 1, 2, ... corresponds to σ-, π-, δ- ... scattering waves. The above equation shows that the number of channels (i.e., the number of coupled differential equations) is very large even when a few vibrational states are considered. It is also desirable to employ an approximate scheme for the exchange effect; the calculation with complete treatment of exchange effects is, in general, unmanageably complex. Two approximations, the orthogonalization approach, or the inclusion of a local exchange potential used in a fixed-nuclei calculation for vibrationally elastic scattering, will be tractable in this case.

The former approach was implicitly employed, however, by Chandra and Temkin[1] in their study of e-N_2 vibrationally inelastic scattering. They performed vibrational close-coupling calculations for the resonant π_g-wave, which gives the dominant contribution to the vibrationally inelastic cross sections, for which the orthogonalization procedures are not necessary because π_g orbitals are not occupied in the target molecule (i.e., the incident π_g-wave is already orthogonal to all orbitals in the target). For all other non-resonant partial waves, which gives mainly vibrationally elastic contributions, the orthogonalization procedures were included with fixed-nuclei close-coupling calculations. They produced reasonable results compared with experimental data.

Our main interest in the present calculation is the vibrational excitation in the low energy region, in particular around the 1.7 eV $^2\Pi$-shape resonance of e-CO scattering. To apply the above orthogonalization approach to our situation is more complex, since π-orbitals are occupied in CO and the numerical procedure in which the incident π-wave is made orthogonal to the orbitals of target for all internuclear separation R in the vibrational close-coupling approximation is very tedious. Therefore, we first performed calculations in a similar manner to the above e-N_2 scattering. That is, the vibrational close-coupling calculation was carried out without inclusion of any exchange effect for the resonant π-wave, and a fixed-nuclei (close-coupling) calculation was made for non-resonant σ, δ ...waves. The input static potential was obtained from the molecular orbital wave functions of McLean and Yoshimine,[3] and the polarization potential was taken partly from experimental data and partly from the above molecular wave functions. All these potentials were computed for each internuclear separation R. The results thus obtained yield much sharper substructures in the vibrational excitation cross section than those of experimental measurements, and the agreements between theory and experiments are generally poor. This implies that the explicit inclusion of exchange effects in vibrationally inelastic scattering for e-CO is more important than it is for e-N_2.

The proper treatment of exchange effects which is of reasonable complexity for the present case seems to be the inclusion of a local exchange potential. We have evaluated the local exchange potential, basically similar to that proposed by Hara[4] and used by others in fixed-nuclei calculations, again from the molecular orbital wave functions of McLean and Yoshimine[3] for each internuclear separation R. With static + local exchange + polarization potentials, the vibrational close-coupling calculations were carried out for all resonant and non-resonant partial waves. Seven vibrational states and ℓ=0, 1, 2, 3, 4 were coupled together. Therefore, the coupled differential equations are of 35, 28, ... channels for σ-, π-, ... waves, respectively. Sample results are presented here. The vibrationally elastic scattering cross section is shown in Figure 1. No comparable experimental (absolute) cross sections exist. The permanent dipole moment contained in the static potential at long range is about 2.4 times bigger than that of the experimental one and thus

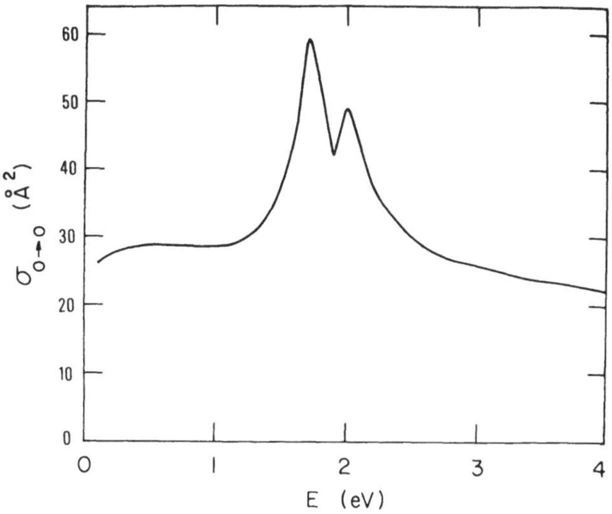

Fig. 1. The vibrationally elastic scattering cross sections obtained from present calculation.

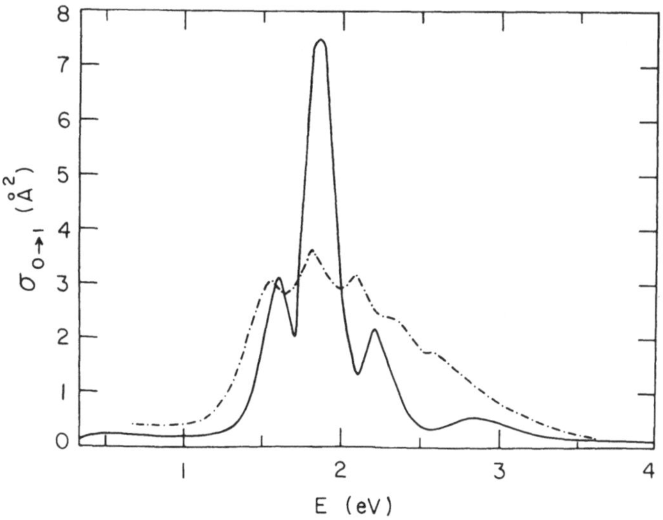

Fig. 2. The 0→1 vibrational excitation cross section, ——: present calculation, -•-•-•: experimental data taken from Ref. 5.

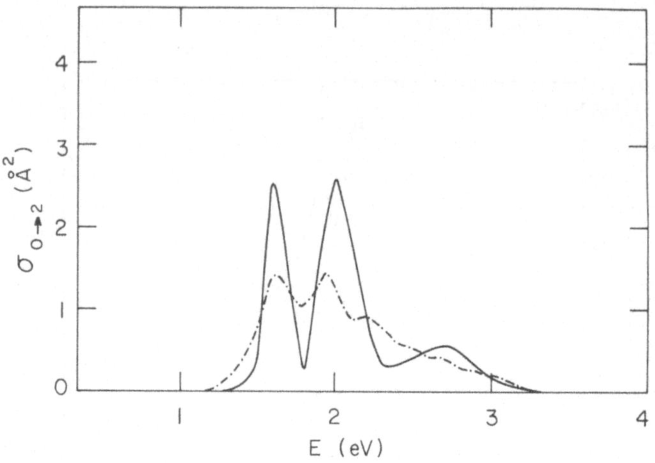

Fig. 3. The 0→2 vibrational excitation cross section, ——: present calculation, -•-•-•: experimental data taken from Ref. 5.

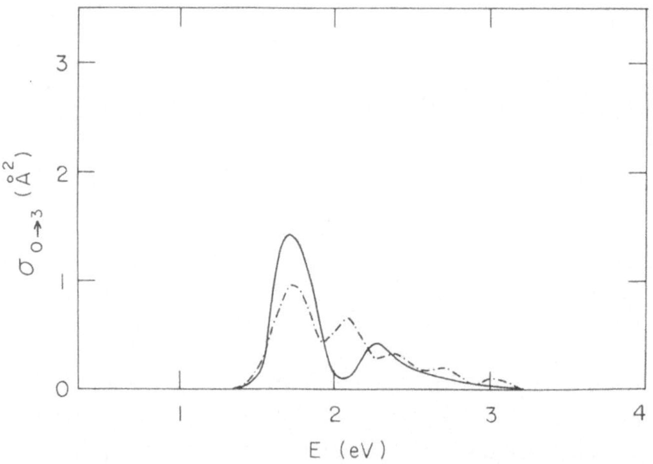

Fig. 4. The 0→2 vibrational excitation cross section, ——: present calculation, -•-•-•: experimental data taken from Ref. 5.

the calculation might overestimate the elastic scattering cross section in the very low energy region. Figures 2-4 shows the $0 \rightarrow 1$, $0 \rightarrow 2$, $0 \rightarrow 3$ vibrational excitation cross sections compared with the experimental data measured by Ehrhardt et al.[5] Although there exist some discrepancies in absolute magnitudes, the theory accounts for all major features in the substructures of these cross sections.

The hybrid theory, a "practicable" _ab initio_ approach, for the calculation of vibrational transition cross sections, is seen to be capable of reasonably good results, when good input potentials, in particular the exchange effect, are properly included. More details of the present calculation will be reported elsewhere.

We thank Dr. A. Temkin for useful discussions.

This work was supported in part by NASA and AF/APL Contract No. F33615-77-C-2011.

REFERENCES

1. N. Chandra and A. Temkin, Phys. Rev. A13, 188 (1976).
2. B. H. Choi and R. T. Poe, Phys. Rev. A16, 1831 (1977).
3. A. D. McLean and M. Yoshimine, IBM J. Res. Dev. Suppl. 12, 206 (1967).
4. S. Hara, J. Phys. Soc. Japan, 22, 710 (1967).
5. H. Ehrhardt, L. Langhans, F. Linder, and H. S. Taylor, Phys. Rev. 173, 222 (1968).

IMPROVED HYBRID THEORY CALCULATION OF $e-N_2$ VIBRATIONAL EXCITATION

A. Temkin

Atomic Physics Office
Laboratory for Astronomy and Solar Physics
Goddard Space Flight Center
Greenbelt, Maryland 20771

I will describe some new calculations on e^--N_2 vibrational excitation using hybrid theory. When we did our original calculations[1] and obtained vibrational excitation cross sections, we were faced with the problem that some upper-atmospheric physicists at Goddard actually wanted to use them; they wanted accurate normalized vibrational excitation cross sections. I would have liked to advise them to use the experimental results; unfortunately, the experimental normalizations are not as yet unique. The situation is demonstrated in Figure 1. The curves on the left are essentially Schulz's original ones.[2] They were differential cross sections measured at 72^0 as a function of energy and the normalization was arbitrary. However, assuming the total cross sections had the same shape as a function of energy (and this is an excellent approximation[1]), Schulz pointed out that if you use the values (in $\AA^2$) given on the ordinate, Figure 1a, then you get agreement with previous experimental results of Haas.[3] Subsequently, Schulz[4] renormalized these curves based on later absolute measurements of Spence et al.[5] which is given in Figure 1b, and you see they are a factor of two larger. In the same article[4] Schulz also showed a figure based on work that S. F. Wong has done which, after conversion to total cross section using our $\frac{d\sigma}{d\Omega}$, gives a normalization yet another factor of two larger. In fact, that normalization in Figure 1c is almost identical to what Chandra and I got in our original hybrid calculation.[1] But in view of the other normalizations, as well as uncertainties in our calculation, I could only advise caution to our atmospheric colleagues and promise them a better calculation. What I'd like to describe here is a preliminary report of this new calculation.

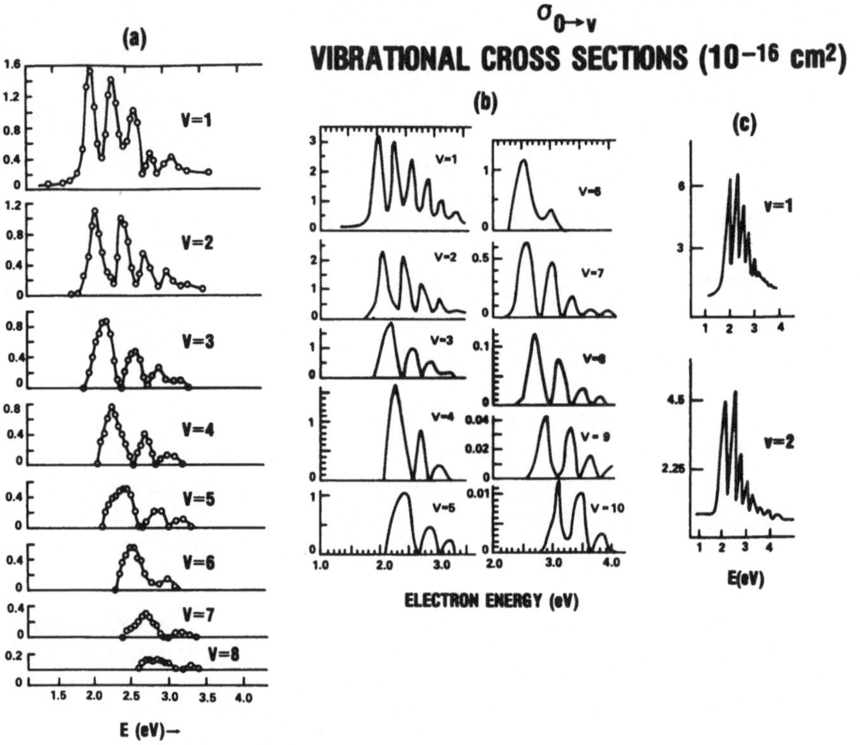

Fig. 1. Various experimental vibrational excitation $\sigma_{0\to v}$ cross sections. The absolute normalization (in $Å^2$) was inferred as described in text from differential cross sections (a) G. J. Schulz, Ref. 2, (b) G. J. Schulz, Fig. 6 of Ref. 4, (c) S. F. Wong, unpublished, displayed in Fig. 4 of Ref. 4.

I think one piece of experimental information that is well known is the total cross section. The cross section as measured by Golden[6] in 1966 is shown in Figure 2; the solid curves are the values which Dube and Herzenberg got.[7] I believe it to be essentially the same calculation that Birtwistle and Herzenberg[8] originally did for the resonant Π_g partial wave; but to their Π_g contribution they added our non-resonant contribution, which is the same as in the fixed-nuclei calculation of Burke and Chandra.[9] In Figure 2 the open circles are our original total cross sections[1]. As I have just said, I believe the experimental normalization is correct. Thus I conclude that the theoretically inferred normalization[1,7] on the basis of present calculations are too large as compared to experiment.

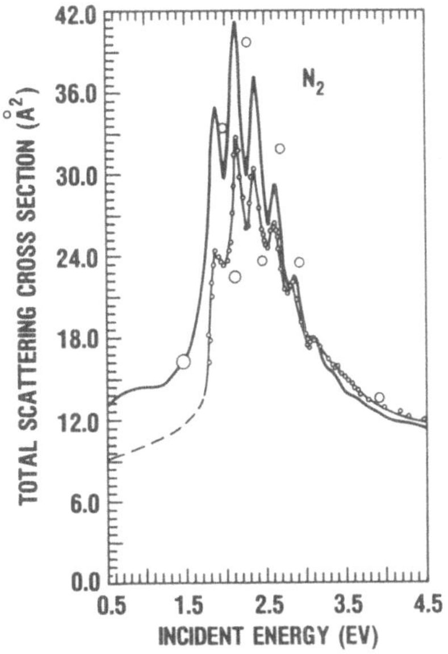

Fig. 2. Total e-N_2 total cross section σ_T thin curve and points experimental results of Golden Ref. 6; heavy curve calculations of Dube and Herzenberg, Ref. 7. Circles are calculated results of hybrid theory of Chandra and Temkin; Ref. 1.

Before presenting our new calculations, the formalism will be reviewed briefly. The idea of the hybrid theory is briefly summarized in Figure 3. The fixed-nuclei theory is such that you have scattering parameters, $a_{\ell\lambda m}$, which are determined from a set of fixed-nuclei equations and which are independent of the orientation The orientation dependence (β_o) is explicit in terms of rotational harmonics ($\mathcal{D}$ functions); the angular dependence of f_{fn} is in terms of spherical harmonics $Y_{\ell m}(\Omega')$. In the adiabatic-nuclei approximation you simply integrate the fixed-nuclei amplitude between vibrational(x_γ) rotational functions on the right and on the left to get a specific vibrational-rotational amplitude. But the essential point is that the only dynamical calculation you do is for the $a_{\ell\lambda m}$, and that is independent of the nuclear orientation. In contrast, in the hybrid theory, we develop an amplitude which we get from a dynamical calculation and which is a matrix in v and v'. However, to get the simultaneous vibration/rotation amplitudes you still do a quadrature with respect to rotation and that remains explicit in the $\mathcal{D}$ functions. That's the idea behind the theory.

What we did to improve the calculation was, firstly, to get a better polarization potential.[10] These potentials are an essential

REVIEW OF HYBRID THEORY

[CHANDRA & TEMKIN, PHYS. REV A 13, 188 (1976)]

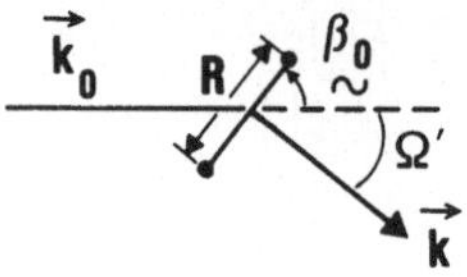

1. FIXED-NUCLEI

$$f_{f.n.}(\underset{\sim}{\beta_0}, R, \Omega') = \Sigma\, a_{\ell\lambda m}(R)\, \mathcal{D}^{(\ell)}_{m'm}(\underset{\sim}{\beta_0})\, \mathcal{D}^{(\lambda)*}_{om}(\underset{\sim}{\beta_0})\, Y_{\ell m'}(\Omega')$$

$a_{\ell\lambda m}$ **DERIVED FROM SET OF FIXED-NUCLEI EQS.**

2. ADIABATIC-NUCLEI

$$f^{(a.n.)}_{j',v';jv}(\Omega') = \langle \mathcal{D}^{(j')}(\beta_0)\chi_{v'}(R) \,|\, f_{f.n.}(\underset{\sim}{\beta_0}, R, \Omega') \,|\, \mathcal{D}^{(j)}(\beta_0)\chi_v(R)\rangle$$

$\rightarrow f^{(a.n.)}$ **OBTAINED BY QUADRATURE FROM** $f_{f.n.}$

3.HYBRID-THEORY: REPLACE $\langle \chi_{v'}, f_{f.n.} \chi_v \rangle$ **BY**

$$f_{HYBRID}(\underset{\sim}{\beta_0}, \Omega') = \Sigma\, a\!f_{v'\ell,v\lambda}\, \mathcal{D}^{(\ell)}_{m'm}(\underset{\sim}{\beta_0})\, \mathcal{D}^{(\lambda)*}_{om}(\underset{\sim}{\beta_0})\, Y_{\ell m'}(\Omega')$$

$a\!f_{v'\ell,v\lambda}$ **DERIVED FROM SET OF VIB C.C. EQS.**

$$f^{(HYBRID)}_{j',v';j,v}(\Omega') = \langle \mathcal{D}^{(j')}(\beta_0) \,|\, f_{HYBRID} \,|\, \mathcal{D}^{(j)}(\beta_0)\rangle$$

Fig. 3. Computation of main formulae of hybrid, adiabatic-nuclei and fixed-nuclei theories, cf. Ref. 1.

part of the vibrational close-coupling equations describing the vibrational dynamics. The R-dependence of the polarizabilities are shown in Figure 4, but it is matrix elements of α between N_2 vibrational functions which actually enter the close-coupling equations. Although we know that this $\alpha(R)$ is not perfect, we feel it is reasonably correct and, in any event, much better than what we previously had,[1] which was completely wrong at R = 0 and increased with a greater slope at R_o. I will come to the solutions later.

To get the non-resonant part, we used another technique, which I will briefly describe. In Figure 5 I present some Σ_u eigenphase sums. The upper solid curve represents the recent Buckley-Burke[11]

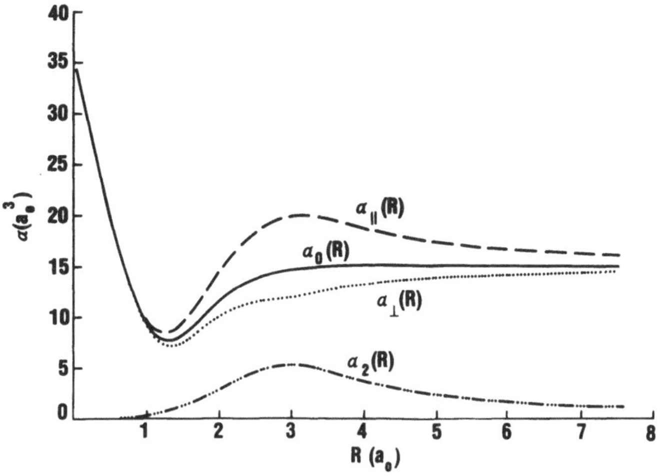

Fig. 4.   Internuclear separation (i.e., R) dependence of polarizabilities of N_2, from Temkin, Ref. 10.

EXCHANGE + POL. (EXCHANGE—ADIABATIC) PHASE
SHIFTS CAN BE WELL SIMULATED BY CHANGING
r_0 IN B.—C. PSEUDOPOTENTIAL PROGRAM IN V_{pol}

$$V_{pol} = \frac{\left[1 - e^{-\left(\frac{r}{r_0}\right)^6}\right]}{r^4}\left\{a_0 + a_2 P_2(\cos\Theta)\right\}$$

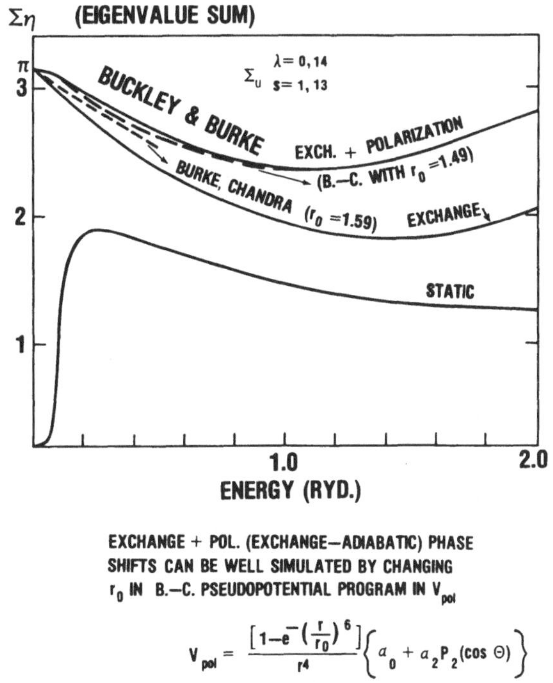

Fig. 5.   Non-resonant Σ_u eigenphase sum in various approximations
(Ref. 11 and B. Buckley private communication). The long
and short dashed curves represent pseudopotential results
using different values of the cut-off parameter r_0.

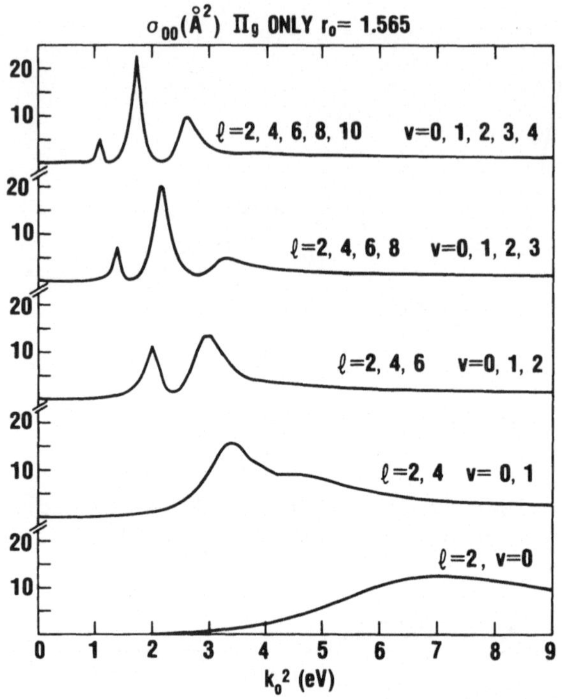

Fig. 6. Vibrational close coupling results for σ_{oo} for increasing number of v's and ℓ's in the c.c. expansion. Note for the value of r_o quoted the first resonance of the highest curve is below the experimental value of E for first resonance.

exchange plus polarization result (we would call this the exchange-adiabatic approximation). We assume that represents the best presently available non-resonant result. The lower short curve is the Burke-Chandra[12] result; they used an orthogonalization procedure to simulate exchange. But the cut-off value $r_o = 1.59$ of their polarization potential was determined by fixing the Π_g resonant partial wave to the experimental value[6] of the resonance energy. It turns out that by slightly changing r_o, you can get a result which is almost identical to that of Buckley and Burke[11] for the non-resonant partial wave. So we use this other value of $r_o = 1.49$ for all non-resonant partial waves. (This r_o also gives better fits for the other non-resonant waves. The difference between Buckley-Burke and B-C ($r_o = 1.49$) in Fig. 5 is exaggerated.)

We then repeated the vibrational close-coupling calculation for π_g, but with several modifications. We included more ℓ-coupling and less v- coupling, but we increased the ℓ and v coupling jointly. What happened was that the substructure started to appear, as it

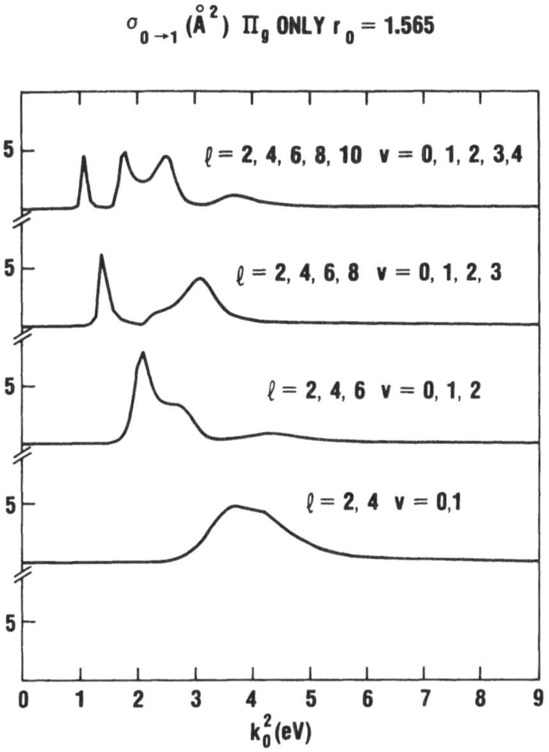

Fig. 7. Same as Fig. 6 but for $\sigma_{\sigma\to 1}$.

did previously,[1] but it showed up at too low an energy (cf. Figs. 6 and 7). What this tells us is that we must use a larger (less attractive) cut-off (r_o) for the resonant (Π_g) partial wave in a close-coupling (c.c.) expansion than the value of r_o that was necessary in a fixed-nuclei calculation.[9] This is illustrated in Fig. 8. Thus we redid the vibrational c.c. calculation with r_o = 1.61. This was our best estimate of the value of r_o that would lead to a converged c.c. calculation with the (first) resonant position at the correct (i.e., experimental energy).

The main result of all this is shown in Figure 9; the total cross section agrees for the first time with the absolute value of the measured value of Golden.[6] Again, I stress that it is only the first peak to which I attribute any reliability of normalization at this stage of our calculation. (The latter must be carried out much further.) Nevertheless, having obtained agreement on the total cross sections, we can similarly have some confidence in our individual cross sections. In Figure 10 we give our current results

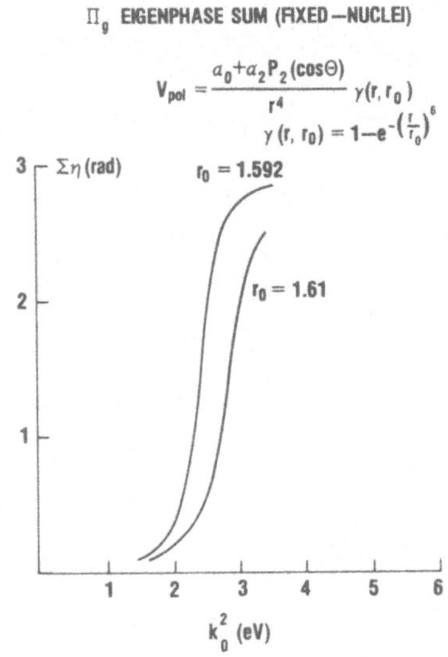

Fig. 8. Fixed-nuclei eigenphase sum for two r_o: the less attractive (larger) r_o gives fixed-nuclei resonance above experimental value (2.4 eV). When used in vib. c.c. calculation, however, this r_o gives manifold resonances about right, c.f. text.

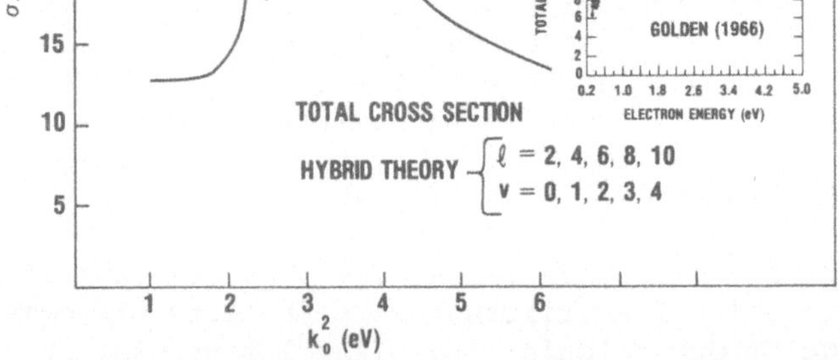

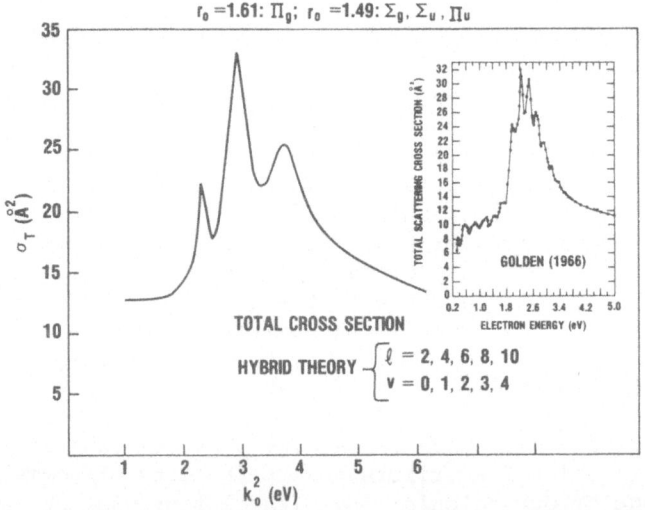

Fig. 9. Calculated total cross section as described in text as compared to experiment of Golden, Ref. 6.

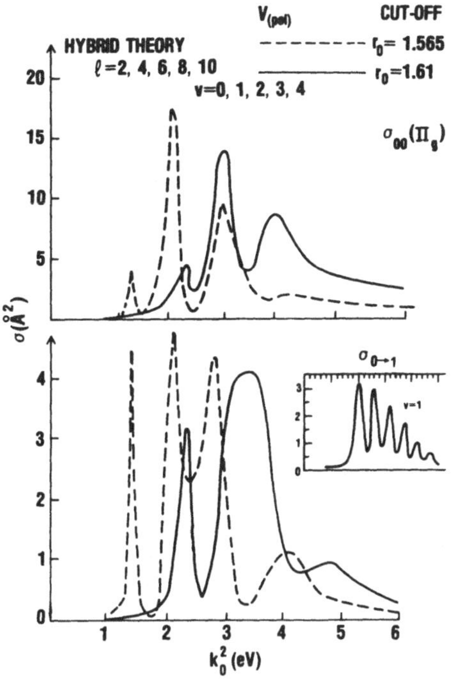

Fig. 10. Absolute cross sections from latest hybrid calculation (solid curves). Note in particular that normalization of first peak agrees with panel (b) of Fig. 1. Note also that if we had used the smaller value of r_o (=1.565) we would have inferred quite a different normalization.

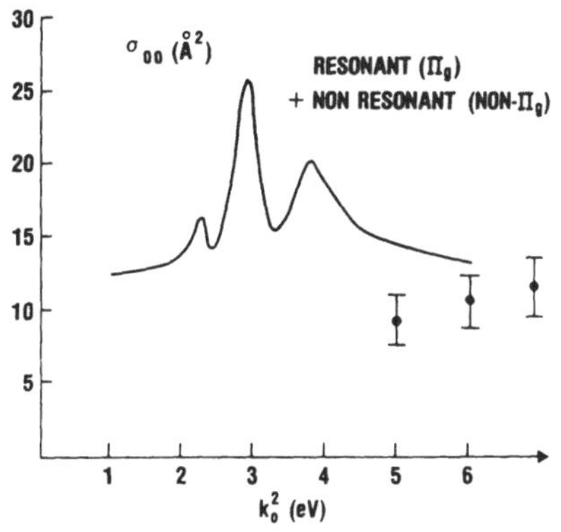

Fig. 11. Calculated σ_{oo}, experimental points are from Ref. 13 and are discussed in text.

for the resonant Π_g contribution to the elastic $(0\to 0)$, and first $(0\to 1)$ vibrational cross sections. The elastic cross section will of course have a large non-resonant contribution, but the inelastic will not. Note that this newer calculation (whose first peak height we believe will not change in a converged calculation) then supports the middle normalization of Fig. 1b. The calculated integrated vibrationally elastic cross section is then shown on Fig. 11, and again it is only the height of the first peak to which we would ascribe quantitative accuracy. An accurate measurement of this cross section at low energies would clearly be invaluable. The present measurement[13] is only at somewhat higher energies, and is based on a e-He normalization,[14] which has now been superceded by a more accurate measurement.[15]

In conclusion, let me thank E. Sullivan for his invaluable contribution to the numerical results displayed here, obtained by a very extensive modification of Chandra's program. Thanks are also due to Dr. Chandra for leaving us a copy of the original program.

REFERENCES

1. N. Chandra and A. Temkin, Phys. Rev. A 13, 188 (1976).
2. G. J. Schulz, Phys. Rev. 135, A988 (1964).
3. R. Haas, Z. Physik 148, 177 (1957).
4. G. J. Schulz, in Principles of Laser Plasmas (John Wiley and Sons, (1976)).
5. D. Spence, J. L. Mauer, G. J. Schulz, J. Chem. Phys. 57, 5516 (1972).
6. D. E. Golden, Phys. Rev. Letters 17, 847 (1966).
7. These results are unpublished; they appear in the Ph.D. thesis of L. Dube, Yale University, Jan. 1978. The results were first presented by the authors at IX ICPEAC, Abstracts of Papers (University of Washington Press, 1975), p. 264. Our own results (Ref. 1) were also presented there ibid, p. 267.
8. D. T. Birtwistle and A. Herzenberg, J. Phys. B 4, 53 (1971).
9. P. G. Burke and N. Chandra, J. Phys. B 5, 16 and 6 (1972).
10. A. Temkin, Phys. Rev. A 17, 1232 (1978).
11. B. D. Buckley and P. G. Burke, J. Phys. B 10, 725 (1975).
12. Cf. Refs. 7 and 8 and L. Dube and A. Herzenberg (to be published).
13. K. Srivastava, A. Chutjian, and S. Trajmar, J. Chem. Phys. 64, 1340 (1976).
14. J. W. McConkey and J. A. Preston, J. Phys. B 8, 63 (1975).
15. R. E. Kennerly and R. A. Bonham, Phys. Rev. A 17, 1844 (1978).

STIELTJES-TCHEBYCHEFF MOMENT-THEORY APPROACH

TO MOLECULAR PHOTOIONIZATION STUDIES

P. W. Langhoff

Department of Chemistry
Indiana University
Bloomington, Indiana 47401

I. INTRODUCTORY REMARKS

Difficult problems often require the introduction of unfamiliar strategies in obtaining solutions. The Stieltjes-Tchebycheff (S-T) moment-theory technique[1-6] for performing atomic and molecular photoionization and related scattering calculations in Hilbert space was devised to provide cross-sectional values[7-30] in the absence of reliable more conventional methods for this purpose. Conventional theoretical approaches to molecular photoionization studies generally require construction in some approximation of scattering eigenfunctions that satisfy appropriate asymptotic boundary conditions. It has proved difficult to construct such scattering functions even in simple static-potential approximations, largely because of the non-central and non-local nature of the molecular field. Consequently, many of the theoretical molecular photoionization studies reported to date involve simplifying approximations designed to avoid or circumvent the latter aspects of the problem. By contrast, finite point-group and electron-exchange symmetries are integral parts of the computational technology of many-electron L^2 bound-state electronic structure calculations,[31-33] providing considerable motivation to devise related Hilbert-space methods for theoretical investigations of molecular electronic continua.

In the present article, a review is given of the S-T approach to photoionization in atoms and molecules. Theoretical aspects of the method are described, recent computational applications are summarized, and some remaining problem areas are indicated. Since the moment strategy is perhaps unfamiliar in the context of scattering and photoionization calculations, an introductory heuristic account of the general approach is given in Section II.

In this, Hilbert-space approximations to scattering functions are investigated, the question of their proper energy normalization is considered, and the relevance of energy moments, distributions, and densities in this connection is indicated.[34-37] An essentially complete and self-contained account of the theory of the S-T technique, giving all the appropriate important theoretical and computational details, is presented in Section III. The convergence of spectral moments in an L^2 basis is established, so-called principal representations of such moments are described, the important Stieltjes and Tchebycheff distributions and densities are defined and their convergence indicated, and the utility of polynomial recurrence coefficients in computational applications is emphasized. Recent applications of the S-T approach to atomic and molecular photoionization[1-30] are described in Section IV. Particular attention is focused on the shape resonances that appear in the partial-channel photoionization cross sections of light diatomic and polyatomic molecules.[21-27] The computational results in these cases indicate that the S-T technique provides a basis for quantitatively reliable studies of the rather structured photoionization cross sections of complex many-electron systems. Some remaining problem areas are described and concluding remarks are made in Section V.

II. THE MOMENT APPROACH TO PHOTOIONIZATION STUDIES

A brief descriptive account of the moment-theory approach to L^2 photoionization calculations is presented in this section. Hilbert-space approximations to scattering functions are described in (A) and the question of their proper energy normalization is considered, in (B) the relevance of energy or spectral moments in this connection is indicated, and in (C) approximations to so-called Stieltjes distributions and densities are described.

A. Pseudospectral Calculations

Linear variational calculations in a finite Hilbert space of L^2 basis functions $\{\chi_i;\ i=1,N\}$ can be made to provide an N-term pseudospectrum of energies and functions $\{\tilde{E}_i,\ \tilde{\Phi}_i;\ i=1,N\}$ satisfying[31-33]

$$\langle \tilde{\Phi}_i | \tilde{\Phi}_j \rangle = \delta_{ij} \tag{1a}$$

$$\langle \tilde{\Phi}_i | H | \tilde{\Phi}_j \rangle = \delta_{ij} \tilde{E}_i, \quad i,j = 1,N, \tag{1b}$$

where H is the target Hamiltonian. In contrast to Eqs. (1), the correct spectrum of eigenvalues and eigenfunctions of an atom or molecule in the Coulomb approximation is comprised of both discrete and continuous intervals $\{E_i,\ \Phi_i;\ i=1,\infty;\ E,\ \Phi_E;\ E_t \leqslant E < \infty\}$, where

E_t is the first threshold energy for ionization.[38] Particular choices of basis functions χ_i can result in $\tilde{E}_i$ values falling in the discrete or continuous (essential) regions of the correct spectrum of H. In the case of the hydrogen atom, for example [Hartree atomic units are employed unless otherwise indicated],

$$H = -\frac{1}{2}\nabla^2 - \frac{1}{r} \tag{2}$$

an L^2 basis of Laguerre functions of order $2\ell + 2$ can be used to span both the discrete and continuous portions of the spectrum of H.[39] In Figure 1 is shown a comparison of the correct p-wave spectrum in atomic hydrogen with a 20-term Laguerre pseudospectrum satisfying Eqs. (1). Since the Laguerre functions are complete in R^3, many of the $\tilde{E}_i$ values obtained are seen to fall in the continuous region of the correct spectrum. By contrast, when the p-type L^2 Laguerre functions of order $2\ell + 1$ are employed,[40] a 20-term calculation in this case simply reproduces the first 20 correct discrete eigenstates in atomic hydrogen, and there are no $\tilde{E}_i$ values in the continuous region of the spectrum.

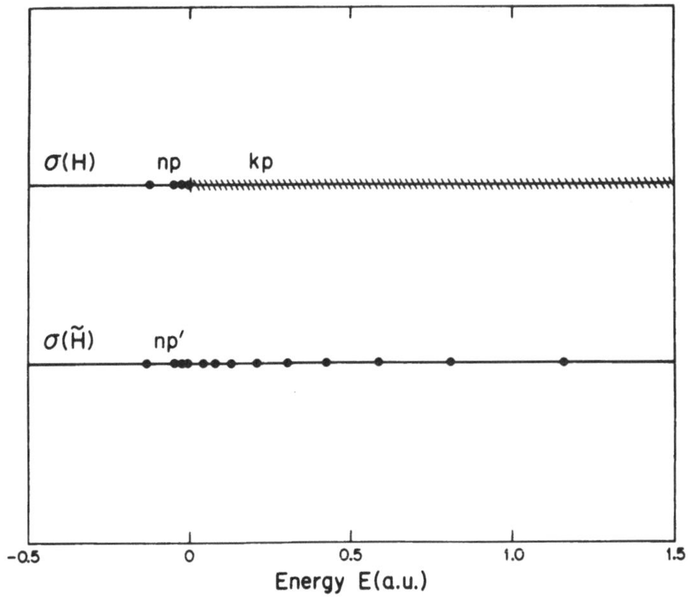

Fig. 1 Comparison of the correct (upper) and a variationally determined (lower) p-wave spectrum in atomic hydrogen. The variationally determined results are obtained from Eqs. (1) employing a twenty-term basis of L^2 Laguerre functions.[29]

The question arises as to the relationship, if any, between the $\tilde{\Phi}_i$ of Eq. (1) and the correct discrete and continuum solutions of the Schrödinger equation for atomic hydrogen. The continuum solutions satisfy appropriate asymptotic scattering boundary conditions and are Dirac delta-function normalized in the momentum or energy variable, whereas the correct bound states are by convention unity normalized.[41] Since the L^2 functions employed in the present case [Figure 1] satisfy the same behavior at the origin as do the bound p states and the regular Coulomb p-waves,[39-41] it is reasonable to suppose the Laguerre pseudospectral functions are closely related to these correct eigenfunctions over finite regions of space.[42] To test this hypothesis, it is helpful to compare the pseudospectrum of transition frequencies and oscillator strengths[43]

$$\tilde{\varepsilon}_i = \tilde{E}_i - E_o , \tag{3a}$$

$$\tilde{f}_i = (2/3)\tilde{\varepsilon}_i |\langle\tilde{\Phi}_i|\mu|\Phi_o\rangle|^2 , \quad i = 1,N \tag{3b}$$

with the discrete,

$$\varepsilon_i = E_i - E_o \tag{4a}$$

$$f_i = (2/3)\varepsilon_i |\langle\Phi_i|\mu|\Phi_o\rangle|^2 , \quad i = 1, \infty, \tag{4b}$$

and continuum,

$$\varepsilon = E - E_o , \tag{5a}$$

$$g(\varepsilon) = (2/3)\varepsilon|\langle\Phi_E|\mu|\Phi_o\rangle|^2, \quad \varepsilon_t \leqslant \varepsilon < \infty, \tag{5b}$$

$$\langle\Phi_E|\Phi_{E'}\rangle = \delta(E-E'), \tag{5c}$$

portions of the correct oscillator-strength profile in atomic hydrogen.[41] Here, μ is the dipole moment operator, and E_o and Φ_o are the correct ground-state eigenvalue and eigenfunction, respectively. Since $\tilde{\Phi}_i$ in Eq. (3b) is normalized to unity [Eq. (1a)], and Φ_E in Eq. (5b) is delta-function normalized in the energy variable [Eq. (5c)], Eq. (3b) and Eq. (5b) for $\varepsilon = \tilde{\varepsilon}_i$ are generally not equal. However, since $g(\varepsilon)$ is known in this case,[41] Eqs. (3b) and (5b) can be made equal by replacing $\tilde{\Phi}_i$ in Eq. (3b) with the renormalized pseudostate $(g(\tilde{\varepsilon}_i)/\tilde{f}_i)^{\frac{1}{2}}\tilde{\Phi}_i$. Since this latter pseudostate is effectively continuum normalized in energy, it is reasonable to suppose that it reproduces Φ_E at $E = \tilde{E}_i$ over a finite region of space, possibly as large as the molecular box provided by the product $\mu\Phi_o$.[42] That is, it is reasonable to expect

$$(g(\tilde{\varepsilon}_i)/\tilde{f}_i)^{\frac{1}{2}}\tilde{\Phi}_i \mu \Phi_o \cong \Phi_{\tilde{E}_i} \mu \Phi_o \,, \tag{6}$$

provided that the L^2 basis set employed is complete in R^3 and sufficient numbers of functions are used in the calculation.

The supposition of Eq. (6) is supported by the results shown in Figure 2, in which are compared the correct discrete and continuum radial transition densities in atomic hydrogen with those obtained from the renormalized Laguerre pseudostates.[29] Evidently, the appropriately renormalized pseudostate transition densities

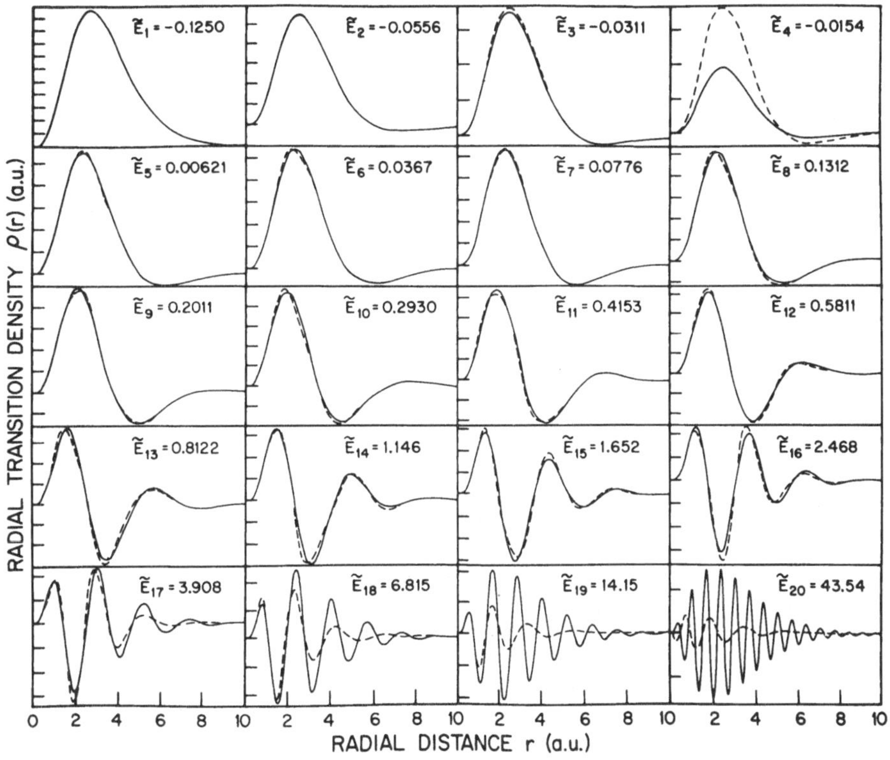

Fig. 2 Comparison of correct discrete 1s → np and continuum 1s → kp radial transition densities (——) in atomic hydrogen with variationally determined approximations (----) obtained from a twenty-term basis of L^2 Laguerre functions.[29] As discussed in the text, the latter for $\tilde{E}_i > 0$ have been renormalized to reproduce the correct continuum oscillator strengths [Eq. (6)]. Hartree atomic units are employed, first row of frames, and 0.05 a.u. for the three other rows of frames.

[Eq. (6)] are in good agreement with the correct results even at very high transition energies. Note in particular that whereas the fourth pseudostate provides a relatively poor approximation to the correct 1s → 5p discrete transition density, the continuum transition densities immediately above the ionization threshold, as well as those at higher energy, are in good accord with the corresponding correct values.

The foregoing observations suggest that the product $\mu\Phi_o$ provides an effective molecular box in which Ritz-principle variational calculations can be employed to obtain pseudostate approximations to the correct continuum eigenfunctions and transition densities. However, the important renormalization factor $(g(\tilde{\varepsilon}_i)/\tilde{f}_i)^{\frac{1}{2}}$ of Eq. (6) is apparently required in this approach, and $g(\varepsilon)$ is, of course, generally not known apriori. Indeed, the photoionization cross section is one of the ultimate aims of the calculation in any event. Clearly, some procedure must be devised to determine $g(\varepsilon)$ directly from the pseudostate calculation, since it is plausible that such calculations can provide the appropriate information in view of the results shown in Figure 2. It is helpful in this connection to examine physical properties that can be expressed as sums or integrals in energy over the spectrum of eigenfunctions.

B. Oscillator-Strength Moments

Although the pseudospectral functions of Eqs. (1) satisfy a different normalization convention than do the regular Coulomb p waves [Eq. (5c)], it is well known that the former can be used in calculations of certain physical properties. Reliable values of dipole polarizabilities,[44] for example,

$$S(-2) = \sum_{i=1}^{\infty} f_i\varepsilon_i^{-2} + \int_{\varepsilon_t}^{\infty} g(\varepsilon)\varepsilon^{-2}d\varepsilon, \tag{7a}$$

are obtained from the variational approximation

$$\tilde{S}(-2) = \sum_{i=1}^{N} \tilde{f}_i\tilde{\varepsilon}_i^{-2}, \tag{7b}$$

where the correct transtion frequencies and oscillator strengths are given by Eqs. (4) and (5) and the variational pseudospectra are given by Eqs. (3). In order to clarify the connection between Eqs. (7a) and (7b), it is helpful to rewrite them in the forms

$$S(-2) = (2/3)\langle\Phi_o|\mu(H-E_o)^{-1}\left\{\sum_{i=1}^{\infty}|\Phi_i\rangle\langle\Phi_i| + \int_{E_t}^{\infty}|\Phi_E\rangle dE\langle\Phi_E|\right\}\mu|\Phi_o\rangle$$

$$= (2/3)\langle\Phi_o|\mu(H-E_o)^{-1}\mu|\Phi_o\rangle, \tag{8a}$$

$$\tilde{S}(-2) \cong (2/3)\langle\Phi_o|\mu(H-E_o)^{-1}\left\{\sum_{i=1}^{N}|\tilde{\Phi}_i\rangle\langle\tilde{\Phi}_i|\right\}\mu|\Phi_o\rangle$$

$$\cong (2/3)\langle\Phi_o|\mu(H-E_o)^{-1}\mu|\Phi_o\rangle, \tag{8b}$$

where the completeness expressions[40]

$$\sum_{i=1}^{\infty}|\Phi_i\rangle\langle\Phi_i| + \int_{E_t}^{\infty}|\Phi_E\rangle dE\langle\Phi_E| = \delta(\underset{\sim}{r}'-\underset{\sim}{r}) \tag{9a}$$

$$\sum_{i=1}^{N}|\tilde{\Phi}_i\rangle\langle\tilde{\Phi}_i| \cong \delta(\underset{\sim}{r}'-\underset{\sim}{r}) \tag{9b}$$

have been employed, and $(H-E_o)^{-1}$ is defined in the subspace orthogonal to Φ_o.[45] Equations (7) to (9) indicate that a physical property such as the dipole polarizability involving summation or integration in energy over the spectrum [Eq. (7a)] can be approximated by a pseudospectrum [Eq. (7b)], as a consequence of the approximate completeness relation of Eq. (9b). Note that the Kronecker-delta normalization of Eq. (1a) is the appropriate one in this connection for the L^2 functions in Eqs. (9), whereas the delta-function normalization in energy of Eq. (5c) is the appropriate one for the scattering states. If these normalization conventions are not employed, appropriate compensating energy density-of-state factors are needed in Eqs. (7) to (9).[41] In the limit $N \to \infty$ Eqs. (8b) and (9b) are exact, in which case the correct polarizability is obtained from Eq. (7b), without explicit construction of the correct spectrum appearing in Eq. (7a).

The polarizability of Eqs. (7) is one of an infinite number of oscillator-strength moments[46] that can be approximated in a pseudospectrum in forms similar to that of Eq. (7b). Perhaps the most familiar of these are the so-called oscillator-strength sum rules, of which the f-sum rule is the prototypical example.[47-49] Comparisons of theoretical values obtained from variational calculations[46] with corresponding experimentally determined quantities[50] indicate that L^2 pseudospectra [Eqs. (1) and (3)] can provide reliable approximations to spectral sums or oscillator-strength moments. This observation suggests that spectral moments can perhaps provide computational or theoretical intermediaries in the construction of approximations to the photoabsorption and photoionization profiles of atoms and molecules from variationally determined pseudospectra. In order to pursue this line of thought, it is helpful to clarify somewhat further connections between the correct and approximate spectra appearing in Eqs. (7a) and (7b), respectively.

C. Distributions and Densities

To clarify procedures for reconstruction of oscillator-strength profiles from variationally detèrmined pseudospectra, employing spectral moments as intermediaries, it is helpful to rewrite Eqs. (7) in the forms

$$S(-2) = \int_0^\infty \varepsilon^{-2} df(\varepsilon) \tag{10a}$$

$$\tilde{S}(-2) = \int_0^\infty \varepsilon^{-2} d\tilde{f}(\varepsilon) \tag{10b}$$

where

$$df(\varepsilon) = \left[\sum_{i=1}^{\infty} f_i \delta(\varepsilon_i - \varepsilon) + g(\varepsilon) \right] d\varepsilon \tag{11a}$$

and

$$d\tilde{f}(\varepsilon) = \left[\sum_{i=1}^{N} \tilde{f}_i \delta(\tilde{\varepsilon}_i - \varepsilon) \right] d\varepsilon \tag{11b}$$

are the correct [Eqs. (4) and (5)] and variationally determined [Eqs. (3)] oscillator strengths, respectively, for transition into the interval ε to $\varepsilon + d\varepsilon$. The quantities in brackets in Eqs. (11) are conveniently referred to as <u>densities</u>, whereas the closely related cummulative oscillator strengths,

$$f(\varepsilon) = \int_0^\varepsilon df(\varepsilon') \tag{12a}$$

$$\tilde{f}(\varepsilon) = \int_0^\varepsilon d\tilde{f}(\varepsilon'), \tag{12b}$$

are referred to as <u>distributions</u>, in accordance with standard usage from the theory of moments.[34-37] Evidently, $\tilde{f}(\varepsilon)$ takes the form of a histogram,

$$\tilde{f}(\varepsilon) = \sum_{i=1}^{j} \tilde{f}_i, \quad \tilde{\varepsilon}_j < \varepsilon < \tilde{\varepsilon}_{j+1}, \tag{13a}$$

when Eq. (11b) is used in (12b), whereas $f(\varepsilon)$ is evidently a histogram in the discrete region of the spectrum $0 \leqslant \varepsilon < \varepsilon_t$, but is a smooth nondecreasing function for $\varepsilon_t < \varepsilon$. When the functions $f(\varepsilon)$ and $\tilde{f}(\varepsilon)$ are differentiated, employing the Dirac convention at the discrete steps, the densities of Eqs. (11) are recovered. However, when the distributions of Eqs. (12a) and (12b) are compared in specific cases it is seen that the latter provide useful histogram

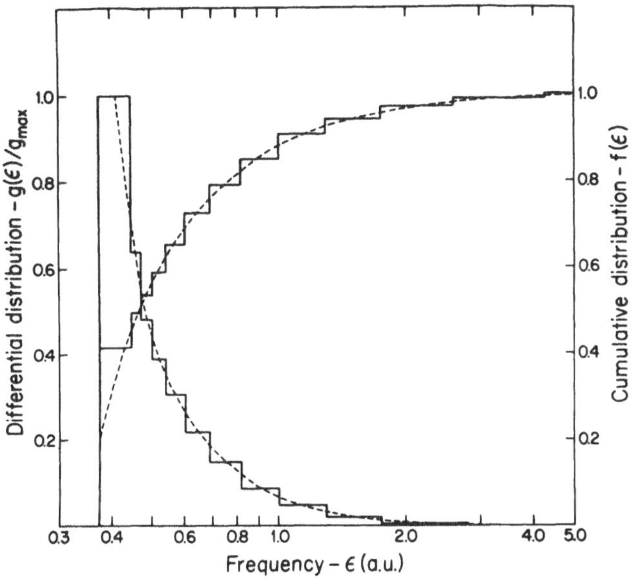

Fig. 3 Comparison of the correct oscillator-strength distribution and density (---) in atomic hydrogen with a fifteen-term variationally determined Stieltjes histogram approximation (——) [Eqs. (13)], obtained employing Laguerre functions.[1] In the discrete portions of the spectrum smooth curves are passed through the correct Stieltjes profiles in each case.

approximations to the former over the entire spectral interval, and this in turn suggests an alternative procedure for differentiating $\tilde{f}(\varepsilon)$.

In Figure 3 is shown the distribution function $f(\tilde{\varepsilon})$ of Eq. (13a) in the case of atomic hydrogen, obtained from a 15-term variationally determined Laguerre pseudospectrum [Eqs. (3)], in comparison with the correct distribution $f(\varepsilon)$.[1] It is evident from the figure that $\tilde{f}(\varepsilon)$ of Eq. (13a) provides a useful histogram approximation to $f(\varepsilon)$ over the entire spectral interval. This result is closely related to the convergence of pseudospectral approximations [Eqs. (7b) and (10b)] to the polarizability [Eqs. (7a) and (10a)], which involve the integrating functions $f(\varepsilon)$ and $\tilde{f}(\varepsilon)$. Moreover, it is also evident from Figure 3 that an appropriate interpretation of the derivative of $\tilde{f}(\varepsilon)$, other than the Dirac interpretation giving the density in Eq. (11b), will provide a suitable approximation to the derivative of $f(\varepsilon)$. Specifically, the so-called Stieltjes derivative of the histogram $\tilde{f}(\varepsilon)$,[35] determined in the form of the histogram

$$\tilde{g}(\varepsilon) = (1/2)(\tilde{f}_{i+1} + \tilde{f}_i)/(\tilde{\varepsilon}_{i+1} - \tilde{\varepsilon}_i), \quad \varepsilon_{i+1} > \varepsilon > \varepsilon_i, \tag{13b}$$

from the slopes of the straight-line segements connecting successive midpoints of the vertical steps of $\tilde{f}(\varepsilon)$, is also shown in Figure 3. The Stieltjes derivative $\tilde{g}(\varepsilon)$ of Eq. (13b) is seen to provide a useful histogram approximation to $g(\varepsilon)$ in the continuous spectral region, and, consequently, gives the renormalization factor of Eq. (6). By contrast, the delta-function sum of Eq. (11b), obtained following the Dirac convention at the steps of $\tilde{f}(\varepsilon)$, does not give a meaningful approximation to the correct density. Since Eqs. (6) and (13) provide approximations to the photoionization continuum at a finite number of frequency points, (Stieltjes) images of $f(\varepsilon)$, $g(\varepsilon)$ and Φ_E in a subspace of the interval $\varepsilon_t \leqslant \varepsilon < \infty$ are thereby obtained.[1-3]

The prescription of Eq. (13b) evidently corresponds to a density obtained by simply averaging the two variationally determined f numbers appearing in the interval $\tilde{\varepsilon}_i$ to $\tilde{\varepsilon}_{i+1}$, and dividing by the appropriate energy interval. In spite of its simplicity, the approach is seen to give a good approximation to $g(\varepsilon)$ in atomic hydrogen. Furthermore, in Figure 4 are shown the experimentally determined photoabsorption distribution $f(\varepsilon)$ and density $g(\varepsilon)$ in atomic helium, in comparison with Stieltjes values [Eqs. (13)] constructed from a variety of previously determined pseudospectra.[51-54] These theoretical results are seen to be in very good mutual accord, and in agreement with the experimental values.[55-57] This suggests that the Stieltjes density is perhaps generally valid, providing useful approximate photoionization cross sections and the factors required for constructing appropriately normalized L^2 scattering functions. It is the case, however, that the bin-smoothing approximation of Eq. (13b) is not entirely satisfactory in general when variationally determined $\tilde{\varepsilon}_i$, $\tilde{f}_i$ values are employed. Indeed, the very encouraging results obtained from Eqs. (13) for atomic hydrogen and helium shown in Figures 3 and 4 are consequences of the fortuitus use of rather special basis functions in the construction of variational pseudospectra. The approach can be extended, refined, and made rigorous, however, and the nature of the L^2 functions required in the calculations clarified, as indicated in the following section.

III. THEORY OF THE STIELTJES-TCHEBYCHEFF TECHNIQUE

In this section theoretical and computational aspects of the S-T technique for molecular photoionization studies are described in detail.[5,6] The convergence of L^2 calculations of spectral moments or oscillator-strength sums is established in (A), so-called principal representations of spectral moments and corresponding principal pseudostates are introduced in (B) and their general properties indicated, the important Stieltjes-Tchebycheff distributions and densities are defined in (C), and computational aspects of the moment problem are treated in (D).

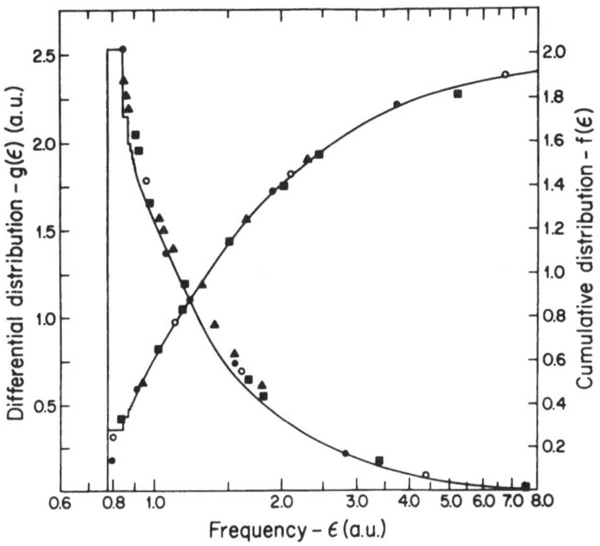

Fig. 4 Comparison of experimentally (——) and theoretically (●, ▲, ■, O) determined oscillator-strength distribution and density in atomic helium. The latter values, corresponding to midpoints of the horizontal portions of the Stieltjes histograms in each case, are obtained from Eqs. (13) and variationally determined pseudospectra.[2]

A. Convergence of Spectral Moments

The polarizability of Eqs. (7) is one of a class of spectral moments that can be written in the forms[46]

$$S(-k) = \int_0^\infty \varepsilon^{-k} df(\varepsilon), \; k = 0, 1, \ldots, \tag{14}$$

where $df(\varepsilon)$ is the oscillator-strength for transition into the interval ε to $\varepsilon + d\varepsilon$ defined by Eq. (11a). These oscillator-strength sums or spectral moments can be written in the alternative forms

$$S(-k) = (2/3)\langle\Phi_0|\mu(H-E_0)^{1-k}\mu|\Phi_0\rangle , \tag{15}$$

following the development of Eqs. (8) and (9). The integrals of Eq. (15) can be evaluated as quadratures for $k = 0, 1,$[48] whereas the operator in the integrand is of the inverse type for $k \geq 2$. In the former case ($k = 0, 1$) the integrals refer to the f sum rule (k = 0), equal to the number of electrons in the target, and to the mean-square size (k = 1) of the target atom or molecule, respectively.[48] Consequently, a knowledge of the ground-state wave function Φ_o is sufficient to determine these moments. In the latter case ($k \geqslant 2$) it is convenient to rewrite Eq. (15) in the form[2,3]

$$S(-k) = (2/3)\langle\Phi_o|\mu|\theta_k\rangle \tag{16}$$

where the θ_k are by definition,

$$\theta_k = (H-E_o)^{1-k}\mu\Phi_o, \tag{17}$$

and they evidently satisfy the inhomogeneous equations[54]

$$(H-E_o)\theta_2 = \mu\Phi_o \tag{18a}$$

$$(H-E_o)\theta_k = \theta_{k=1}, \ k \geqslant 3. \tag{18b}$$

The particular solutions of Eqs. (18) required are those in the subspace of H orthogonal to Φ_o.[45] Finally, from the definition of the θ_k [Eq. (17)],

$$S(1-k-\ell) = (2/3)\langle\theta_k|\theta_\ell\rangle \ . \tag{19}$$

The development of Eqs. (14) to (19) indicates that the spectral moments S(-k) can be calculated in Hilbert space. Equation (19) shows that the moments are the elements of the (Gramian) metric tensor corresponding to the functions θ_k of Eqs. (16) to (18).[58] For $k = \ell$ Eq. (19) shows that S(1-2k) is determined by the norm of θ_k, which consequently can be calculated in an L^2 basis when the moments are finite. Introducing an N-term primitive L^2 many-electron basis $\{\chi_i; \ i=1, N\}$, Eq. (15) has the approximate

$$\tilde{S}(-k) = (2/3) \ \underset{\sim}{d}^{\dagger} \cdot \underset{\sim}{\varepsilon}^{1-k} \cdot \underset{\sim}{d} \tag{20}$$

where

$$(\underset{\sim}{d})_i = \langle\chi_i|\mu|\Phi_o\rangle \tag{21a}$$

$$(\underset{\sim}{\varepsilon})_{ij} = \langle\chi_i|H-E_o|\chi_j\rangle, \tag{21b}$$

are vector and matrix representatives of the transition moment and Hamiltonian operators, respectively. It is generally convenient, but not necessary, to introduce a unitary transformation among the χ_i, under which Eq. (20) is invariant, that provides a representative of in diagonal form. The resulting linear combinations of primitive functions constitute an N-term pseudospectrum of L^2

eigenfunctions, as discussed in Section II(A). In terms of these pseudospectral functions [Eqs. (1)], Eq. (20) takes the form

$$\tilde{S}(-k) = \sum_{i=1}^{N} \tilde{\varepsilon}_i^{-k} \tilde{f}_i, \tag{22}$$

where the $\tilde{\varepsilon}_i$ and $\tilde{f}_i$ are the discrete pseudo-transition frequencies and oscillator strengths of Eqs. (3). Equation (22) provides a straight-forward generalization of Eq. (7b) to arbitrary k.[46]

In view of Eqs. (15), (16), and (19), and the assumed completeness of the L^2 basis, we have the important result

$$\text{Lim } (N \to \infty)\tilde{S}(-k) \to S(-k). \tag{23}$$

Although this observation provides a formal basis for L^2 convergence of the spectral moments, and of the subsequent Stieltjes-Tchebycheff analysis described below, it is important to note that the moments need not be calculated explicitly in the ensuing development. Rather, when pseudospectral calculations are performed, the pseudospectra $\{\tilde{E}_i, \tilde{\Phi}_i;\ i=1, N\}$ and $\{\tilde{\varepsilon}_i, \tilde{f}_i;\ i=1, N\}$ provide appropriate and sufficient computational information from which the underlying oscillator-strength density is obtained. Of course, if spectral moments are available in closed analytic forms, not necessarily obtained from pseudospectral calculations, these can be employed directly in the Stieltjes-Tchebycheff analysis given below.

B. Principal Representations

In Section II(C) pseudo-transition frequencies and oscillator strengths obtained from L^2 variational calculations were employed in constructing so-called Stieltjes approximations to distributions and densities [Eqs. (13)]. Although these can be satisfactory in specific cases, as indicated above, it is necessary to note that the results so obtained are generally basis-set dependent, and that different results can be obtained employing different basis sets. The Stieltjes histograms constructed from pseudospectra corresponding to nonlocal potentials, or those obtained from configuration-interaction calculations, for example, are found to be particularly unsatisfactory.[3,14]

Although variationally determined pseudospectra are generally basis-set dependent, it is anticipated from Section III(A) that they can provide suitably converged approximations to finite numbers of corresponding invariant spectral moments. While it is by no means obvious, and will be demonstrated explicitly below, [Section IV(A)], it is the lower-order moments ($0 \leqslant k \leqslant 2n$, $n < N$) and quantities derived from them that are rapidly convergent in an N-term L^2 basis;

$$\tilde{S}(-k) \to S(-k),\ 0 \leqslant k \leqslant 2n,\ n < N. \tag{24}$$

The higher-order spectral moments ($k > 2n$, $n < N$), or, more precisely, specific quantities derived from them that are described below [Sections III(D) and IV(A)], are generally more slowly convergent in an L^2 basis. As might be expected from their inverse-power nature, however, the higher-order moments and quantities derived from them rapidly approach known asymptotic values with increasing k, and consequently do not require precise calculations.

Given spectral moments $S(-k)$, $0 \leqslant k \leqslant 2n$, obtained from variational calculation in an N term basis ($n < N$), or by some other means, it is possible to construct discrete transition frequencies and oscillator strengths that satisfy

$$S(-k) = \sum_{i=1}^{n} \varepsilon_i^{-k} f_i,\ 0 \leqslant k < 2n-1. \tag{25}$$

The ε_i, f_i so obtained are generalized Gaussian quadrature points and weights, and are said to provide a <u>principal representation</u> of the defining spectral moments.[35] This principal pseudospectrum $\{\varepsilon_i, f_i, i = 1, n\}$ exhibits certain useful properties that the variational pseudospectrum $\{\tilde{\varepsilon}_i, \tilde{f}_i, i = 1, N\}$ generally does not exhibit. Specifically, the principal transition frequencies and oscillator strengths satisfy[35]

$$\varepsilon_1 \leqslant \varepsilon_i < \infty, \tag{26a}$$

$$f_i \geqslant 0,\ i=1, n, \tag{26b}$$

where ε_1 is the correct lower end of the spectrum, and they provide a basis for constructing convergent (Stieltjes) approximations to the associated oscillator-strength distribution and density, described in the following section.

In addition to the princiapl representation of Eqs. (25), it is useful to also consider generalized Gaussian quadrature points and weights that provide representations having one frequency fixed at a prespecified point ε in the spectrum and satisfying $[\varepsilon_0(\varepsilon) \equiv \varepsilon]$

$$S(-k) = \sum_{i=0}^{n} \varepsilon_i(\varepsilon)^{-k} f_i(\varepsilon),\ 0 \leqslant k \leqslant 2n. \tag{27}$$

The principal representation $\{\varepsilon_i(\varepsilon), f_i(\varepsilon);\ i=0, n\}$ so obtained satisfies[35]

$$-\infty < \varepsilon_i(\varepsilon) < +\infty, \tag{28a}$$

$$f_i(\varepsilon) \geqslant 0, \; i=0, n, \tag{28b}$$

and provides a basis for the construction of convergent (Tchebycheff) approximations to the associated distribution and density, described in the following section. Note that the functional dependences, $f_i(\varepsilon)$, $\varepsilon_i(\varepsilon)$ in Eqs. (27) and (28) simply imply that the indicated values are dependent upon where in the spectrum the prespecified frequency point ε is placed.

In the particular case that the S(-k) are calculated in an N-term L^2 basis, it is helpful to note that the defining equations for the principal pseudospectra can be written

$$\sum_{i=1}^{N} \tilde{\varepsilon}_i^{-k}\tilde{f}_i = \sum_{i=1}^{n} \varepsilon_i^{-k} f_i, \; 0 \leqslant k \leqslant 2n-1, \; n < N, \tag{29a}$$

$$\sum_{i=1}^{N} \tilde{\varepsilon}_i^{-k}\tilde{f}_i = \sum_{i=0}^{n} \varepsilon_i(\varepsilon)^{-k} f_i(\varepsilon), \; 0 \leq \varepsilon \leq 2n, \; n < N . \tag{29b}$$

Equations (29) suggest that the principal pseudospectra provide a smoothing in some sense of the generally larger variationally determined pseudospectrum, and they suggest that the former can be constructed from the latter without explicit reference to the intermediate spectral moments [Eq. (22)]. Computational procedures for the solution of Eqs. (25), (27), and (29) are described in Section III(D).

In certain special cases, variationally determined N-term pseudospectra can reproduce the first 2N spectral moments of Eq. (14) exactly in the form

$$\tilde{S}(-k) = \sum_{i=1}^{N} \tilde{\varepsilon}_i^{-k}\tilde{f}_i = S(-k), \; 0 \leqslant k \leqslant 2N-1, \tag{30}$$

as opposed to Eq. (23), in which the limit $N \to \infty$ is required. In these cases the pseudospectrum $\{\tilde{\varepsilon}_i, \tilde{f}_i;\ i=1, N\}$ provides a principal representation of the correct spectral moments, in accordance with Eq. (25) [n = N], and the corresponding pseudospectrum $\{\tilde{E}_i, \tilde{\phi}_i;\ i=1, N\}$ is said to be comprised of <u>principal pseudostates</u>.[1-3] These pseudostates provide an optimal N-term L^2 representation of the correct excitation spectrum under study in that they reproduce the first 2N spectral moments of Eq. (14) exactly. As a consequence, the corresponding $\tilde{\varepsilon}_i$ and $\tilde{f}_i$ satisfy certain useful theorems from the theory of moments, indicated in the following section, without requiring explicit solutions of the moment problem of Eq. (25).

The first n principal pseudostates of an arbitrary system can be determined from the θ_k of Eqs. (17) and (18), expressed as expansions

$$\theta_k = \sum_{i=1}^{\infty} \tilde{\varepsilon}_i^{\,1-k} \langle \tilde{\Phi}_i | \mu | \Phi_o \rangle \tilde{\Phi}_i, \quad 0 \leqslant k \leq n-1, \tag{31}$$

in terms of an arbitrary nonprincipal pseudospectrum. Principal pseudostates are then obtained by employing the n θ_k of Eq. (31) as nonorthogonal basis functions in the diagonalization of the Hamiltonian matrix.[3] Consequently, it is seen that any basis set of functions that spans the space of n of the functions θ_k, and hence provides solutions of the first n of Eqs. (18), can be used in the construction of a finite number of principal pseudostates. In the case of atomic hydrogen, the Laguerre functions employed in Section II(A) span the appropriate space,[53,54] and, consequently, principal pseudostates and the exact spectral moments are obtained from them.[1,2] It is largely because of this that the histograms shown in Figures 3 and 4 provide useful approximations to the photoionization distributions and densities in atomic hydrogen and helium, as is clarified more fully in the following section.

C. Stieltjes-Tchebycheff Distributions and Densities

When sufficiently accurate spectral moments have been determined, and the principal representations of Eqs. (25) and (27) obtained, so-called Stieltjes distributions can be constructed in the form[35]

$$f_s^{(n)}(\varepsilon) = 0, \quad 0 < \varepsilon < \varepsilon_1, \tag{32a}$$

$$f_s^{(n)}(\varepsilon) = \sum_{i=1}^{j} f_i, \varepsilon_j < \varepsilon < \varepsilon_{j+1}, \tag{32b}$$

$$f_s^{(n)}(\varepsilon) = \sum_{i=1}^{n} f_i = S(0), \quad \varepsilon_n < \varepsilon, \tag{32c}$$

where the superscript n indicates an n-term principal pseudospectrum is employed in the development. Such distributions satisfy the so-called Tchebycheff inequalities,[35]

$$f_s^{(n)}(\varepsilon_i - 0) < f_s^{(n+1)}(\varepsilon_i - 0) \leqslant f(\varepsilon_i) \leqslant f_s^{(n+1)}(\varepsilon_i + 0) < f_s^{(n)}(\varepsilon_i + 0), \tag{33}$$

at their points of increase ε_i, where $f(\varepsilon_i)$ is the correct cumulative oscillator-strength distribution [Eq. (12a)] evaluated at the point ε_i. Equation (33) indicates that the Stieltjes distributions converge monotonically in n at their points of increase to the correct distribution, which provides the motivation for their construction. Most important, the corresponding Stieltjes densities, obtained from the slopes of line segments connecting successive midpoints of the vertical portions of the Stieltjes distribution histogram [Eqs. (32)],

$$df_s^{(n)}/d\varepsilon = 0, \; 0 < \varepsilon < \varepsilon_1 \quad (34a)$$

$$df_s^{(n)}/d\varepsilon = (1/2)(f_{i+1}+f_i)/(\varepsilon_{i+1}-\varepsilon_i), \; \varepsilon_i < \varepsilon < \varepsilon_{i+1}, \quad (34b)$$

$$df_s^{(n)}/d\varepsilon = 0, \; \varepsilon_n < \varepsilon, \quad (34c)$$

also converge (in the Stieltjes sense) to the correct oscillator-strength density in the limit of large n.[6,59] Consequently, a convergent histogram approximate to the photoionization oscillator-strength density is obtained in the continuous spectral interval (ε_t, ∞). Moreover, the Stieltjes histogram in the discrete region of the spectrum provides information from which approximations to the discrete oscillator strengths are obtained if the corresponding transition energies are known.[1-3] The development consequently provides a generalization of the familiar quantum-defect or Coulomb approximation in the discrete spectral interval.[60]

Evidently, the Stieltjes distributions [Eq. (32)] and densities [Eq. (34)] are similar in form to the elementary variational approximations [Eqs. (13)] described in Section II(C). However, unless principal pseudostates are employed, variationally determined distributions [Eq. (13a)] do not necessarily satisfy the Tchebycheff inequalities of Eq. (33), and, consequently, the associated densities [Eq. (13b)] need not provide useful approximations to the correct oscillator-strength density. The smoothing of the variational pseudospectrum inherent in Eq. (29a) insures that the principal representations satisfy Eq. (33), and that they are comprised of relatively uniformly spaced and slowly varying effective oscillator strengths, in which case Eqs. (32) and (34) provide useful approximations. Evidently, the final expression of Eq. (34) corresponds to a simple bin-smoothing of the two effective principal f numbers present in the interval ε_i to ε_{i+1}. By contrast, bin smoothing of variationally determined pseudospectra can give rise to highly irregular Stieltjes oscillator-strength densities, as has been indicated above.

Since distributions having one frequency point at a preassigned value also satisfy the Tchebycheff inequalities [Eq. (33)],[35] a useful continuous approximation to the cumulative oscillator strength is obtained from the representation of Eqs. (27) and (29b) in the form

$$f_t^{(n)}(\varepsilon) = \sum_i f_i(\varepsilon) + \frac{1}{2} f_o(\varepsilon), \tag{35}$$

where the sum is over all $f_i(\varepsilon)$ corresponding to frequency points $\varepsilon_i(\varepsilon) < \varepsilon$. The so-called Tchebycheff distribution of Eq. (35) refers to the midpoint of the vertical rise of the corresponding Stieltjes histogram at the point ε, regarded as a function of the prespecified point ε. The properties of the Tchebycheff distribution are not obvious, but have been studied in considerably detail.[6] It is sufficient to note here that $f_t^{(n)}(\varepsilon)$ is a nondecreasing function of ε that is continuous on the real line, is generally smooth in the appropriate photoionization interval, exhibits step-like structures at the appropriate transition frequencies in the discrete region of the spectrum, and converges to $f(\varepsilon)$ in the limit of large n.[4-6]

The density corresponding to the distribution of Eq. (35) is obtained simply from the customary derivative in the form

$$df_t^{(n)}/d\varepsilon = \sum_i df_i(\varepsilon)/d\varepsilon + \frac{1}{2} df_o(\varepsilon)/d\varepsilon. \tag{36}$$

The Tchebycheff density so defined is real, nonnegative, and continuous on the real axis, is generally smooth in the photoionization interval, exhibits delta-function-like structures in the discrete region, and is convergent to the correct density $df/d\varepsilon$ [Eq. (11a)] in the limit of large n.[4-6]

The Stieltjes-Tchebycheff distributions and densities are seen to provide a general means for the construction of photoexcitation and ionization cross sections from L^2 variational calculations, or from spectral moments obtained following theoretical or semiempirical procedures. In order to construct the Stieltjes and Tchebycheff distributions and densities it is evidently necessary to solve Eqs. (25) and (27), or Eqs. (29), and to evaluate the derivatives appearing in Eq. (36).

When principal representations of the Stieltjes [Eq. (25)] or Tchebycheff [Eq. (27)] type have been constructed from variational or other spectral moments, it is desirable to make optimal use of the resulting principal transition frequencies and oscillator strengths. As an alternative to the Stieltjes [Eqs. (34)] and Tchebycheff [Eq. (36)] densities, it is useful to also fit the Stieltjes distributions [Eq. (32)] of various orders with explicit analytical forms, and to compare derivatives of these with the results of Eqs. (34) and (36). Polynomials in $1/\varepsilon$[3,15] and cubic splines[11] have proved particularly useful for this purpose. The Stieltjes procedure can also be made to provide information over a potentially greater range of frequency values than those values provided by the principal representations [Eq. (25)] through introduction of variable transformations and appropriate moment calculations.[12] Mutually consistent results are generally obtained when the variational or other moment calculations are satisfactory, and sufficient numbers of recurrence coefficients have been employed to achieve convergence from the various imaging procedures. In these cases, the Stieltjes and Tchebycheff densities are, of course also mutually consistent, providing a number of closely related moment-theory procedures for determining the unknown spectral density from L^2 variational or other calculations.

D. Computational Aspects of the Moment Problem

In order to construct solutions of Eq. (25) in an efficient manner it is useful to consider the Stieltjes integral[35]

$$I(z) = \int_0^\infty \frac{\varepsilon(df/d\varepsilon)d\varepsilon}{\varepsilon - z} = S(0) + S(-1)\,z + S(-2)\,z^2 + \ldots, \tag{37}$$

which has the indicated power-series expansion in terms of the spectral moments $S(-k)$. Such integrals have convergent ($n \to \infty$) representations in terms of Padé approximants $[n, n-1](z)$[61] or equivalent J-type continued-fraction approximants $A_n(z)$.[34] The former are conveniently written

$$[n, n-1](z) = P_{n-1}(z)/Q_n(z) \tag{38a}$$

where

$$P_{n-1}(z) = \sum_{i=0}^{n-1} a_i z^i \tag{38b}$$

$$Q_n(z) = 1 + \sum_{i=1}^{n} b_i z^i \tag{38c}$$

are appropriate polynomials, whereas the latter are of the form

$$A_n(z) = \frac{\beta_0}{/1/z - \alpha_1} \Big/ - \frac{\beta_1}{/1/z - \alpha_2} \Big/ \ldots - \frac{\beta_{n-1}}{/1/z - \alpha_n} \Big/, \tag{39}$$

employing standard notation from the theory of continued fractions.[34] The coefficients a_i, b_i of Eqs. (38) and the α_i, β_i of Eq. (39) are found by requiring that the power-series expansions of Eqs. (38) and (39), respectively, equal that of the defining integral Eq. (37) up to sufficient order;

$$[n,\ n-1](z) = I(z) = (1/z)A_n(z),\ O(z^{-2n-1}). \tag{40}$$

The left-hand equality of Eq. (40) provides an algorithm for the a_i, b_i in terms of the S(-k) in the form of linear inhomogeneous equations,[62] whereas the right-hand equality of Eq. (43) gives so-called quotient-difference or product-difference algorithms for the α_i, β_i in terms of S(-k).[63-65] Both $[n,\ n-1](z)$ and $(1/z)A_n(z)$ have the partial-fraction representation

$$[n,\ n-1](z) = (1/z)A_n(z) = \sum_{i=1}^{n} \frac{\varepsilon_i f_i}{\varepsilon_i - z}\ . \tag{41}$$

Introducing Eq. (41) into Eq. (40) shows that the ε_i, f_i are precisely those of Eq. (25). That is, Eq. (40) using the partial-fraction representation of Eq. (41) gives the defining moment relations of Eq. (25). Consequently, Eqs. (37) to (41) indicate there are three ways in which the moment problem can be solved for principal representations associated with Stieltjes distributions and densities. First, the nonlinear equations [Eqs. (25)] for the ε_i, f_i can be solved directly. Second, the Padé approximants of Eqs. (38) can be constructed by solving linear equations [Eq. (40)] for the a_i, b_i using the S(-k). Finally, the continued-fraction approximant $A_n(z)$ can be constructed using the product-difference algorithm [Eq. (40)] for the α_i, β_i in terms of the S(-k). In the latter two cases the ε_i, f_i are obtained from the appropriate roots and residues of Eqs. (38) and (39).[62-66] These algorithms and closely related variants[67] are generally satisfactory when appropriate spectral moments are available. In the case that the available spectral moments are expressed in terms of a variationally determined pseudospectrum [Eq. (22)], an alternative particularly stable algorithm described further below is available for calculations of the necessary α_i, β_i coefficients.[5]

Convenient, computationally useful recurrence relations for the $P_{n-1}(z)$ and $Q_n(z)$ are obtained directly from the equality of $[n,\ n-1](z)$ and $(1/z)A_n(z)$ [Eq. (40)] in the forms,

$$Q_n(z) = (1-\alpha_n z)Q_{n-1}(z) - z^2\beta_{n-1}Q_{n-2}(z) \tag{42a}$$

$$Q_{-1}(z) = 0,\ Q_o(z) = 1 \tag{42b}$$

$$P_{n-1}(z) = (1-\alpha_n z)P_{n-2}(z) - z^2\beta_{n-1}P_{n-3}(z) \tag{42c}$$

$$P_{-1}(z) = 0, \; P_o(z) = \beta_o. \tag{42d}$$

These three-term recurrence relations can be cast into alternative perhaps more familiar forms by introducing the closely related polynomials[68,69]

$$p_n(x) = x^{n-1} P_{n-1}(1/x) \tag{43a}$$

$$q_n(x) = x^n Q_n(1/x), \tag{43b}$$

which are defined with respect to the density $F(x) \equiv S(0)-f(1/x)$. In the following, the polynomials $Q_n(z)$ will be referred to as those orthogonal with respect to $f(\varepsilon)$ since they satisfy[5]

$$\int_0^\infty Q_n(\varepsilon)Q_m(\varepsilon)\varepsilon^{-m-n}df(\varepsilon) = \beta_o\beta_1\ldots\beta_n\delta_{nm}, \tag{44}$$

whereas the $P_{n-1}(z)$ are referred to as the associated numerator polynomials [Eq. (30)]. Equation (44) follows directly from the defining relations for the a_i, b_i in terms of the $S(-k)$ [Eq. (40)].

The development of Eqs. (42) to (44) indicates that the α_n, β_n are recurrence coefficients for the $P_{n-1}(z)$, $Q_n(z)$, the roots of which in turn determine the required ε_i, and which also provide the f_i from residues of $[n, n-1](z)$. Although the product-difference algorithm for the α_n, β_n in terms of the $S(-k)$ is generally satisfactory for small numbers of moments,[63,64] an alternative algorithm has proved useful for large numbers of moments. This is obtained directly from the recurrence relations [Eqs. (42)] in the form[66]

$$\alpha_n = \frac{Y_{n-1,m}}{Y_{n-1,n-1}} - \frac{Y_{n-2,n-1}}{Y_{n-2,n-2}}, \; \alpha_o = 0, \; \alpha_1 = \frac{S(-1)}{S(0)} \tag{45a}$$

$$\beta_n = \frac{Y_{n,n}}{Y_{n-1,n-1}}, \; \beta_{-1} = 0, \; \beta_o = S(0), \tag{45b}$$

where the matrix elements

$$Y_{n,\ell} \equiv \int_0^\infty \varepsilon^{-\ell-n} Q_n(\varepsilon)df(\varepsilon) \tag{45c}$$

satisfy

$$Y_{n,\ell} = Y_{n-1,\ell+1}, \; -\alpha_n Y_{n-1,\ell} - \beta_{n-1} Y_{n-2,\ell} \tag{45d}$$

$$Y_{n,\ell} = 0, \ \ell < n \tag{45e}$$

$$Y_{o,\ell} = S(-\ell) \tag{45f}$$

$$Y_{n,1} = Y_{-1,\ell} = 0. \tag{45g}$$

Equations (45) are a variant of the so-called Stieltjes-Tchebycheff algorithm.[34]

Although Eqs. (37) to (45) are satisfactory when spectral moments are available, a more stable algorithm can be employed when variationally determined pseudospectra are calculated. In this case it is convenient to write the α_n and β_n in the forms[5,11]

$$\alpha_n = \frac{1}{\beta_o\beta_1\ldots\beta_{n-1}} \int_o^\infty (1/\varepsilon)^{2n-1} Q_{n-1}(\varepsilon)^2 df(\varepsilon) \tag{46a}$$

$$\beta_n = \frac{1}{\beta_o\beta_1\ldots\beta_{n-1}} \int_o^\infty (1/\varepsilon)^{2n} Q_n(\varepsilon)^2 df(\varepsilon), \tag{46b}$$

which are obtained directly from Eqs. (42). In terms of variationally determined pseudospectra [Eqs. (3)], the integrals of Eqs. (46) are given by the quadratures

$$\alpha_n = \frac{1}{\beta_o\beta_1\ldots\beta_{n-1}} \sum_{i=1}^{N} (1/\tilde{\varepsilon}_i)^{2n-1} Q_{n-1}(\tilde{\varepsilon}_i)^2 \tilde{f}_i \tag{47a}$$

$$\beta_n = \frac{1}{\beta_o\beta_1\ldots\beta_{n-1}} \sum_{i=1}^{N} (1/\tilde{\varepsilon}_i)^{2n} Q_n(\tilde{\varepsilon}_i)^2 \tilde{f}_i. \tag{47b}$$

Since the values $Q_n(\tilde{\varepsilon}_i)$, $Q_{n-1}(\tilde{\varepsilon}_i)$ can be obtained from the recurrence relation of Eq. (42) employing α_m, β_m values of lower order than the α_n, β_n being calculated, Eqs. (47) provide a particularly convenient and stable algorithm. When the α_n and β_n have been determined in this manner, or from Eqs. (45) if only moments are available, the ε_i and f_i are obtained from the defining equations

$$Q_n(\varepsilon_i) = 0 \tag{48a}$$

$$f_i = P_{n-1}(\varepsilon_i)/\varepsilon_i Q_n'(\varepsilon), \tag{48b}$$

employing matrix diagonalization and polynomial recurrence techniques.[63-66]

The foregoing development is sufficient to construct the Stieltjes distributions and densities of Eqs. (25), (29a), (32), and (34). In the case of the Tchebycheff distributions and densities of Eqs. (27), (29b), (35), and (36), additional considerations are required. Specifically, the roots and residues of the rational fractions,[5,66]

$$\{n, n-1\}(z) = \tilde{P}_{n-1}(z)/\tilde{Q}_n(z) \tag{49a}$$

$$\tilde{P}_{n-1}(z) = Q_n(\varepsilon)P_{n-2}(z)-Q_{n-1}(\varepsilon)P_{n-1}(z) \tag{49b}$$

$$\tilde{Q}_n(z) = Q_n(\varepsilon)Q_{n-1}(z)-Q_{n-1}(\varepsilon)Q_n(z), \tag{49c}$$

obtained from the expressions

$$\tilde{Q}_n(\varepsilon_i(\varepsilon)) = 0 \tag{50a}$$

$$f_i(\varepsilon) = \tilde{P}_{n-1}(\varepsilon_i(\varepsilon))/\varepsilon_i(\varepsilon)\tilde{Q}_n'(\varepsilon_i(\varepsilon)), \tag{50b}$$

provide the necessary pseudospectrum of Eqs. (27) and (29b).[5,6] Since the so-called quasiorthogonal polynomials $\tilde{P}_{n-1}(z)$, $\tilde{Q}_n(z)$ of Eqs. (49) are defined in terms of the orthogonal polynomials $P_{n-1}(z)$, $Q_n(z)$, the development of Eqs. (37) to (48) is sufficient to determine the Tchebycheff distributions and densities of Eqs. (35) and (36). Matrix diagonalization and polynomial recurrence relations[6] are employed in the solution of Eqs. (50).[62-66]

IV. APPLICATIONS OF THE STIELTJES-TCHEBYCHEFF TECHNIQUE

In this section applications of the S-T technique are made to the photoionization cross sections of atoms and molecules.

A. Atomic Photoionization Studies

As an illustration of the general development given in the preceding section, the so-called Bethe-Ohmura approximation to the oscillator-strength density and distribution in H^- is considered.[4,5] In this case the correct density and distribution are[70,71]

$$df/d\varepsilon = (8/3\pi)\left[(\varepsilon-1)^{\frac{1}{2}}/\varepsilon\right]^3, \quad 1 \leqslant \varepsilon < \infty, \tag{51a}$$

$$f(\varepsilon) = (2/\pi)\left[\frac{(\varepsilon-1)}{3}^{\frac{1}{2}}\left(\frac{2}{\varepsilon^2} - \frac{5}{\varepsilon}\right) + \tan^{-1}(\varepsilon-1)^{\frac{1}{2}}\right], \quad 1 \leqslant \varepsilon < \infty, \tag{51b}$$

respectively, where the dimensionless variable $\varepsilon = \omega/\varepsilon_t$ is employed, and the density is normalized to unit area. The corresponding recurrence coefficients of Eqs. (39) to (46) can be written in the explicit forms,

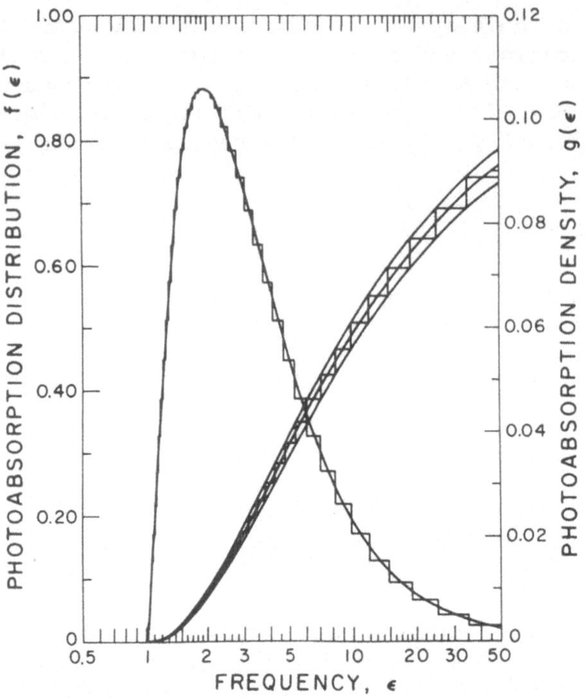

Fig. 5 Comparison of the oscillator-strength distribution and density in H^- in the so-called Bethe-Ohmura approximation of Eqs. (51)[70,71] with corresponding Stieltjes-Tchebycheff results for n = 50, obtained from the recurrence coefficients of Eqs. (52) and the development of Section III(D).[4,5]

$$\alpha_n = \frac{1}{2}(4n^2-3)/(4n^2-1), \ n = 1,2, \ldots, \tag{52a}$$

$$\beta_n = \frac{1}{4}(2n+3)(2n-1)/(4n+2)^2, \ \beta_o = 1, \ n = 1,2, \ldots, \tag{52b}$$

obtained from solution of Eqs. (45),[4,5] making variational or other calculations unnecessary. These expressions are used in the development of the preceding section in constructing Stieltjes and Tchebycheff distributions and densities, which are found to converge to the correct values of Eqs. (51) for sufficiently large n. The Stieltjes and Tchebycheff results for n = 50, for example, shown in Figure 5 are seen to be convergent to $df/d\varepsilon$ and $f(\varepsilon)$, although the Tchebycheff bounds on $f(\varepsilon)$ have not closed over the entire spectral integral. The mean values of these bounds, however – the Tchebycheff distribution [Eq. (35)] – are seen to be indistinguishable from the results of Eq. (51b). Similarly, although the steps of the

Stieltjes density [Eqs. (34)] are clearly discernable, the histogram values at the midpoints of the various energy intervals, as well as the Tchebycheff density [Eq. (36)], are indistinguishable from the results of Eq. (51a). Moreover, as few as five α_n and β_n values also provide convergent Stieltjes-Tchebycheff distributions and density [not shown], although the Tchebycheff bounds are weaker in this case, and the steps of the Stieltjes density are larger than those of Figure 5.

In the foregoing example the important recurrence coefficients α_n, β_n are available in explicit analytical forms [Eqs. (52)]. Consequently, the development of Eqs. (32) to (50) can be implemented when necessary α_n, β_n values are known. More generally, it is possible to obtain α_n, β_n values from variational calculations in Hilbert space and appropriate asymptotic information.[4,5]

The H^- ion treated as a two-electron problem, as opposed to the approximation of Eqs. (51), provides a useful illustrative example of determinations of convergent recurrence coefficients.[5] Two different pseudospectra (a and b) of transition frequencies and oscillator strengths, constructed from different basis sets of L^2 functions, are used in the variational calculation of α_n and β_n values for H^-.[5] Spectrum a is constructed from an 80 term $^1S_o{}^e$ ground-state function, comprised of configurations of (7s, 2p) Slater orbitals multiplied by Hylleraas correlation factors, and a 109-term $^1P_1{}^o$ pseudospectrum obtained from configurations of (6s, 11p) Slater orbitals multiplied by Hylleraas factors.[72] Spectrum b is constructed from a 135-term $^1S_o{}^e$ ground-state function, made up of basis functions of the Pekeris type, and a 110-term $^1P_1{}^o$ pseudospectrum made up of correlated functions of the Breit type.[73] In addition, the spectrum b basis contains the special functions required to satisfy the S(2), S(1), and S(0) sum rules.[74]

The first ten of the recurrence coefficients for H^- obtained from the a and b spectra are shown in Table I. Evidently, the lower-order values are in good mutual accord, in spite of the generally different natures of quantum-mechanical calculations involved. By contrast, for $n > 8$ there are significant differences between the respective results. This suggests that the lower-order recurrence coefficients converge more rapidly than do the higher-order ones, as intimated in Section III(A). In particular, the spectrum b coefficients do not approach the indicated correct asymptotic values. These latter are obtained from the general and important relations[5,34]

$$\alpha_\infty = 1/(2\varepsilon_t) \tag{53a}$$

$$\beta_\infty = 1/(4\varepsilon_t)^2, \tag{53b}$$

where ε_t is the threshold excitation energy for photoionization.

Table I. Polynomial recurrence coefficients for the dipole spectrum of the negative hydrogen ion.[a]

n	Spectrum a α_n	Spectrum a β_{n-1}	Spectrum b α_n	Spectrum b β_{n-1}
1	7.48208	1.99806	7.48373	2.00000
2	16.17492	47.16656	16.17802	47.03106
3	17.54044	64.74453	17.51533	64.26250
4	17.81489	71.04796	17.85973	70.47542
5	17.78246	75.46819	17.95406	74.51359
6	17.73632	78.44970	17.95009	76.27444
7	17.81685	79.78305	17.94283	77.95384
8	18.04287	79.25322	17.20083	80.63289
9	18.18408	77.92561	9.88699	72.49630
10	18.24869	77.56184	1.23642	1.51100
∞[b]	18.01768	81.15916	18.01768	81.15916

[a]Values in Hartree atomic units obtained from Eqs. (42) and (47) and two independently determined variational pseudospectra.[5]

[b]Asymptotic limits obtained from Eqs. (53) as discussed in the text, where $\varepsilon_t = 0.027751$ a.u. is the threshold frequency for photoionization in this case.[5]

Construction of the Stieltjes and Tchebycheff densities and distributions using the coefficients of Table I indicates that convergence is achieved with increasing numbers of coefficients for both spectra until values for $n > 8$ are employed. At this point the spectrum b Tchebycheff profiles exhibit structures apparently associated with the underlying discrete pseudospectrum. Similarly, when sufficient numbers of spectrum a recurrence coefficients for $n > 10$ (not shown) are employed, the resulting Tchebycheff profiles also exhibit structures associated with the corresponding underlying discrete pseudospectra. It is reasonable to conclude on the basis of these observations that the failure to converge to the asymptotic values of Eqs. (53) of variationally determined recurrence coefficients, and the related structures in the corresponding Stieltjes-Tchebycheff profiles, are related to the use of finite L^2 basis sets in the calculations. Indeed, solutions of Eqs. (25), (27) and (29) for n=N will result in Tchebycheff profiles that simply reproduce the variationally determined delta-function density of Eq. (11b).

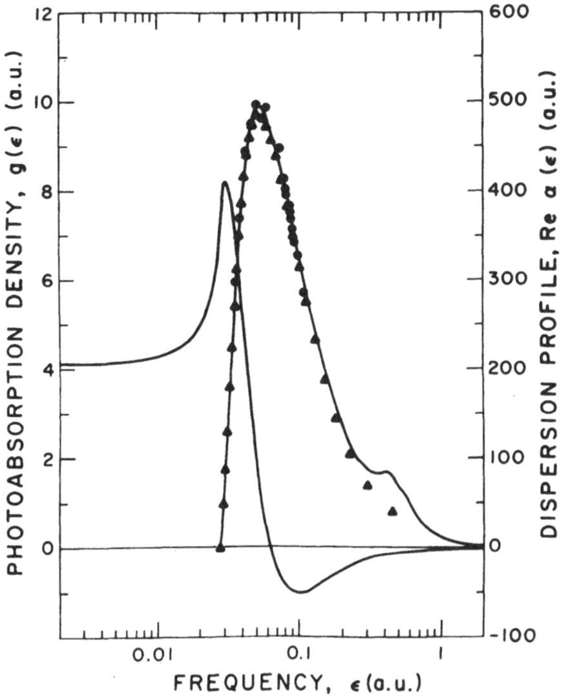

Fig. 6 Photoabsorption and dispersion profiles in H^- obtained from the S-T technique employing large-basis-set L^2 variational calculations and the recurrence-coefficient extension procedure of Eq. (54) (——),[5] in comparison with experimental (●) and previously reported continuum wave-function calculations (▲).[75,76]

The foregoing observations suggest that a complete set of appropriate recurrence coefficients can be obtained employing variational calculations in a sufficiently large L^2 basis set for the low-order coefficients, and a suitable extrapolation procedure to the known asymptotic expressions Eqs. (53) for the higher-order coefficients. Plausible forms for the extrapolated coefficients are given by the expressions[5]

$$\alpha_n = \alpha_\infty(1 + \delta_1/n + \delta_2/n^2 + \ldots), \tag{54a}$$

$$\beta_n = \beta_\infty(1 + \gamma_1/n + \gamma_2/n^2 + \ldots), \; n \to \infty, \tag{54b}$$

which correspond to asymptotic expansions in $1/n$. The parameters δ_i, γ_i are obtained from the variationally determined lower-order recurrence coefficients, or from the known analytic behavior in ε of the photoionization cross section at threshold.

In Figure 6 is shown an illustration of the use of the coefficients of Table I and Eqs. (54) for H^-. A four-term interpolation formula is used in this case, employing calculated α_1 α_2, β_1 β_2 values to determine δ_1 δ_2, γ_1 γ_2. The results shown are for $n = 20$, making use of ten of the coefficients of Table I and ten of those obtained from Eqs. (54). The profiles of Figure 6 are convergent in the sense that they are invariant to the use of additional recurrence coefficients [Eqs. (64)] in the S-T analysis. Evidently, the photoionization profile for H^- obtained from this procedure is generally in very good agreement with previous continuum eigenfunction calculations,[75] and with the available experimental values.[76] The structure in the S-T result in the figure is a consequence of the use of configurations in the calculations that correspond to resonance effects associated with the 2s final hydrogen-atom state.[5] Although the calculated corresponding dispersion profile in H^- is also shown in Figure 6, there are apparently no experimental values available for comparison.

Table II. Polynomial recurrence coefficients in atomic lithium.[a]

n	α_n	β_{n-1}
1	0.69336	1.15304
2	4.04781	1.23440
3	4.22279	3.06532
4	3.73326	3.68506
5	3.28640	3.64210
6	3.22493	3.03250
7	3.31091	2.56230
8	2.99237	1.97390
9	3.35833	2.81429
10	2.17304	0.94942
∞[b]	2.52209	1.59024

[a] Values in Hartree atomic units obtained from Eqs. (42) and (47) employing a 119-term variationally determined pseudospectrum.[13]

[b] Asymptotic limits obtained from Eqs. (53) employing the experimental ionization energy $\varepsilon_t = 0.198142$ a.u..

As a third illustrative application of the S-T approach, an (8s, 6p) Slater basis and Hylleraas correlation factors are used in the calculation of a 150-term $^2S^e$ ground-state function, 120-term $^2P^o$ pseudospectrum, corresponding transition frequencies and oscillator strengths, and recurrence coefficients in atomic lithium.[13] The various configurations employed include both K-shell and valence-shell excitations, allowing for core polarization and configuration interaction in both ground and excited states.[77-78] The resulting polynomial recurrence coefficients for Li shown in Table II evidently exhibit small oscillations with n in the interval $n \leq 10$, and larger oscillations are evident for $n > 10$ (not shown). When the extension procedure described above is applied in the case of Li, the 20th order (convergent) profile shown in Figure 7 is obtained. Also shown in the figure are Stieltjes points obtained in seventh- and eight-orders for comparison with the Tchebycheff result. These results are evidently in good mutual agreement, except perhaps at the highest excitation energy shown, where the Stieltjes results are seen to fall somewhat below the Tchebycheff profile.

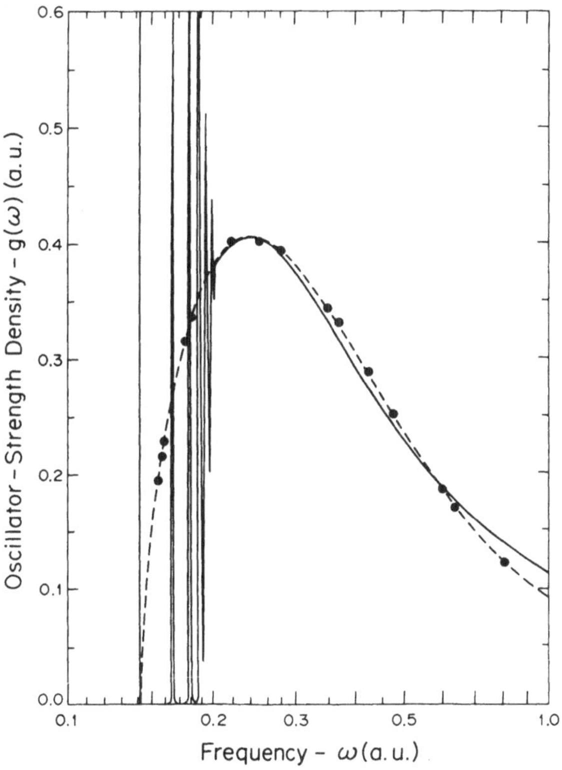

Fig. 7 Tchebycheff (——) and Stieltjes (●) oscillator-strength densities in Li, obtained from large-basis-set L^2 variational calculations and the S-T technique.[13]

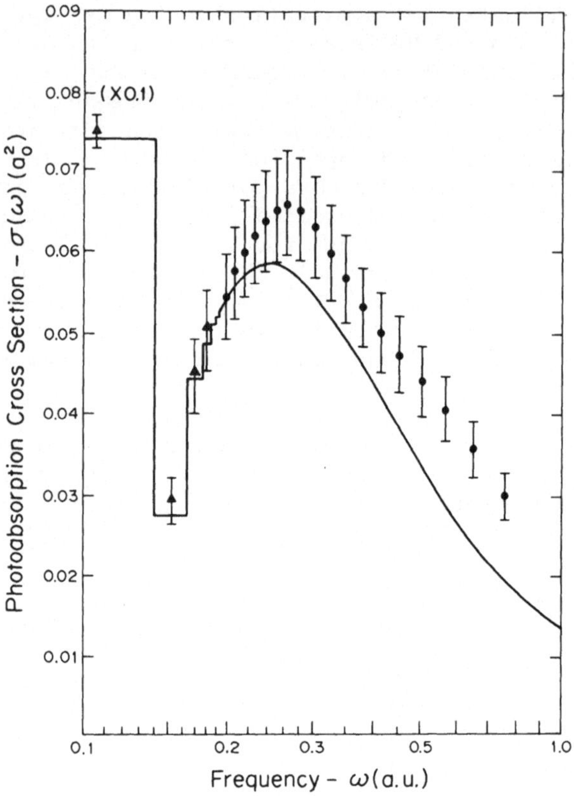

Fig. 8 Photoexcitation and ionization cross sections in Li obtained from the S-T technique (——)[13] in comparison with experimental values (▲, ●).[79,80]

In Figure 8 the calculated S-T photoabsorption cross section in Li is compared with available measured values. The discrete region of the calculated spectrum, presented in the Stieltjes sense, is in good agreement with the measured f numbers,[79] whereas the calculated photoionization continuum cross section is in good agreement at threshold with measured values,[80] but falls below these at higher energies. However, the calculated values shown in Figure 8 are in very good agreement with recent polarized-orbital[81] and many-body calculations[82] (not shown), and the discrepancy between theory and experiment can be attributed to systematic errors in the measured values in this case.[13]

In addition to the atomic and ionic photoionization calculations reported here, S-T determinations of detachment cross sections in atomic anions and of the cross sections of metastable atoms have been reported employing various computational approximations.

Model potential studies in Li^- and Na^-[9] and ab initio calculations in H^-[12] and F^-[18] have provided Stieltjes detachment cross sections in good accord with available experimental determinations and previous theoretical investigations. Photoionization of the metastable ^3P state in Ne has been studied employing extensive CI calculations and S-T procedures,[15] and approximate separations of the partial-channel photoionization cross section in B have been attempted employing CI and Stieltjes techniques.[11] Very recent studies of the 1s and 2s photoionization cross sections in Ne employing the Stieltjes approach[30] have made necessary re-evaluations of earlier continuum wave function calculations in this case, suggesting that the S-T approach can be computationally preferrable to continuum wave function calculations even for atomic systems.

B. Molecular Photoionization Studies

Conventional molecular photoionization calculations require the construction of an appropriate ground- or target-state eigenfunction, and of the continuum functions associated with the photoejection of electrons over a range of ionization energies. Computational difficulties largely related to the non-central and non-local nature of the potential have made necessary the introduction of simplifying approximations in construction of molecular continuum functions. The various computational approaches employed include the additive atomic approximation, use of plane-wave or orthogonalized-plane-wave states, one- and two-center Coulomb wavefunctions, one-center partial-wave expansions, and the scattered-wave X_α method. Although these approaches can provide useful information in specific cases, the resulting cross sections are generally of unknown quality, and it is generally not possible to further refine the calculations without abandoning the model adopted. The S-T approach to photoionization described in the preceding sections is particularly well suited to studies of molecular continua, since it can make use of the technology of computational quantum chemistry devised for L^2 bound-state problems.[31,32] As illustrations, the total and partial-channel vertical-electronic photoionization cross sections of diatomic molecules calculated in the separated-channel static-exchange approximation, employing the customary adiabatic treatment of nuclear motion, are described here.[21-28]

In the separated-channel static-exchange approximation an appropriate Hartree-Fock function is first constructed near the equilibrium internuclear molecular configuration.[31] The resulting canonical molecular orbitals are then used in constructing appropriate one-electron Hamiltonians for describing the vertical ionization of individual electrons. These are of the form

$$h_\Gamma = T + V + V_\Gamma \tag{55}$$

where T and V are the kinetic- and nuclear-framework-potential-energy operators, respectively, and V_Γ is the one-electron channel potential. For excitations of configurations of electrons involving doubly-occupied shells only (σ^2, $a_1{}^2$, $b_1{}^2$, $b_2{}^2$, etc.), V_Γ takes the form[21]

$$V_\Gamma = \sum_i (2\hat{J}_i - \hat{K}_i) + \hat{J}_\Gamma + \hat{K}_\Gamma, \tag{56}$$

where $\hat{J}_i$, $\hat{K}_i$ are the Coulomb and exchange operators for the doubly-occupied unexcited orbitals, and $\hat{J}_\Gamma$, $\hat{K}_\Gamma$ are the corresponding operators for the excited shell.[31] For open-shell situations, and those configurations requiring symmetry restrictions upon excitation (π^2, δ^2, etc.), alternative channel potentials are employed.[26] The potential of V_Γ is a noncentral and non-local one that has Coulombic form, corresponding to an ionic target, for sufficiently large distance from the center of mass. Consequently, the spectrum of h_Γ obtained from

$$(h_\Gamma - \varepsilon)\phi_\varepsilon = 0 \tag{57}$$

will exhibit an infinite number of bound states, corresponding to so-called improved virtual orbitals (IVO), as well as a photoionization continuum.[83-87]

It is anticipated that the separated-channel static-exchange approximation, using the customary adiabatic treatment of the nuclear motion, will provide reliable vertical-electronic molecular photoionization cross sections at ionization energies sufficiently above threshold, and in those cases in which the effects of core relaxation are negligible. Since the photoelectron spectra of the outer valence shells of light diatomic and polyatomic molecules are generally in one-to-one correspondence with the corresponding HF molecular-orbital binding-energy spectrum, and shake-up lines are generally weak, the separate-channel static-exchange approximation should be appropriate in these cases.[88] By contrast, inner valence-shell and K-shell photoelectron spectra indicate the presence of significant contributions from final-ionic-state correlation effects in certain cases.[89-92] Consequently, it will be necessary to exercise caution in interpreting the results of separated-channel static-exchange calculations in these situations.

In Figure 9 are shown the S-T IVO partial-channel vertical-electronic photoionization cross sections for removal of the $3\sigma_g$, $1\pi_u$, $2\sigma_u$ valence orbitals in N_2, and the corresponding cross section for the production of parent N_2^+ ions.[22] These results are obtained using large basis sets of Gaussian functions of appropriate symmetry[93] in variational solutions of Eq. (57). The S-T procedure is then applied to the resulting pseudospectra for ionization of the three

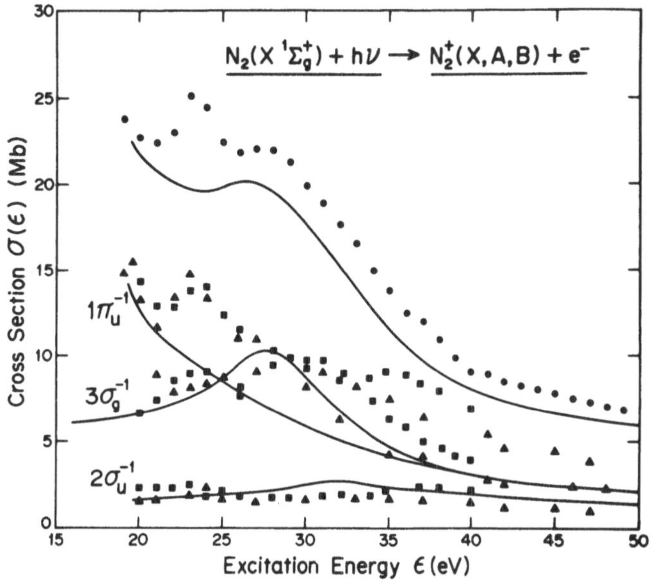

Fig. 9 Partial-channel and total photoionization cross sections in N_2 for the removal of $3\sigma_g$, $1\pi_u$, and $2\sigma_u$ electrons, and for the production of N_2^+ parent ions, obtained from the IVO S-T approach (——),[22] in comparison with recent line-source (◆),[96] synchrotron-radiation (■),[97,98] (e, 2e) coincidence (▲),[99] and (e, e + ion) coincidence measurements (•).[100]

indicated valence orbitals. The dimensionalities of the pseudospectra employed are 22, 28, 22, 27, and 19 for σ_u, σ_g, π_u, π_g, and δ_g final-state symmetries, respectively. Finally, the cross section for parent N_2^+ ion production is obtained from summation of the three vertical valence-electron cross sections, under the assumption that the appropriate Franck-Condon regions do not extend into the dissociative continua, and that contributions from higher-lying ionic states are negligible. Inspection of the potential curves for the three lowest-lying ionic states indicates that the appropriate Franck-Condon regions are entirely in the bound-state regions in these cases.[94] However, a small contribution to N_2^+ production can be expected in the ~20 - 40 eV interval from the neglected $C^2\Sigma_u^+$ ionic state, which is largely predissociated by the $B^2\Sigma_u^+$ and a $^4\Pi_u$ ionic state.[95]

Evidently, the IVO calculations shown in Figure 9 are in generally good agreement with the recent line-source,[96] synchrotron-radiation,[97,98] and (e, 2e)[99] measurements of the partial-channel

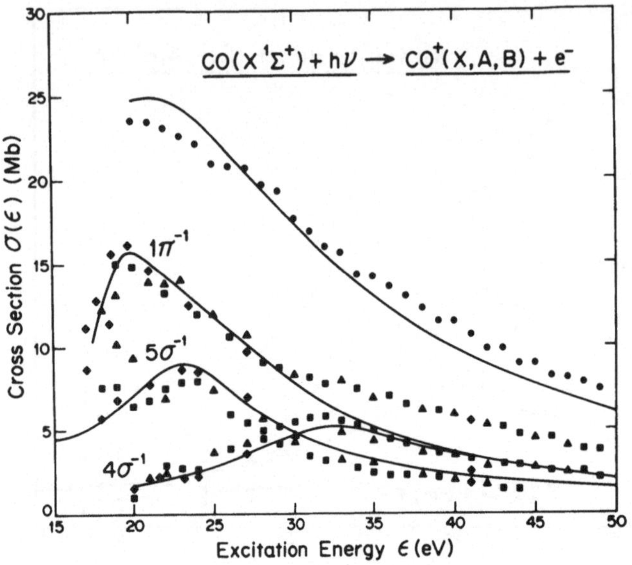

Fig. 10 As in Figure 9, for the removal of 5σ, 1π, and 4σ outer valence electrons in CO.[23]

photoionization cross sections. Similarly, the sum of these cross sections is in generally good agreement with the recent (e, e + ion) photoionization cross section measurements corresponding to N_2^+ production.[100] Note that the prominent resonance-like feature in the $3\sigma_g$ cross section approximately 10 eV above threshold appears as only a small inflection in the N_2^+ profile, emphasizing the importance of the partial-channel measurements in identifying such features.[96-99]

In Figure 10 are shown the results of vertical-electronic photoionization cross-section calculations in the separated-channel static-exchange approximation for the 5σ, 1π, and 4σ valence orbitals in CO, and for the production of parent CO^+ ions.[23] Large basis sets of Gaussians similar to those employed in the case of N_2 are used,[22] providing pseudospectra having dimensionalities of 41, 28, and 22 for σ, π, and δ final-state symmetry, respectively. These results are seen to be in good agreement with recent line-source,[101] synchrotron-radiation[97] and (e, 2e)[99] partial-channel photoionization cross-section measurements. Similarly, the sum of the three calculated vertical electronic cross sections is in good accord with the corresponding (e,e + ion) measurements for CO^+ production.[100] As in N_2, inspection of the potential-energy curves for the three ionic states involved shows that the appropriate Franck-Condon regions are limited to the lowest few vibrational levels.[102]

Consequently, the vertical electronic cross sections for removal of the three outer valence electrons correspond only to production of CO^+ ions, with ionic fragments (O^+, C^+) arising from higher-lying electronic states.[100] In contrast to the situation in N_2 [Fig. 9], in which case a resonance appears in the $3\sigma_g$ channel but not in the $2\sigma_u$ channel, both the 5σ and 4σ photoionization channels in CO exhibit resonance-like features approximately 10 eV above threshold.

In addition to the IVO S-T photoionization studies cited here, IVO calculations of photoionization cross sections in CH,[19] O_2,[26] H_2CO,[24] and H_2O [28] have been reported recently. Large scale configuration-interaction studies of photoionization in H_2 [20] and metastable Ar_2^* [17] have also been carried out. Configuration-interaction studies of photoionization intensities and assignments in H_2CO [14] are in generally good agreement with the more recent IVO S-T calculations.[24] Photoionization cross sections in a large number of additional molecules, including NO, CO_2, [25] CH_4, NH_3, F_2[27] and O_3 to mention some representative examples, are presently under study in various computational approximations. Although a detailed discussion is inappropriate here, it is the case that the partial channel photoionization cross sections of diatomic and polyatomic molecules are found to generally exhibit resonance-like features that can be attributed to the presence of compact valence-like molecular orbital contributions to the corresponding electronic continua. The results indicate that the S-T theory provides a highly-reliable technique for applying conventional L^2 molecular-orbital and configuration-interaction methods to studies of molecular electronic photoionization continua.

V. Discussion and Concluding Remarks

The present review attempts to provide an overall descriptive account of the Stieltjes-Tchebycheff moment-theory technique for photoioniztion and related scattering studies. The qualitative discussion given in Section II, and the detailed theoretical considerations presented in Section III, provide sufficient information for performing S-T studies similar to those described in Section IV. These illustrative results indicate that a variety of quantum-mechanical calculational approaches can be employed in conjunction with the S-T method in obtaining reliable atomic and molecular photoionization cross sections. The clarifications of resonance features in the photoionization cross sections of diatomic and polyatomic molecules provided by the S-T IVO results are particularly satisfying.

Although the S-T technique is generally satisfactory for the problems described here, and further applications to photoionization studies in diatomic and polyatomic molecules are currently underway, additional theoretical and computational developments are desirable.

The properties of so-called principal pseudostates discussed in Section III(B) have yet to be established in detail, particularly in connection with many-electron systems and the contributions of various atomic and molecular shells to spectral moments and corresponding recurrence coefficients. It is also necessary to investigate further the recurrence-coefficient extension procedure of Section IV(A), and to devise convenient expressions that incorporate the appropriate threshold and asymptotic cross section behaviors in various situations. Investigations of the broad, flat densities associated with K-shell ionization, for example, require introduction of correspondingly appropriate asymptotic recurrence coefficients. Procedures for investigations of the angular distributions of photoejected electrons in the context of the S-T approach require clarification, and methods for L^2 variational calculations in the separate-channel static-exchange approximation must be extended to coupled-channel investigations. The latter approximation is required particularly in connection with the ionization of inner-valence electrons in molecules, in which cases relatively strong satellite lines are generally present. Although there are no formal difficulties in incorporating vibrational and rotational degrees of freedom in S-T investigations of dissociative photoionization, it is apparently the case that ionic fragment production in diatomic molecules proceeds in many cases via inner valence electron ionization, necessitating theoretical studies at least at the coupled-channel static-exchange level. In connection with further theoretical developments, the failure of the adiabatic approximation in regions just above ionization thresholds, and the many autoionizing states appearing in these regions, have yet to be dealt with satisfactorily in the context of the L^2 approach. Finally, although the various computational approaches described in Section IV employed in conjunction with the S-T approach are generally satisfactory, additional developments are clearly desirable. Most important, the various available integral programs employing Cartesian Gaussian functions can be profitably modified to include s- and p- and d-functions having additional nodes, to insure that oscillations in molecular scattering functions can be conveniently represented in modest sized L^2 basis sets, and possibly avoiding the relatively large expansions presently employed in S-T calculations.

It is a pleasure to acknowledge the important contributions made by a great many colleagues and coworkers in developments and applications of the Stieltjes-Tchebycheff method. The theoretical developments were made in close collaboration with C. T. Corcoran, and particular benefit in this connection derived from conversations at the Joint Institute for Laboratory Astrophysics with W. P. Reinhardt, S. V. ONeil, and P. M. Johnson. Early computational applications were made possible by collaborations with J. T. Sims, F. Weinhold, R. M. Glover, and S. R. Langhoff, and the more recent partial-channel molecular photoionization studies were done with groups at Lawrence Livermore Laboratory (T. N. Rescigno, A. U. Hazi, C. F. Bender, A. E. Orel), California Institute of Technology (B. V. McKoy, A. Gerwer, and C. Asaro), University of Campinas (G. N. Csanak and N. Padial), and University of Sydney (G. R. J. Williams). Financial assistance was kindly provided by the Donors of the Petroleum Research Fund, administered by the American Chemical Society, and by the Proctor and Gamble Corporation, for which the author is most grateful.

VI. REFERENCES

1. P. W. Langhoff, Chem. Phys. Letters 22, 60 (1973).
2. P. W. Langhoff and C. T. Corcoran, J. Chem. Phys. 61, 146 (1974)
3. P. W. Langhoff, J. S. Sims, and C. T. Corcoran, Phys. Rev. A 10, 829 (1974).
4. P. W. Langhoff and C. T. Corcoran, Chem. Phys. Letters 40, 367 (1976).
5. P. W. Langhoff, C. T. Corcoran, J. S. Sims, F. Weinhold, and R. M. Glover, Phys. Rev. A 14, 1042 (1976).
6. C. T. Corcoran and P. W. Langhoff, J. Math. Phys. 18, 651 (1977)
7. G. W. F. Drake, Astrophys. J. 184, 145 (1973).
8. P. W. Langhoff and S. L. Seidman, Chem. Phys. Letters 27, 195 (1974).
9. R. F. Stewart, C. Laughlin, and G. A. Victor, Chem. Phys. Letters 29, 353 (1974).
10. P. W. Langhoff, Int. J. Quantum Chem. S8, 347 (1974).
11. R. K. Nesbet, Phys. Rev. A 14, 1065 (1976).
12. J. T. Broad and W. P. Reinhardt, Chem. Phys. Letters 37, 212 (1976).
13. P. W. Langhoff, C. T. Corcoran, and J. S. Sims, Phys. Rev. A 16, 1513 (1977).
14. P. W. Langhoff, S. R. Langhoff, and C. T. Corcoran, J. Chem. Phys. 67, 1722 (1977).
15. A. U. Hazi and T. N. Rescigno, Phys. Rev. A 16, 2376 (1977).
16. P. W. Langhoff, Int. J. Quantum Chem. S11, 301 (1977).
17. T. N. Rescigno, A. U. Hazi, and A. E. Orel, J. Chem. Phys. 68, 5283 (1978).
18. T. N. Rescigno, C. F. Bender, and B. V. McKoy, Phys. Rev. A 17, 645 (1978).
19. J. Barsuhn and R. K. Nesbet, J. Chem. Phys. 68, 2783 (1978).
20. S. V. ONeil and W. P. Reinhardt, J. Chem. Phys. 69, 2126 (1978).
21. T. N. Rescigno and P. W. Langhoff, Chem. Phys. Letters 51, 65 (1978).
22. T. N. Rescigno, C. F. Bender, B. V. McKoy, and P. W. Langhoff, J. Chem. Phys. 68, 970 (1978).
23. N. Padial, G. Csanak, B. V. McKoy, and P. W. Langhoff, J. Chem. Phys. 69, 2992 (1978).
24. P. W. Langhoff, A. E. Orel, T. N. Rescigno and B. V. McKoy, J. Chem. Phys. 69, 4689 (1978).

25. N. Padial, G. Csanak, B. V. McKoy, and P. W. Langhoff, (to be published)
26. A. Gerwer, C. Asaro, B. V. McKoy, and P. W. Langhoff, J. Chem. Phys. xx, xxxx (1979).
27. A. E. Orel, T. N. Rescigno, B. V. McKoy, and P. W. Langhoff, J. Chem. Phys. xx, xxxx (1979).
28. G. R. J. Williams and P. W. Langhoff, Chem. Phys. Letters 60, 201 (1979).
29. D. J. Margoliash and P. W. Langhoff, Phys. Rev. xx, xxxx (1979).
30. R. L. Martin and E. R. Davidson, J. Chem. Phys. A xx, xxxx (1979).
31. H. F. Schaefer III, The Electronic Structure of Atoms and Molecules (Addison Wesley, Reading, MA, 1972); Ann. Rev. Phys. Chem. 27, 261 (1976).
32. S. D. Peyerimhoff and R. J. Buenker, Adv. Quantum Chem. 9, 69 (1975).
33. S. T. Epstein, The Variation Method in Quantum Mechanics (Academic, N.Y., 1974).
34. H. S. Wall, Analytic Theory of Continued Fractions (Van Nostrand, N.Y., 1948).
35. J. A. Shohat and J. D. Tamarkin, The Problem of Moments, Mathematical Surveys 1 (American Mathematical Society, Providence, R. I., 1950), 2nd ed.
36. N. I. Akhiezer, The Classical Moment Problem (Oliver and Boyd, London, 1965).
37. Yu V. Vorobyev, Method of Moments in Applied Mathematics (Gordon and Breach, New York, 1965).
38. E. C. Titchmarsh, Eigenfunction Expansions (Oxford, U.P., Oxford, 1958).
39. P.-O. Löwdin and H. Shull, J. Chem. Phys. 23, 1362 (1955); 30, 617 (1959).
40. L. I. Schiff, Quantum Mechanics (McGraw-Hill, N.Y., 1968).
41. H. A. Bethe and E. E. Salpeter, Quantum Mechanics of One- and Two-Electron Atoms (Springer, Berlin, 1957).
42. A. U. Hazi and H. S. Taylor, Phys. Rev. A 1, 1109 (1970).
43. U. Fano and J. W. Cooper, Rev. Mod. Phys. 40, 441 (1968).
44. A. Dalgarno, Advan. Phys. 11, 281 (1962).
45. P. W. Langhoff, S. T. Epstein, and M. Karplus, Rev. Mod. Phys. 44, 602 (1972).
46. P. W. Langhoff, J. Chem. Phys. 57, 2604 (1972).
47. A. Dalgarno, Rev. Mod. Phys. 35, 522 (1963).
48. J. O. Hirschfelder, W. Byers Brown, and S. T. Epstein, Advan. Quantum Chem. 1, 256 (1964).
49. R. Jackiw, Phys. Rev. 157, 1220 (1967).
50. P. W. Langhoff and M. Karplus, J. Opt. Soc. Am. 59, 863 (1969).
51. Y. M. Chan and A. Dalgarno, Proc. Phys. Soc. (London) 86, 777 (1965).
52. R. J. Bell and A. E. Kingston, Proc. Phys. Soc. (London) 88, 901 (1966).
53. R. E. Johnson, S. T. Epstein, and W. J. Meath, J. Chem. Phys. 47, 1271 (1967).

54. P. W. Langhoff and M. Karplus, J. Chem. Phys. 52, 1435 (1970).
55. J. F. Lowry, D. H. Tomboulian, and D. L. Edere, Phys. Rev. 137, A1054 (1965).
56. J. A. R. Sampson, Adv. At. Mol. Phys. 2, 187 (1966).
57. J. B. West and G. V. Marr, Proc. Roy. Soc. (London) A 349, 397 (1976).
58. F. Weinhold, J. Chem. Phys. 46, 2448 (1967); 50, 4136 (1969).
59. J. Deltour, Physica 39, 413, 424, 431 (1968).
60. A. Burgess and M. J. Seaton, Mon. Not. R. Astron. Soc. 120, 121 (1960).
61 G. A. Baker, Jr., Essentials of Padé Approximants (Academic, N. Y., 1975).
62. P. W. Langhoff and M. Karplus, in The Padé Approximant in Theoretical Physics, G. A. Baker, Jr., and J. L. Gammel, Editors, (Academic, N. Y. 1970), pp. 41-95.
63. R. G. Gordon, J. Math. Phys. 9, 655 (1968).
64. J. C. Wheeler and R. G. Gordon, in The Padé Approximant in Theoretical Physics, G. A. Baker, Jr., and J. L. Gammel, Editors, (Academic, N. Y., 1970), pp. 99-128.
65. P. W. Langhoff, R. G. Gordon, and M. Karplus, J. Chem. Phys. 55, 2126 (1971).
66. C. Blumstein and J. C. Wheeler, Phys. Rev. B 8, 1764 (1973).
67. R. A. Sack and A. F. Donovan, Numer. Math. 18, 465 (1972).
68. G. Szego, Orthogonal Polynomials (American Mathematical Society, Providence, 1959).
69. U. W. Hochstrasser, in Handbook of Mathematical Functions, edited by M. Abramowitz and I. A. Stegan (U. S. GOP, Washington, D. C., 1964), Chap. 22.
70. B. H. Armstrong, Phys. Rev. 131, 1132 (1963).
71. M. Inokuti and Y.-K. Kim, Phys. Rev. 173, 154 (1968).
72. J. S. Sims, S. A. Hagstrom, and J. R. Rumble, Int. J. Quantum. Chem. 10, 853 (1976).
73. R. M. Glover and F. Weinhold, J. Chem. Phys. 66, 191 (1977).
74. A. Dalgarno and S. T. Epstein, J. Chem. Phys. 50, 2837 (1969).
75. S. Geltman, Astrophys. J. 136, 935 (1962).
76. S. J. Smith and D. S. Burch, Phys. Rev. 116, 1125 (1959).
77. J. S. Sims and S. A. Hagstrom, Phys. Rev. A 11, 418 (1975).
78. J. S. Sims, S. A. Hagstrom, and J. R. Rumble, Phys. Rev. A 13, 242 (1976).
79. C. A. Martin and W. L. Wiese, Phys. Rev. A 13, 699 (1976).
80. R. D. Hudson and V. L. Carter, Phys. Rev. 137, A 1648 (1965); J. Opt. Soc. Am. 57, 651 (1967).
81. A. K. Bhatia, A. Temkin, and A. Silver, Phys. Rev. A 12, 2044 (1975).
82. T. N. Chang and R. T. Poe, Phys. Rev. A 11, 191 (1975).
83. H. Lefebvre-Brion, C. M. Moser, and R. K. Nesbet, J. Chem. Phys. 35, 1702 (1961); J. Mol. Spectro. 13, 418 (1964).
84. H. P. Kelly, Phys. Rev. 136, B896 (1964).
85. P. W. Langhoff, M. Karplus, and R. P. Hurst, J. Chem. Phys. 44, 505 (1966).

86. W. J. Hunt and W. A. Goddard, Chem. Phys. Letters 3, 414 (1969); 24, 464 (1974).
87. P. W. Langhoff and S. W. Chan, Mol. Phys. 25, 345 (1973).
88. L. S. Cederbaum and W. Domcke, Adv. Chem. Phys. 36, 205 (1977).
89. L. S. Cederbaum, J. Schirmer, W. Domcke, and W. von Niessen, J. Phys. B 10, L549 (1977).
90. J. Schirmer, L. S. Cederbaum, W. Domcke, and W. von Niessen, Chem. Phys. 26, 149 (1977).
91. J. Schirmer and L. S. Cederbaum, J. Phys. B 11, 1889 (1978).
92. J. Schirmer, W. Domcke, L. S. Cederbaum, J. Phys. B 11, 1901 (1978).
93. T. H. Dunning, Jr., and P. J. Hays, in Modern Theoretical Chemistry, H. F. Schaefer, III, editor (Plenum, N. Y., 1976), Vol. 3, Chap. 1.
94. A. Lofthus and P. H. Krupenie, J. Phys. Chem. Ref. Data 6, 113 (1973).
95. A. L. Roche and H. Lefebvre-Brion, Chem. Phys. Letters 32, 155 (1975).
96. J. A. R. Samson, G. N. Haddard, and J. L. Gardner, J. Phys. B 10, 1749 (1977).
97. E. W. Plummer, T. Gustafsson, W. Gudat, and D. E. Eastman, Phys. Rev. A 15, 2339 (1977).
98. P. R. Woodruff and G. V. Marr, Proc. Roy. Soc. (London) A 358, 87 (1977).
99. A. Hamnett, W. Stoll, and C. E. Brion, J. Electron Spectrosc.-Relat. Phenom. 8, 367 (1976).
100. G. R. Wigth, M. J. Wan der Wiel, and C. E. Brion, J. Phys. B 9, 675 (1976).
101. J. A. R. Samson and J. L. Gardner, J. Electron Spectrosc. Relat. Phenom. 8, 35 (1976).
102. P. H. Krupenie, Natl. Stand. Ref. Data Ser. Natl. Bur. Stand. 5 (1966).

DISCUSSION

Nesbet: I would like to point out that the Stieltjes technique is, among other things, very useful in describing the oscillator strength distribution in the discrete region of the spectrum--which is something Dalgarno and others have exploited in the past. It turns out that one can turn the procedure around, using the experimentally known positions of the transition energies, to get reasonable estimates of entire Rydberg series of oscillator strengths.

Kelly: A question that has been coming up several times is the question of the interweaving oscillator strengths in a Stieltjes imaging problem. It was a problem for us in our HF calculations. And so I'm wondering if people have seen this problem.

Nesbet: Do you actually compute the principal moments, that's the most important question?

Kelly: Yes. And we wanted to break them up of course into series. We decided that maybe if we did a careful population analysis, that might help, but we couldn't do it by inspection.

Nesbet: We do need good working criteria to separate series that are superimposed like this so we won't get into false structure or oversimplify the problem by just drawing smooth curves where there might be some structure. That's the difficulty.

Taylor: I have a question. Say you have some testing function, ϕ, and some continuum function, ψ_{ε}. Now Stieltjes imaging generally solves for this:

$$|\langle\phi|\psi_{\varepsilon}\rangle|^2 .$$

Now if you're doing photoionization ϕ is $r\psi_o$. What are the limitations on ϕ? Does the testing function have to be square integrable? Can it be a delta-function? What kind of problems can you apply this to?

Langhoff: We've looked at a problem where the moments don't exist formally. That is to say, you can justify this procedure mathematically using what mathematicians call regularization.

Taylor: Does that mean you can let ϕ be anything you want?

Langhoff: I'm not saying that. I'm saying you can make the moment theory work in a number of ways even when you're dealing with moments that are not formally defined. A regularization is provided simply by the fact that when you calculate moments in a square-integrable basis set, you just have a finite sum and you always get a finite number in a square-integrable basis. So you can say that a finite Hilbert space representation of the spectrum itself provides a regularization.

Taylor: This is turning out to be a very important question.

Langhoff: One more point about positive moments. The method is very important in connection with the way one formulates the procedure for calculating the recurrence coefficients. Bob Nesbet talks about multiplying the inverse of the Hamiltonian matrix into a vector.

Nesbet: There's no doubt it's easier to multiply by a matrix than by its inverse.

Langhoff: Yes, that's the essential point. When you have positive moments, you deal with the Hamiltonian itself.

Nesbet: Just one more point. Would it be possible to put in exponential factors in calculating moments and wouldn't that make it better behaved?

Langhoff: Yes, we have done it both ways in test cases on H and H^-, so we have looked at this in detail.

THE CONTINUUM MULTIPLE-SCATTERING APPROACH TO ELECTRON-MOLECULE SCATTERING AND MOLECULAR PHOTOIONIZATION*

J. L. Dehmer

Argonne National Laboratory
Argonne, Illinois 60439

Dan Dill

Department of Chemistry
Boston University
Boston, Massachusetts 02215

I. INTRODUCTION

Properties of the electronic continuum of molecules such as cross sections for photoionization and electron scattering are both scientifically interesting and important for practical applications. Nevertheless, this subject is largely unexplored owing to the computational difficulty associated with the nonseparable nature of the multicenter wave equation, for which no practical, comprehensive method of solution exists. This volume summarizes the most promising approaches designed to cope with this dilemma, e.g., R-matrix, single-center expansion, L^2, and continuum multiple-scattering methods. Each shares the same general goal and has met with substantial success, but always with limitations in scope or accuracy dictated by the compromises required in developing a workable scheme.

In this talk we describe the multiple-scattering approach to the electronic continua of molecules. The continuum multiple-

---

*) Work performed under the auspices of the U. S. Department of Energy and National Science Foundation Grant CHE78-08707. Acknowledgment is made to the Donors of The Petroleum Research Fund, administered by the American Chemical Society, for the partial support of this research.

scattering model (CMSM) was developed[1,2] as a survey tool and, as such was required to satisfy two primary requirements: First, it had to have a very broad scope, by which we mean (i) molecules of arbitrary geometry and complexity containing any atom in the periodic system, (ii) continuum electron energies from 0-1000 eV, and (iii) capability to treat a large range of processes involving both photoionization and electron scattering. Second, the structure of the theory was required to lend itself to transparent, physical interpretation of major spectral features such as shape resonances, which are proving to be a dominant feature in the continuum dynamics of molecules. In particular, such interpretive analysis is greatly aided by easy access to a partial wave representation of the continuum wavefunction in the vicinity of the atomic cores and at large distances from the molecule. In Sec. II we collect together the CMSM and model-independent formulation upon which this work is based. In Sec. III applications to electron-molecule scattering are reviewed with special emphasis on the role of shape resonances in several electron-molecule systems. And in Sec. IV applications to molecular photoionization are summarized with a similar emphasis.

II. THEORETICAL FRAMEWORK

The purpose of this section is to collect together in one place the key equations which have been developed for this project in various papers over the last several years. This should be viewed as a theoretical framework rather than an exhaustive review, and reference will be made to original papers for details. The scope of this section includes the multiple-scattering approach to the molecular continuum and applications to electron- and photon-molecule interactions. A review of bound-state studies is given by Johnson.[3] We'd like to emphasize at the outset that, although the CMSM has been the vehicle of this work, most of the formulation that follows is model independent, e.g., expression of cross sections in the angular-momentum transfer (j_t) basis. In fact, following Eq. (4) all expressions are applicable to other methods of treating the molecular continuum.

A. The Molecular Potential

The crux of the CMSM is the partitioning of the molecular field into spherical regions as indicated schematically in Fig. 1. First a sphere is constructed around each atom and then around the entire cluster in close-packed fashion. Usually the potential in each atomic sphere (regions I_i) and in the spherical region surrounding the molecule (region III) is spherically symmetric, although this limitation has already been relaxed[4] to include a general multipole potential in these regions with only modest additional effort. (This generalization is presently being used to study molecules with permanent dipole moments.) In the interstitial region (region II), the potential is approximated by a constant. This leads to the great economy of the model since the solution of the wave equation for region II is expressable in terms of Bessel functions, whose closed-form, efficient re-expansion about different sites is used to propagate the molecular wavefunction among the various spherical regions. This approximation cannot be relaxed without a major jump in computer costs, and, in any case, seems to be justified as a first approximation by the results obtained in numerous examples discussed below.

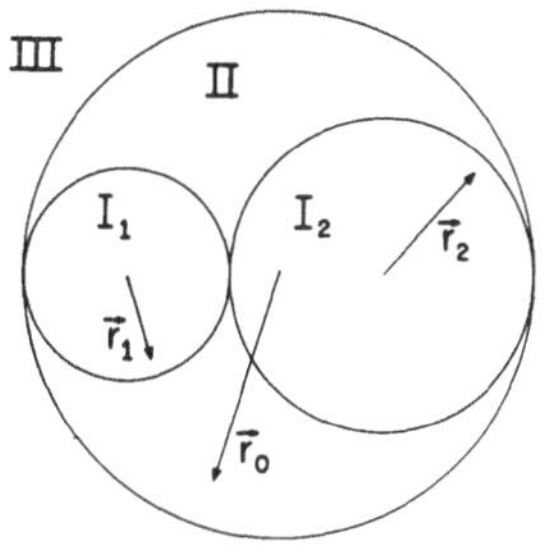

Fig. 1. Partitioning of the molecular field for a heteronuclear diatomic molecule.

In general, the potential in each region of space is comprised of three parts - a direct and an exchange electrostatic potential plus a polarization potential. The direct and exchange terms are determined directly from an SCF ground-state charge distribution of the molecule and the polarization potential, important mainly for electron-molecule scattering, is added to region III, which corresponds rather well to the region in which polarization is important and has assumed its asymptotic form. We have studied[5] various local exchange approximations, including the Slater,[6] Hara,[7] and semiclassical[8,9] forms. We now know that Slater exchange is most natural and successful for bound-state studies and Hara exchange is most natural and effective for electron-molecule scattering. The best approach to molecular photoionization has not yet been conclusively established, although both the Slater and Hara forms have been successfully used in several calculations.

Molecular potentials, constructed in this way, describe the field near the nuclei very well, lead to rapid convergence of the partial-wave expansion of the molecular wavefunction in each region, and are easily constructed for molecules of arbitrary geometry, containing any atomic constituents.

B. Form of the Wavefunction

The CMSM molecular wavefunctions are constructed from partial-wave expansions in solutions of the radial-wave equation for each region.[1,2] In each atomic sphere the molecular wavefunction has the form

$$\psi_i = \sum_L A_L^i f_\ell^i (kr_i) Y_L (\hat{r}_i) , \quad (1)$$

where L represents the double index (ℓ,m), i is shorthand for I_i which represents a particular atomic sphere, r_i is the distance from the ith nucleus, f_ℓ^i is the ℓth radial solution (regular at $r_i = 0$) of the Schroedinger equation for region I_i, and k is the continuum electron's wave number $k = \sqrt{E/Ry}$. In region II, the form is

$$\psi_{II} = \sum_L A_L^{II_0} j_\ell (\kappa r_0) Y_L(\hat{r}_0) + \sum_i \sum_{L'} B_{L'}^{II_i} n_{\ell'} (\kappa r_i) Y_{L'} (\hat{r}_i) , \quad (2)$$

where j_ℓ and n_ℓ are regular and irregular spherical Bessen functions, κ is the local wave number in region II, and r_0 is distance measured from the molecular center of mass. This form, when re-expanded on any spherical boundary in the CMSM field, is complete and therefore capable of matching with the wavefunctions forms in all spherical regions.[1,2] Finally, in region III, we use the form

$$\psi_{III} = \sum_L \left[A_L^{III} f_\ell^{III} (kr_0) + B_L^{III} g_\ell^{III} (kr_0) \right] Y_L (\hat{r}_0) , \quad (3)$$

where f_ℓ^{III} and g_ℓ^{III} are regular and irregular solutions of the wave equation in region III that approach $r_0 \to \infty$ 90^0 out of phase. This flexibility is necessary to represent the phase-shifted asymptotic wave in the electronic continuum. For bound state problems, of course, a single set of functions which vanish at $r_0 \to \infty$ would be used.

Besides converging rapidly in L, molecular wavefunctions constructed with these regional expansions facilitate mapping, e.g., the distribution of asymptotic partial waves into which a particular partial wave near an atomic nucleus is scattered by the anisotropic molecular field. This has proven to be a very illuminating way of looking at certain physical processes, most notably inner-shell photoionization.

C. The CMSM Linear System

Requiring the regional wavefunctions, Eqs. (1)-(3), and their derivatives to match at all spherical boundaries, $r_i = \rho_i$, yields the linear system of equations[1,2]

$$\sum_{L'}\left[{}^{II}A^{0}_{L'}J^{0j}_{L'L} + \sum_i (1-\delta_{ij})\,{}^{II}B^{i}_{L'}N^{ij}_{L'L}\right] + \frac{[n_\ell(\kappa\rho_j), f^j_\ell(k\rho_j)]}{[j_\ell(\kappa\rho_j), f^j_\ell(k\rho_j)]}\,{}^{II}B^{j}_{L} = 0 \,, \tag{4a}$$

$$\sum_i \sum_{L'} {}^{II}B^{i}_{L'}J^{i0}_{L'L} + \frac{[j_\ell(\kappa\rho_{III}), g^{III}_\ell(k\rho_{III})]}{[n_\ell(\kappa\rho_{III}), g^{III}_\ell(k\rho_{III})]}\,{}^{II}A^{0}_{L} = \frac{[f^{III}_\ell(k\rho_{III}), g^{III}_\ell(k\rho_{III})]}{[n_\ell(\kappa\rho_{III}), g^{III}_\ell(k\rho_{III})]}\,A^{III}_{L} \,, \tag{4b}$$

where we use the notation $[X,Y] = XY' - YX'$, and $N^{ij}_{L'L}$ and $J^{ij}_{L'L}$, called structure factors, are re-expansion coefficients which transform region-II basis functions of angular momentum L' on site i to those with angular momentum L on site j. These re-expansion coefficients are known in closed form and are very efficient to evaluate, thus making all quantities in Eq. (4) easy to compute. Note that whereas the linear system for the bound state problem is homogeneous,[1-3] Eq. (4) is inhomogeneous. Specification of the source term in Eq. (4b), discussed in the next section, allows solving this linear system for the coefficients $\{A^{II}_L\}$ and $\{B^{II}_L\}$. Then use of auxiliary conditions, derived elsewhere,[1,2] determines the coefficient A^{I}_{L} and B^{III}_{L}, which were originally eliminated to form the linear system. Hence, once A^{III}_{L} is specified, all expansion coefficients for the regionwise wavefunction forms can be determined, thereby completely specifying the entire molecular wavefunction. Finally, we note that, although Eq. (4) is expressed in the complex angular momentum (ℓ,m) basis, in actual calculation the full linear system is reduced to block-diagonal form by transforming to a real symmetry-adapted basis. This operation both saves computer time and automatically classifies each continuum wavefunction according to the irreducible representations of the molecule's point group.

D. Standing-Wave (R-Matrix) Solution

It is customary to solve Eq. (4) for a complete set of standing-wave or K-matrix-normalized wavefunctions at each energy. In this way, the bulk of the numerical work involves only real quantities and any alternative set of functions, e.g., an S-matrix normalized set can be subsequently obtained by linear transformation. To see how this is done, one need only inspect the asymptotic forms of the region-III and the K-matrix-normalized wavefunctions. The asymptotic form of ψ_{III} in Eq. (3) is

$$\psi_{III} \sim (kr_0)^{-1} \sum_{L'} \left[A_{L'}^{III} \sin\theta_{\ell'} - B_{L'}^{III} \cos\theta_{\ell'}\right] Y_{L'}(\hat{r}_0) \quad , \qquad (5)$$

where $\theta_\ell \equiv$ asymptotic phase for the long-range part of the molecular potential. The asymptotic form of the K-matrix-normalized wavefunction is

$$\psi_{III,L} \sim (\pi k)^{-\frac{1}{2}} r_0^{-1} \sum_{L'} \left[\delta_{LL'} \sin\theta_{\ell'} + K_{LL'} \cos\theta_{\ell'}\right] Y_{L'}(\hat{r}_0) . \qquad (6)$$

where each member L of this set of functions has only one regular (sin) component, namely that with orbital momentum quantum number L, and irregular (cos) components with all values of L'. The coefficients $K_{LL'}$ comprise the real symmetric K-matrix and each matrix element $K_{LL'}$ indicates how the regular wave L is coupled to each irregular wave L' by the anisotropic molecular field. The K-matrix is built up by successive solutions of the linear system with the inhomogeneity chosen as

$$A_{L'}^{III} = \delta_{LL'} (k/\pi)^{\frac{1}{2}} , \qquad (7)$$

that is with a regular part in only one region III partial wave. Each solution with this condition yields a row of the K-matrix according to

$$-K_{LL'} = (\pi/k)^{\frac{1}{2}} B_{L'}^{III} . \qquad (8)$$

We end up then with N linearly independent solutions of the linear system (4), with the asymptotic form (6).

E. Alternative Boundary Conditions

Converting K-matrix normalization to S-matrix normalization is the first step toward matching the molecular wavefunction to the physical boundary conditions of interest. S-matrix normalization has an incoming-wave (-) and outgoing-wave (+) form defined by the large r behavior

$$\psi_{III,L}^{\pm} \sim (\pi k)^{-\frac{1}{2}} (2ir_0)^{-1} \quad \times$$

$$\times \sum_{L'} [\delta_{LL'} e^{\mp i\theta_{\ell'}} - S^{\pm}_{LL'} e^{\pm i\theta_{\ell'}}] Y_{L'}(\hat{r}_0), \qquad (9)$$

where the coefficients $S_{LL'}$ comprise the unitary, symmetric S matrix. The notation $S^{\pm}_{LL'}$ indicates that $S_{LL'}$ and $S^{*}_{LL'}$ should be used for the ψ^{+} and ψ^{-} forms, respectively. The S and K matrices are related by the transformation

$$\sum_k S_{ik}(\delta_{kj} - iK_{kj}) = \delta_{ij} + iK_{ij}, \qquad (10)$$

and the S- and K-matrix-normalized basis sets are related by the linear transformation

$$\psi^{\pm}_{III,L} = \sum_{L'} c^{\pm}_{LL'} \psi_{III,L'}, \qquad (11)$$

where

$$c^{\pm}_{LL'} = \mp (I \mp iK)^{-1}_{LL'} . \qquad (12)$$

Note that up to this point, all quantities have been expressed in a coordinate system fixed to the molecule - the so-called body frame - whereas the coordinate system more natural for describing most experimental measurements is the laboratory-fixed frame (or simply the lab frame). We write the continuum molecular wavefunction in the lab frame as a linear combination of the functions obtained from solution of the body-frame Schroedinger equation,

$$\psi^{\pm}_{III}(\text{LAB FRAME}) = \sum_L a^{\pm}_L \psi^{\pm}_{III,L}(\text{BODY FRAME}) . \qquad (13)$$

The transformation coefficients $a^{\pm}_L$ are derived by matching the asymptotic form of the RHS of Eq. (13) to alternative asymptotic forms appropriate to particular physical processes in the lab frame. Thus for e^{-}-molecule scattering, an outgoing-wave-normalized plane wave is used, with the asymptotic form

$$\psi^{+}_{III} \sim [(2\pi)^{-3}k/2]^{\frac{1}{2}} [\exp(i\vec{k}' \cdot \vec{r}'_0) + f^{+}(\hat{k}' \cdot \hat{r}'_0) \times \exp(ikr_0)/r_0], \qquad (14)$$

where primed coordinates indicate the lab frame. The first term on the RHS is the incident plane wave repsenting the incoming electron, and the second term represents the scattered part as an outgoing spherical wave. The quantity $f^{+}(\hat{k}' \cdot \hat{r}'_0)$ is the scattering amplitude. As described elsewhere,[1,2] matching the incoming-wave parts of the RHS and LHS of the outgoing-wave form of Eq. (13) yields the expansion coefficients

$$a_L^+ = -i^\ell \left[(2\ell+1)/4\pi\right]^{\frac{1}{2}} D_{0m}^{\ell *}(\hat{R}) \tag{15}$$

where $\hat{R}$ represents the rotation taking the lab frame into the body frame. In most cases, cross section expressions derived from lab frame wavefunctions such as Eq. (13) are averaged over $\hat{R}$, e.g., molecular orientation in the lab frame. This corresponds to typical experiments in which rotational structure is not resolved. More will be said about this in Sec. II.H.

For photoionization, an incoming-wave-normalized, lab-frame wavefunction is sought. Its asymptotic form, appearing more complicated because of the Coulomb field, is given by

$$\psi_{III}^- \quad \left[(2\pi)^{-3} k/2\right]^{\frac{1}{2}} (\exp\{i[\vec{k}' \cdot \vec{r}_0' + (Z/k)\ln k(r_0 - Z_0)]\}$$

$$+ f^-(\hat{k}' \cdot \hat{r}_0')\exp\{-i[kr_0 - (Z/k)\ln k(r_0 + z_0)]\}/r_0) \ . \tag{16}$$

By matching the outgoing-wave parts of the RHS and LHS of the incoming-wave form of Eq. (13), one obtains[2] the expansion coefficients

$$a_L^- = i^\ell e^{-i\sigma_\ell} \sum_{m'} D_{m'm}^{\ell *}(\hat{R}) Y_{\ell m'}^*(\hat{k}') \ . \tag{17}$$

Thus, once the real K-matrix-normalized solutions of the molecular field are computed, successive transformations given by Eqs. (11) and (13) produce molecular wavefunctions normalized to alternative physical boundary conditions in the lab frame.

F. Cross Sections for Electron-Molecule Scattering

The cross section for electron-molecule scattering is given in terms of the scattering amplitude f^+ by

$$\frac{d\sigma}{d\hat{r}'_0} = \int d\hat{R} |f^+(\hat{r}'_0, \hat{R})|^2 \ , \tag{18}$$

where $\hat{r}'_0$ is measured relative to the incident electron direction $\hat{k}'$. The scattering amplitude is obtained from the outgoing-wave-normalized wavefunction ψ_{III}^+ in (13) by subtracting the plane wave part, as indicated by the asymptotic form in (14). This produces an expression for f^+ in terms of the S matrix,

$$f^{+}(\hat{r}_0', \hat{R}) = (i\pi^{\frac{1}{2}}/k) \sum_{LL'm''} i^{\ell-\ell'} (2\ell+1)^{\frac{1}{2}} (\delta_{LL'} - S_{LL'})$$

$$\times D^{\ell *}_{0m}(\hat{R})\, Y_{\ell' m''}(\hat{r}_0') D^{\ell'}_{m''m'}(\hat{R}) \; . \tag{19}$$

Now, $d\sigma/d\hat{r}_0'$ can be straightforwardly calculated by evaluating the <u>coherent sum</u> over L, L', L", L'" which arises in the squared modulus of Eq. (19); however, this proves to be extremely inefficient. Rather we transform to the so-called j_t basis (see, e.g., Ref. 10) in which components of the scattering amplitude are classified by the angular momentum transferred during the collision,

$$\vec{j}_t = \vec{\ell}' - \vec{\ell} \; . \tag{20}$$

The expression for the electron-molecule scattering cross section[11] which results from this change of representation can now be cast as an <u>incoherent sum</u> over j_t and is more efficient to evaluate, as indicated below. Thus,

$$\frac{d\sigma}{d\Omega} = \frac{\pi}{k^2} \sum_{j_t=0}^{j_t(\max)} \sum_{m_t m_t'} (2j_t + 1)^{-1} \left| B^{j_t}_{m_t m_t'}(\Omega) \right|^2 , \tag{21}$$

$$B^{j_t}_{m_t m_t'}(\Omega) = \sum_{LL'} (-1)^{m} i^{\ell - \ell'} (2\ell + 1)^{\frac{1}{2}} T_{LL'} \tag{22}$$

$$\times (\ell - m, \ell' m' | j_t m_t')(\ell 0, \ell' m_t | j_t m_t) Y_{\ell' m_t}(\Omega),$$

where m_t and m_t' are projections of j_t along the laboratory and molecular axes, respectively, and the T matrix is related to the S matrix by

$$T_{LL'} \equiv \delta_{LL'} - S_{LL'} \; . \tag{23}$$

A significant property of B^{j_t} is its rapid convergence with increasing j_t, which reflects the physical fact that small angular momentum changes are more probable than large ones. This permits the truncation of the sum of Eq. (21) when its integral over scattering angle Ω agrees with the integrated cross section,

$$\sigma = \frac{\pi}{k^2} \sum_{LL'} \left| T_{LL'} \right|^2 , \tag{24}$$

to some tolerance. This truncation can result in great numerical savings when calculating the differential cross section $d\sigma/d\Omega$.

G. Cross Sections for Molecular Photoionization

Before dealing directly with the general form of the molecular photoionization cross section, we will introduce the "electron-optical" description of inner-shell photoionization - a great simplification in which energy-dependent and energy-independent parts of the dipole amplitude are identified and analyzed separately.

In deep inner-shell photoionization, the photon is absorbed near an atomic nucleus in an "atomic-like" environment. There the final state wavefunction has the form

$$\psi_L^- = \sum_{L'} A_{LL'}^- \bar{f}_{L'}(r) Y_{L'}(\hat{r}) \, , \tag{25}$$

where

$$A_{LL'}^- = \sum_{L''} (I+iK)^{-1}_{LL''} A_{L''L'} \tag{26}$$

are the incoming-wave-normalized region I expansion coefficients (the superscript denoting the region has been dropped here to simplify notation), which carry an index L' denoting the partial wave in region I and an index L denoting a row of the K or S matrix. We denote by $\bar{f}_L$ that the radial waves are chosen to be energy-independent at $r \to 0$ so that the small-r energy dependence ψ_L^- is isolated in the coefficients $A_{LL'}^-$. The integrated cross section is

$$\sigma = \frac{4}{3} \pi^2 \alpha h\nu \sum_{Lm_\gamma} |D_{Lm_\gamma}^-|^2 \, , \tag{27}$$

$$D_{Lm_\gamma}^- = -\sum_{\ell' m'} (\ell_0 m_0, 1m_\gamma | \ell' m')(10, \ell' 0 | \ell_0 0) A_{LL'}^{-*} \bar{R}_{\varepsilon\ell', n\ell_0} \, , \tag{28}$$

$$\bar{R}_{\varepsilon\ell', n\ell_0} = \int r^2 dr \bar{f}_{\varepsilon\ell'} r f_{n\ell_0} \, , \tag{29}$$

where $f_{n\ell_0}$ denotes the initial state and $\bar{f}_{\varepsilon\ell'}$ is the dipole-allowed partial wave component(s) of the final state of energy ε. Note that in an atomic core in a molecule, the field is roughly spherically symmetric, so that the dipole selection rules limit the allowed final-state partial-wave components to $\ell' = \ell_0 \pm 1$. For the K-shell case, Eq. (27) becomes

$$\sigma(ns_0) = \frac{4}{3} \pi^2 \alpha h\nu \sum_{Lm'} \frac{1}{3} |A_{L,1m'}^-|^2 \bar{R}^2_{\varepsilon p, ns_0} \, , \tag{30}$$

so that only the p-wave component of the final state in region I is allowed. For a general initial state, the cross section has region I terms for $\ell_0 \pm 1$;

$$\langle \sigma(n\ell_o)\rangle = \frac{4}{3}\pi^2\alpha h\nu\,(2\ell_o+1)^{-1}\{(\ell_o+1)\sum_{Lm'}|A^-_{L,\ell_o+1\,m'}|^2$$

$$\times\,(2\ _o+3)^{-1}\bar{R}^2_{\varepsilon\ell_o+1,n\ell_o}$$

$$+\ell_o\sum_{L,m'}|A^-_{L,\ell_o-1m'}|^2\,(2\ell_o-1)^{-1}\bar{R}^2_{\varepsilon\ell_o-1,n\ell_o}\}\ . \qquad (31)$$

where $\langle\sigma\rangle$ denotes averaging over initial state projections m_o.

The crux of the electron-optical interpretation hinges on the following considerations: The overlap between the initial and final state is in a strong field near an atomic nucleus. Hence, when normalized to be energy-independent at r=0, $\bar{f}_L$, and therefore $\bar{R}$, will be approximately energy-independent over a spectral range comparable to the initial state binding energy. Any significant energy variation is then isolated in the factor $|A_{LL'}|^2$. This factor embodies the relative probability of a partial wave L' at r=0 escaping to $r\to\infty$ in the Lth asymptotic partial wave, and therefore can be interpreted as a transmission coefficient for the anisotropic molecular field. This point of view simplifies the analysis of the striking energy variation in the inner-shell cross sections by focussing attention on the coefficients $|A^-_{LL'}|^2$, which at the same time are interpretable in physical terms as an electron-optical property of the molecular field. Applications to the σ shape resonances in N_2 and CO K-shell photoionization[12-15] are given below.

More generally, the cross section for molecular photoionization is written[13]

$$\frac{d\sigma}{d\Omega} = \sum_{j_t} 4\pi^2\alpha h\nu(2j_t+1)^{-1}\sum_{\ell\ell'}T(j_t;\ell\ell';\theta) \qquad (32)$$

where is measured relative to the electric vector of the light and where, again, we label ionization channels by j_t, the angular momentum transferred to the target during photoionization. The quantity T is defined by

$$T(j_t;\ell\ell';\theta) = i^{\ell'-\ell}e^{i(\sigma_\ell-\sigma_{\ell'})}\,I(j_t;\ell\ell')\,\Theta(j_t;\ell\ell';\theta)\ . \qquad (33)$$

The factor I represents the dynamical part of the photoionization process

$$I(j_t;\ell\ell') = \sum_{\substack{mm_\gamma\\ m'm'_\lambda}} D^{-j_t*}_{\ell'm'm'_\gamma}D^{-j_t}_{\ell mm_\gamma}\,\delta_{m-m_\gamma,m'-m'_\gamma}\ , \qquad (34)$$

$$D^{-j_t}_{Lm_\gamma} = (-1)^{m_\gamma} (j_t m - m_\gamma | 1 - m_\gamma, \ell m) D^{-}_{Lm_\gamma} \quad , \tag{35}$$

$$D^{-}_{Lm_\gamma} = \sum_{L'} (I - iK)^{-1}_{LL'} D_{L'm_\gamma} \quad , \tag{36}$$

$$D_{L'm_\gamma} = (\tfrac{4\pi}{3})^{\frac{1}{2}} \langle \psi_{L'} | r Y_{1m_\gamma}(\hat{r}) | \psi_0 \rangle . \tag{37}$$

The angular factor[10,16]

$$\Theta\,(j_t;\ell\ell';\theta) = (-1^{j_t}) 4\pi)^{-1} (2j_t+1)[(2\ell+1)(2\ell'+1)]^{\frac{1}{2}}$$

$$\times \sum_K \begin{Bmatrix} 1 & 1 & K \\ \ell & \ell' & j_t \end{Bmatrix} (\ell 0, \ell' 0 | K0) P_K(\cos\theta) \;\; (K0|10,10) \tag{38}$$

embodies the purely geometric aspects of the process. The motivation[10] for writing Eqs. (32)-(38) in this way is threefold. First, the evaluation of the cross section is separated into a geometrical part, which is easily evaluated for all cases, and a dynamical part which will reflect the unique aspects of a particular process. Second, organization of the ionization channels according to j_t results in an incoherent summation over photoionization amplitudes for alternative j_t values, so that interference phenomena are confined within each j_t component. Third, in certain cases, only one j_t may contribute to a process, leading to a completely specified angular distribution independent of dynamics.[16,17]

The final step in developing the photoionization formulae is to cast Eq. (32) in the familiar form

$$\frac{d\sigma}{d\Omega} = \frac{\sigma}{4\pi} \left[1 + \beta P_2(\cos\theta)\right] \equiv \sum_{j_t} \frac{d\sigma(j_t)}{d\Omega} \tag{39}$$

where β is the photoelectron asymmetry parameter and P_2 is the second Legendre polynomial. Explicit evaluation of Θ yields

$$\beta = (1/\sigma) \sum_{j_t} \sigma(j_t)\beta(j_t) \tag{40}$$

with

$$\sigma = \sum_{j_t} \sigma(j_t) \quad , \tag{41}$$

$$\sigma(j_t) = (4\pi^2/3)\alpha h\nu \sum_\ell I(j_t;\ell\ell) \; , \tag{42}$$

$$\beta(j_t) = \left[(2j_t + 1)\sum_\ell I(j_t;\ell\ell)\right]^{-1}$$

$$\times\{(j_t+2)\ I(j_t;j_t + 1,\ j_t + 1) + (j_t -1)I(j_t;j_t-1,j_t-1)$$

$$+ 6\left[j_t(j_t+1)\right]^{\frac{1}{2}}\ \mathrm{Re}\left[e^{i(\sigma_{j_t+1}-\sigma_{j_t-1})} I(j_t;j_t+1,j_t-1)\right]\}. \tag{43}$$

H. Effects of Nuclear Motion

The preceding formulation implicitly assumes that rotational states remain unresolved in the process and that the nuclei are frozen, e.g., in their equilibrium geometry. We will briefly discuss these two assumptions in turn.

In most experimental situations involving electron-molecule scattering or molecular photoionization, rotational structure is not resolved. In the related theoretical formulation, this corresponds to analytically averaging the cross section formulae over the molecular orientation $\hat{R}$. This was done for all the expressions given above. When specific rotational transitions are observed, the corresponding cross section formula is obtained by simply specifying particular weights in the average over $\hat{R}$. Another interesting case arises when the molecular orientation is fixed. This can arise for molecules adsorbed on surfaces, oriented in a molecular beam, or for molecules whose orientation in a particular event is determined, e.g., by coincidence with a rapid recoil fragment. In this case, the photoionization cross section has the form[18]

$$\frac{d\sigma}{d\Omega} = \sum_{K=0}^{2\ell_{max}} \sum_M A_{KM} Y_{KM}(\theta\phi) \quad . \tag{44}$$

This expression has been evaluated[19-21] in molecular photoionization to reveal angular distribution patterns which dramatically exhibit the partial-wave character of shape resonances, as illustrated below. Analyses for electron-molecule scattering follow similar lines.[22]

Consideration of vibrational motion within the context of the adiabatic nuclei approximation is accomplished (at least to a first approximation) by averaging transition amplitudes over vibrational wavefunctions. We can define a general transition amplitude

$$\Delta(v_f \leftarrow v_i) \propto \int dQ_j \chi_{v_f}(Q_j)\Delta(Q_j)\chi_{v_i}(Q_j) \quad , \tag{45}$$

where $\chi(Q_j)$ represents a vibrational wavefunction for a particular normal coördinate Q_j. Clearly for vibrationally elastic transitions, $v_i = v_f$, the effect of Eq. (45) is to simply average the processes occurring at different nuclear separations, weighted by the probability of the molecule being in each configuration. For $v_i \neq v_f$ the probability amplitude for the corresponding vibrational transition is obtained. As a specific example, in terms of the T matrix, Eq. (23), for electron-molecule scattering, the vibrational excitation cross section, in this so-called adiabatic nuclei approximation,[23] is given by

$$\sigma^j_{f\leftarrow i} = \pi\left(\frac{E_f}{E_i}\right)^{\frac{1}{2}} E_i^{-1} \Sigma_{L,L'} \left|\int dQ_j \chi_{v_f}(Q_j) T_{LL'}(Q_j) \chi_{v_i}(Q_j)\right|^2 \quad (46)$$

This expression is evaluated in connection with results discussed below.

Going beyond the adiabatic nuclei approximation involves taking into account coupling between electronic and vibrational modes caused by interactions which depend upon the velocity of the nuclei. This has been done using vibrational close coupling by Chandra and Temkin in their "hybrid" treatment[24] of e^--N_2 scattering, and accounts fairly well for vibrational substructure on the broad π_g shape resonance at 2.4 eV. This level of refinement has not been used in CMSM calculations to date, but we plan to explore nonadiabatic effects in terms of Nesbet's "energy-modified adiabatic approximation"[25] in the near future.

III. ELECTRON-MOLECULE SCATTERING RESULTS

Within the last year, survey calculations of electron-molecule scattering have begun in earnest. Following an extensive study[5,11,26-28] of the prototype system e^--N_2 over the range 0-1000 eV, the molecules CO_2,[29,30] OCS,[29] CS_2,[29] and SF_6[31] have been investigated. Although this body of work has hardly scratched the surface, it has already characterized many new resonance phenomena and has shown the CMSM to be an effective survey tool. Here we will survey the highlights of these initial studies. During the course of this survey, we will present examples of integrated, differential, and vibrationally inelastic cross sections as alternative manifestations of the collision process. Moreover, examples of collisions at high impact energies (up to 1000 eV) and involving large molecules (e.g., SF_6) will test the scope of the method.

A. e^--N_2 Scattering

Owing to the large body of data and availability of accurate alternative calculations, e^--N_2 was chosen as the prototype system to test the CMSM. During the course of this work various local exchange approximations, e.g., Slater,[6] Hara,[7] and semiclassical[8,9] were analyzed and tested. We will not focus here on details of this analysis and comparison, but will merely state our general conclusion, namely, for electron-molecule scattering calculations using the CMSM, Hara and semiclassical exchange are clearly more useful approximations than Slater exchange, with the Hara form proving to be slightly more realistic than the semiclassical form, particularly in differential scattering cross sections. For a detailed discussion of this conclusion, see Refs. 5 and 27. Nevertheless, our earliest work using Slater exchange will be used in some cases to illustrate certain general characteristics of e^{-}-N_2 scattering which persist regardless of the exchange approximation used.

The eigenphase sum is the generalization for anistropic potentials of the single-channel phase shift for central potentials. It is defined in units of π radians (modulo 1) by

$$\mu_{SUM} = \pi^{-1} \delta_{SUM} = \pi^{-1} \sum_i \tan^{-1} (U^{\dagger} KU)_{ii} , \tag{47}$$

where U is the unitary matrix that diagonalizes the real, symmetric K matrix. The eigenphase sum is a fingerprint in multichannel scattering processes and its rapid increase indicates the presence of a shape resonance which will be reflected by a local maximum in the scattering cross section.

Figure 2 gives the eigenphase sums for the $\sigma_{g,u}$, $\pi_{g,u}$, and $\delta_{g,u}$ channels in e^{-}-N_2 scattering computed with the Slater exchange approximation.[26] The most prominent feature in Fig. 2 is the rapid rise in the eigenphase sum for the π_g channel at $\sim$2.4 eV. Inspection of the K matrix indicates the $\ell = 2$ component of the π_g wavefunction is responsible for this rapid rise by $\sim\pi$ radians over the narrow energy range. This sharp rise constitutes a shape resonance and indicates that the incident electron will be temporarily trapped in a quasidiscrete d-like orbit at $\sim$2.4 eV incident energy. This feature is the best known aspect of e^{-}-N_2 scattering and has been studied by numerous authors.[11,24,26,27,32] Not widely recognized, however, are less dramatic increases in the eigenphases of the δ_g and σ_u channels at somewhat higher energies. The weak σ_u shape resonance has recently been observed in total e^{-}-N_2 scattering,[33] and will be shown below to produce enhanced vibrational excitation.

These features in the eigenphase sum are reflected in the total elastic scattering cross section[26] shown in Figure 3, again computed with the Slater exchange approximation. Here we see the strong π_g resonance at 2.4 eV, as well as broad weak features at higher energies due to the weak resonances in the δ_g and σ_u

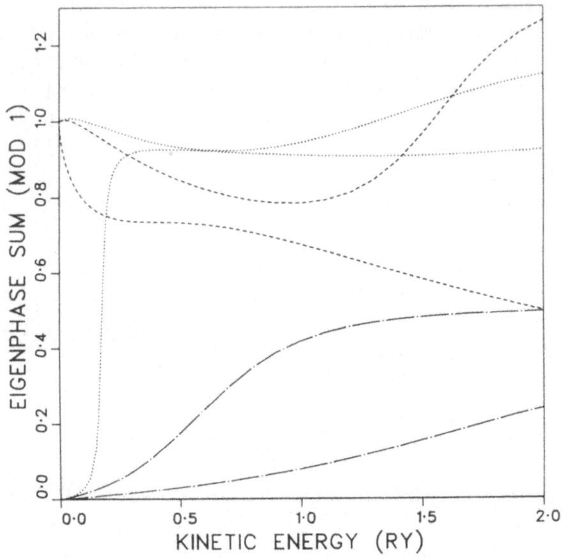

Fig. 2. Eigenphase sums for the λ=0 (dashed lines), 1 (dotted lines), and 2 (dash-dot lines) channels in electron-nitrogen scattering. The σ_u, π_u, and δ_g lie above σ_g, π_g and δ_u curves, respectively.

channels. A quantitative comparison with experiment will be made below.

A four-way comparison between experiment and CMSM calculations using the three exchange approximations mentioned above is shown in Figure 4 over the range 0-1000 eV. The dashed curve is our earliest calculation[26] using the Slater exchange approximation with $\alpha = 1$, plus the term $-12/r^4$ Ry a_0^4 in region III to approximate the polarization potential. The resonance position is in good agreement with experiment in this case, but the cross section is overestimated throughout most of the energy range. The dotted curve was an attempt to make adjustments (lower α to 2/3 and remove the polarization term) to approximate the conditions of high-energy scattering. While

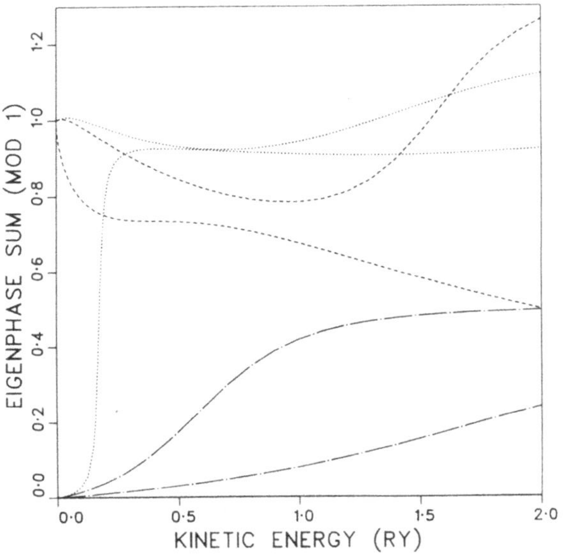

Fig. 3. Total electron-nitrogen elastic scattering cross sections (in $\AA^2$) for λ=0 (dashed lines), 1 (dotted lines), 2 (dash-dot lines), and their sum (solid line). For each set of λ channels the gerade component (e.g., σ_g, π_g, δ_g) lies above the ungerade component, at least from 0 to 0.4 Ry.

the agreement with experiment was thereby improved, the importance of incorporating the energy dependence in a more natural way was recognized. This was accomplished[5,27] by employing two alternative exchange approximations, Hara free-electron gas exchange[7] and semi-classical exchange.[8,9] Both of these are designed for scattering calculations and explicitly depend on the kinetic energy of the incident electron. The calculations employing Hara exchange with polarization and semiclassical exchange with polarization are represented in Figure 4 by the solid and dash-dot curves, respectively. Both curves position the π_g resonance very well and represent a good approximation to the energy dependence of the cross section over the broad energy range shown in Figure 4. It is still necessary to reduce the polarization interaction to achieve perfect agreement with experiment at high energy. This is believed to be a satisfactory state of affairs until an energy-dependent polarization interaction can be developed for complex systems. Key partial cross sections for the latter two exchange approximations are shown in Figure 4 and generally agree well with one another. Although these two exchange treatments are roughly equivalent, in this context, we have found that differential cross sections[27] for e^--N_2 and unpublished integrated cross sections for e^--CO_2, OCS, CS_2 favor the Hara exchange approximation.

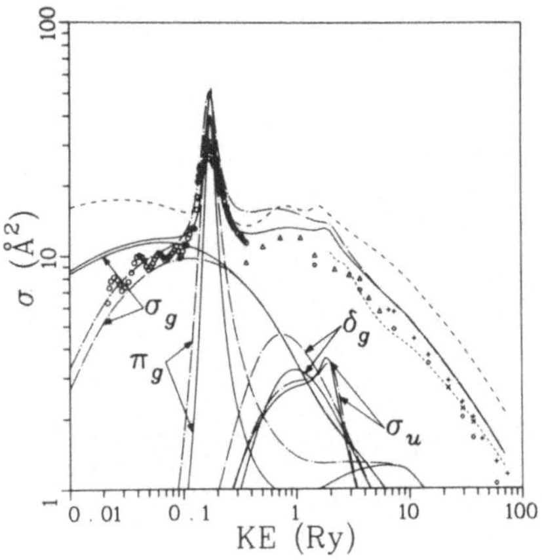

Fig. 4. Comparison of theoretical and experimental elastic e^--N_2 scattering cross sections from 0 to 1000 eV. Theoretical calculations differ mainly in the treatment of exchange interactions: Hara (——), semiclassical (— · — ·), Slater potential A of ref. 26 (----), and Slater potential B of ref. 25 (....). Experimental data include absolute measurements by Golden[34] (o), Bromberg[35] (x), DuBois and Rudd[36] (◇), and the normalized measurements of Srivastava et al.[37] (Δ), and Hermann et al.[38] (+). Note that Golden's measurements include vibrationally inelastic processes.

A more detailed view of the scattering process is reflected in the differential scattering cross section (DCS).[11,27] An overview[11] of the DCS for e^--N_2 scattering is given in Figure 5, utilizing the Slater exchange approximation. Near zero kinetic energy, the DCS is rather isotropic as the centrifugal forces exclude all except the $(\ell,\lambda)=(0,0)$ partial wave. As the energy increases, backscattering dominates briefly, until the π_g shape resonance is reached at 2.4 eV. The d-wave (ℓ=2) character of the 2.4 eV resonance stands out dramatically in this DCS surface. Then, above 0.5 Ry the spectral variation becomes more gradual and the angular distribution becomes progressively more peaked in the forward direction, as the electron transfers less and less momentum to the target. Note the strong resemblance to the experimental DCS surface in Figure 30 of Ref. 39.

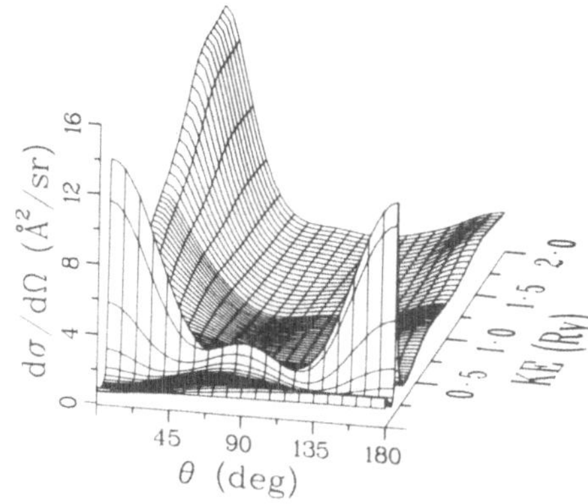

Fig. 5. Electron-nitrogen differential scattering cross sections from 0.001 to 2.0 Ry.

In Figure 6 slices of this DCS surface are compared with experiment and more recent DCS calculations,[27] employing Hara and semiclassical exchange approximations. It is clear in Figure 6 that the overall agreement between a variety of experimental data sets and all three DCS calculations is very good. Note that the linear plots in the second row for 15-30 eV are expanded on a log-linear plot in the third row to bring out the minor differences. Upon close inspection, a systematic differentiation between the theoretical curves does emerge. Namely, the Slater-exchange-based curves (---) at higher energy (15-30 ev) tend to have oscillations near $\theta = 90$ which are not present in experimental data. Furthermore, although the Hara (——) and semiclassical (-·-·) curves are usually close together, the Hara curve is systematically more free of extraneous structure near $\theta = 90^0$ at high energies, and is in better agreement at low scattering angles at lower impact energies. Hence, the DCS study points to the Hara exchange approximation as the most realistic of the three for electron-molecule scattering calculations. Further tests are necessary, of course, to firmly establish this tentative conclusion. Comparison of these results with other calculations is detailed elsewhere.[27] We can summarize these comparisons by saying that very good agreement is observed, on the average, with the single-center expansion calculations of Chandra and Temkin[24] and Buckley and Burke,[32d] both of which use more accurate exchange approximations. We have also investigated the DCS at high energy, 300-500 eV, with similar good agreement with the absolute experiments of Bromberg,[35] and

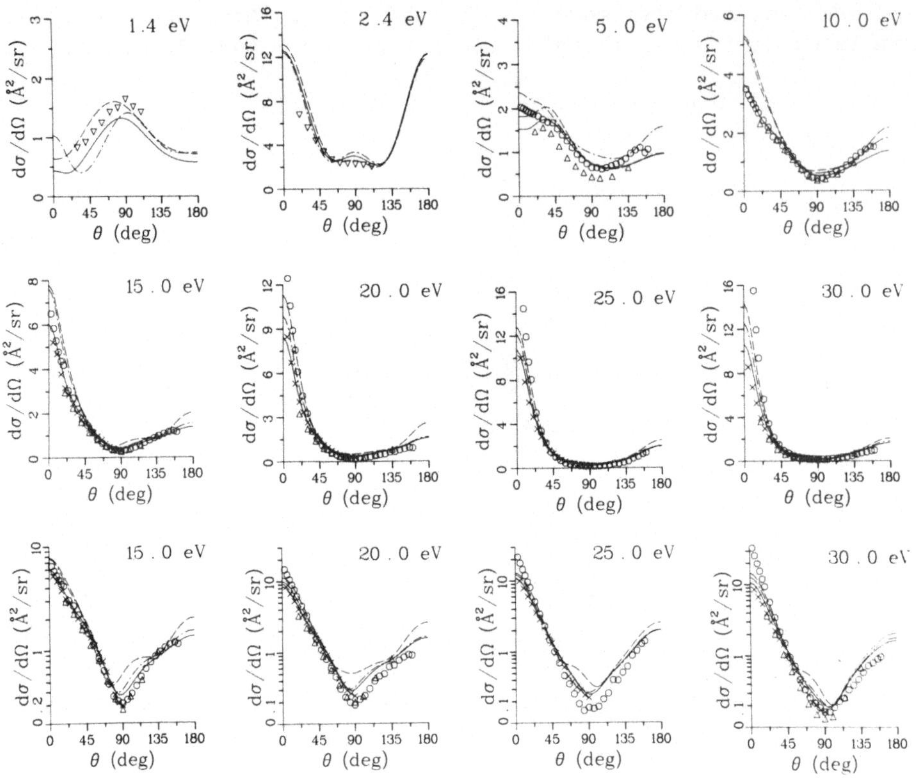

Fig. 6. Comparison of experimental and theoretical results for the e^{-}-N_2 differential cross section between 1.4 and 30.0 eV. Theoretical results are denoted by smooth curves using the same convention as Fig. 5. Experimental results include ∇, Ehrhardt and Willmann (Ref. 40); O, Shyn et al. (Ref. 41); Δ, Srivastava et al. (Ref. 37); and X, Finn and Doering (Ref. 42).

refer the reader to the original article for discussion of these high-energy DCS's.

Another major facet of electron-molecule scattering is the effect of nuclear motion. We have studied this for e^{-}-N_2 scattering between 0-50 eV within the adiabatic-nuclei approximation (see II.H). Hence, we do not introduce effects which depend upon the velocity of the nuclei, such as the vibrational substructure on the π_g resonance (see, e.g., Ref. 24). The R-averaged cross section in the vicinity of the π_g resonance is shown in Figure 7, using the Hara exchange approximation. Averaging resonance profiles for different values of R over the probability distribution in the ground vibrational state has the effect of smearing out the fixed-nuclei results, leading to a lower, broader resonance peak, in very good agreement with experiment. In all cases studied so far, the on-resonance results are similarly affected. This stands to reason since shape resonances have large amplitude in the molecular core, owing to the quasi-discrete nature of the state, and are therefore very sensitive to changes in molecular geometry. Another way of looking at it is that the delicate balance between centrifugal forces and electrostatic forces, which result in the potential barrier which traps the shape resonance, is sensitive to shifts in the electrostatic potential brought on by changes in molecular geometry. Note that, in general, the off-resonance cross section is relatively insensitive to this R-averaging procedure. This is clearly seen in Figure 7.

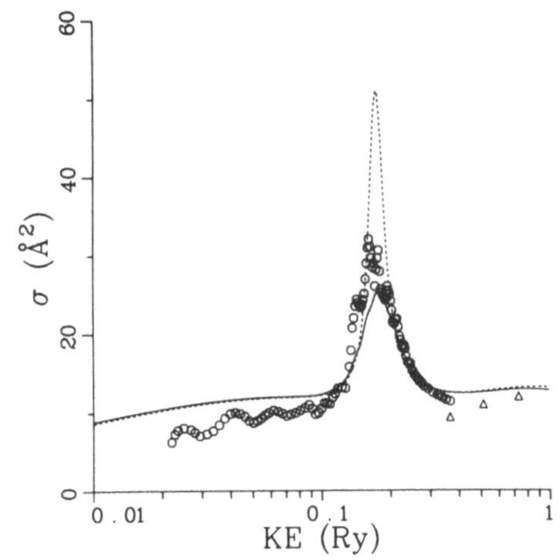

Fig. 7. CMSM calculations using the Hara exchange approximation at fixed internuclear distance (····) and R-averaged (——). Experimental data given in same convention used in Fig. 5.

The vibrationally elastic and inelastic cross sections[28] between 0-50 eV are shown in Figure 8. The 0 → 0 cross section has been discussed above, but is included here to emphasize how weak resonances can be barely detectable in the 0 → 0 channel and yet stand out prominently in the vibrational excitation channels. Thus, only the π_g resonance stands out in the 0 → 0 curve, whereas the weak σ_u resonance is prominent in all the vibrational excitation channels. The failure of the weak δ_g resonance to stand out in vibrational excitation is attributed to its orientation away from the nuclear axis and its weak resonant character, both of which cause it to couple only weakly with the nuclear motion. The π_g state is also oriented perpendicularly to the axis, but, in contrast to the δ_g, it is strongly

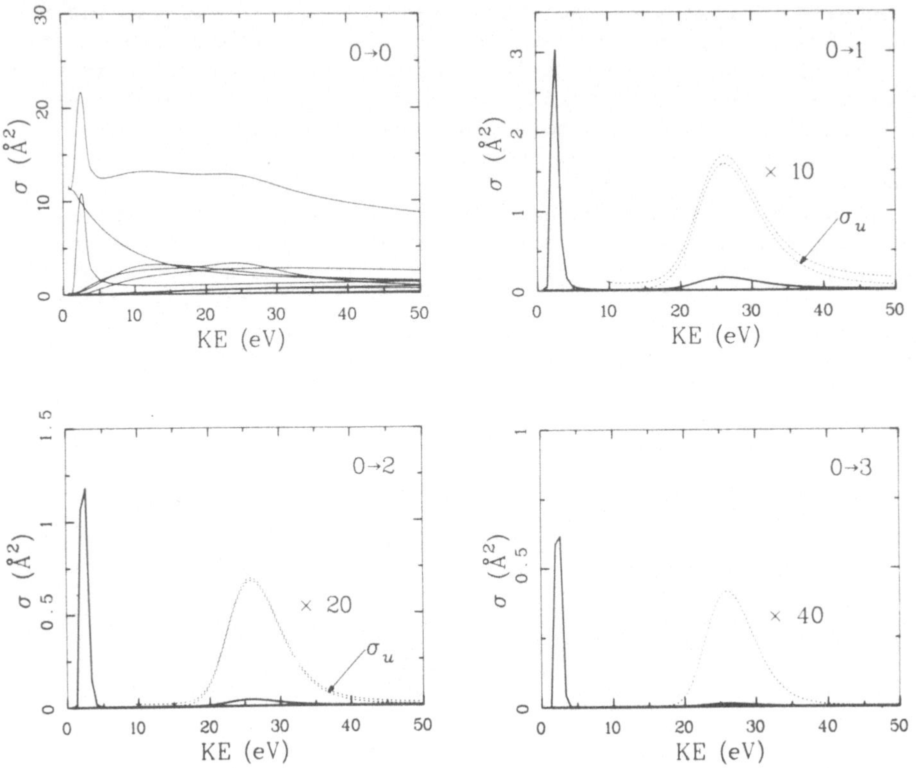

Fig. 8. Vibrational excitation spectra for e^--N_2 scattering using the CMSM with the Hara exchange approximation. Cross sections $\sigma_{v_f \leftarrow v_i}$ are evaluated in the adiabatic nuclei approximation using Eq. (46).

localized in the molecular core and hence couples strongly with nuclear motion.

As stated above, our model presently neglects vibrational close-coupling, so that we must compare our π_g results with an average of the peaks in the experimental spectrum.[9] Our peak values of 3.1, 1.3, and 0.6 $\mathring{A}^2$ for the (0,1), (0,2), and (0,3) transitions, respectively, agree fairly well with a gross average of the vibrational substructure presented by Chandra and Temkin,[24] and lie somewhat above experimental values. Note, however, that normalization of the experimental values remains in doubt by as much as a factor of two.[24] This is discussed in the accompanying article by Temkin.

The novel aspect of the calculated vibrational spectrum is the broad feature extending from 15-40 eV, centered at 26 eV. This feature is due wholly to the weak but axially oriented σ_u shape resonance, and corresponds reasonably well with the broad hump observed by Pavlovic et al.[43] This feature was originally interpreted by Pavlovic et al. in terms of a large manifold of overlapping compound states above 20 eV, including possible shape resonances and singly and multiply core-excited Feshbach resonances. In view of the calculation in Figure 8, we have recently proposed the alternative simple explanation of the experimental spectrum in terms of the σ_u shape resonance. This conclusion is still tentative as discussed in more detail elsewhere.[28]

It is instructive to make the connection with a related phenomenon - photoionization of N_2. In this case, there is one less electron in the molecular field to screen the attractive nuclear charge. This causes all features discussed above to appear lower in the spectrum (relative to a photoionization threshold). Hence, in K-shell photoionization, we observe an intense, dominant π_g peak below threshold in the discrete part of the spectrum.[12-15] The final state for this peak strongly resembles the quasidiscrete resonance state in the e^--N_2 system. In addition, the very weak σ_u feature in e^--N_2 scattering appears in K-shell photoionization as a prominent shape resonance,[12-15] ~0.8 Ry above the ionization threshold. The δ_g feature is unobservable in K-shell photoionization due to dipole selection rules, but may be discernible in valence-shell spectra. Finally, we re-emphasize the point made extensively in the study[12,13] of photoionization of N_2 - namely, that the large contribution of high angular momentum components to continuum processes in first row diatomics is due to the influence of the molecular field on the process. For example, ℓ=2,3, etc. partial waves would not cause resonant behavior in the near-threshold photoionization of either atomic nitrogen or the united atom-silicon; hence, the high-ℓ character of the dominant structure in the K spectrum of N_2 is due not to the amount of nuclear charge in the target but rather to the distribution of this nuclear charge over multiple sites separated by distances on a molecular scale.

B. e^--CO_2 Scattering

Only very recently has any attempt been made to calculate cross sections for electron scattering from polyatomic molecules. CO_2 was the first molecule treated in detail, with both single-center-expansion methods[44] and the CMSM. The CMSM results summarized here are described in detail by Lynch et al.[29] and Welch et al.[30] The eigenphase sums for e^--CO_2 from 0-100 eV are shown in Figure 9. Significant rises in the eigenphase sums are observed for the π_u, δ_g, σ_g and σ_u channels, in order of increasing energy. The p-like π_u resonance at 3.4 eV is well known, both experimentally[45-49] and theoretically.[29,30,44,50] The higher-energy features, however, were observed in the CMSM calculation for the first time.

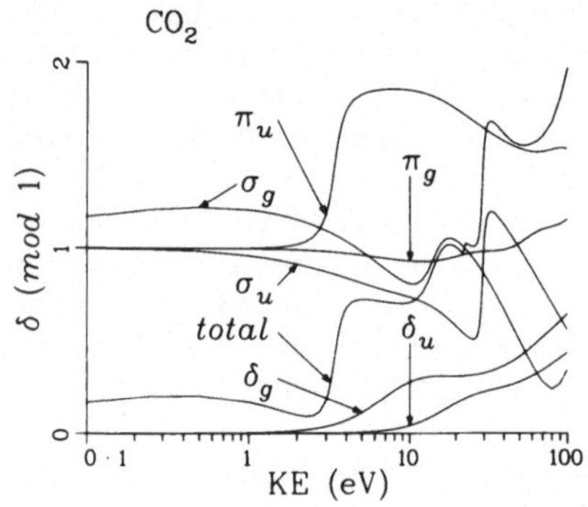

Fig. 9. Eigenphase sums for the λ=0,1,2 channels and the total eigenphase sum for e^--CO_2 elastic scattering from 0-100 eV.

The integrated elastic scattering cross sections for each channel are shown in Figure 10. Here the labels A-D mark the positions of local maxima in the π_u, δ_g, σ_g, and σ_u channels, respectively. The well known π_u resonance stands out clearly; however, the resonant features at higher energy form only weak undulations on the total cross section. As we shall see below, the σ_u peak (D) is further washed out by R averaging. However, the σ_g and σ_u features emerge clearly in the vibrational excitation spectrum. In Figure 11 our results are compared with experiment, and the single-center-expansion results of Morrison et al.[44] Reasonably good agreement is observed among all the curves in Figure 11, although the single-center-expansion result is in better agreement with experiment at low incident energies, and the CMSM result is in better agreement at higher energy.

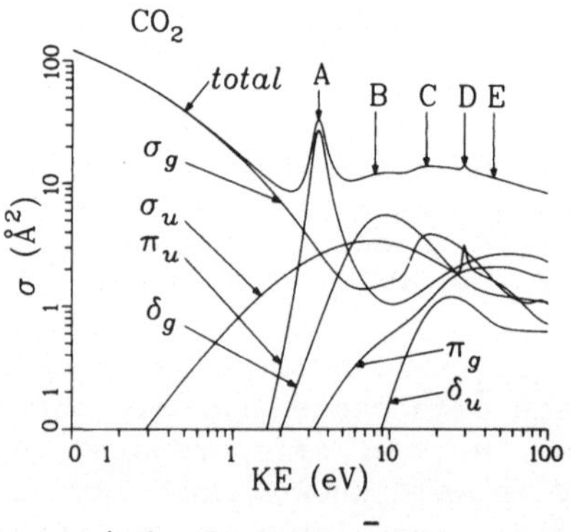

Fig. 10. Total and partial elastic e^--CO_2 scattering cross sections from 0-100 eV.

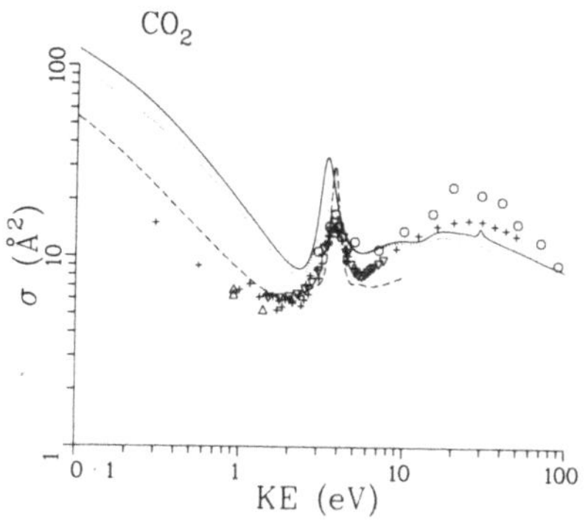

Fig. 11. Total e^--CO_2 elastic scattering cross sections from 0 to 100 eV. CMSM results are denoted by a solid line (equilibrium internuclear separation) and a dotted line (R-averaged). Single-center expansion results from ref. 44 are denoted by dashed line. Experimental data are taken from ref. 45 (Δ); ref. 46 (+); ref. 47 (o); and ref. 48 (∇).

C. e^--CS_2, OCS Scattering

In the spirit of survey calculations, cross sections for e^--CS_2 and e^--OCS scattering were calculated in order to establish the extent to which this family of linear triatomics is similar. The comparison of the three systems has been discussed extensively by Lynch et al.[29] and will only be summarized here. Briefly, each system was found to have a substantially similar pattern in their eigenphase sum curves, i.e., a sequence of rises in the π_u, δ_g and σ_g channels (π, δ, and σ for OCS). The main difference was that, with increasing molecular size, the ℓ's participating in these resonances shifted to larger values, even though the pattern of features retained the same order and roughly the same energy. The relative strength of higher-energy features also changed in the sequence CO_2-OCS-CS_2, e.g., the σ_u is prominent in CO_2 but very weak in OCS and CS_2. The integrated cross section for CS_2 is shown in Figure 12, where A, B, and C refer to π_u, δ_g, and σ_g, respectively. The σ_u feature at higher energy no longer stands out, even in the fixed nuclei calculation. The strong π_u resonance has only been observed in back scattering,[51] but was located at 1.15 eV, in very good agreement with the calculation. The e^--OCS integrated elastic cross section is shown in Figure 13, and compared with the data of Syzmtkowski and Zubek.[48] Clearly the π shape resonance (labeled A) agrees well with the experimental resonance position, although the magnitudes differ significantly. The difference is attributed to the neglect of the molecule's permanent dipole field in the theoretical calculation. The experimental curve rises at ~2 eV in agreement with the theory which indicates enhanced scattering in the δ(B) and σ(C) channels. These results speak for the general similarity of resonance structures in closely related molecules, but show significant systematic trends at a more detailed level, e.g., shifts in resonance energies, intensities, widths, and asymptotic ℓ character. To complete this set of studies, it is most important to have experimental data on e^--CS_2 and take into account scattering by the dipole field in e^--OCS. Moreover, just as in e^--N_2 and e^--CO_2, it is important to investigate the vibrational excitation channels in order to verify the existence of weak resonances at higher kinetic energy which are too weak to observe in the total scattering cross section.

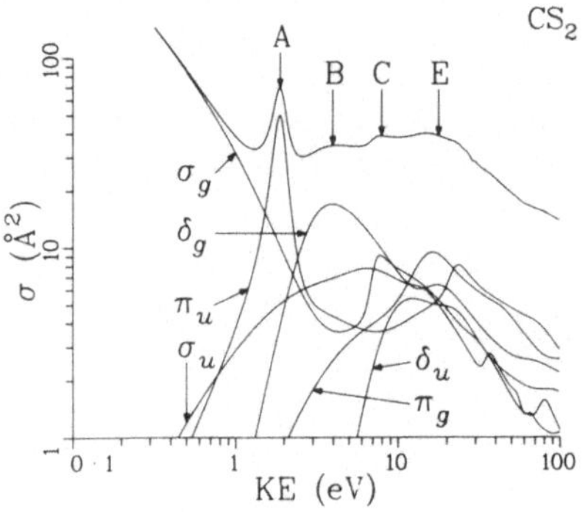

Fig. 12. Total and partial elastic e^--CS_2 scattering cross sections from 0-100 eV.

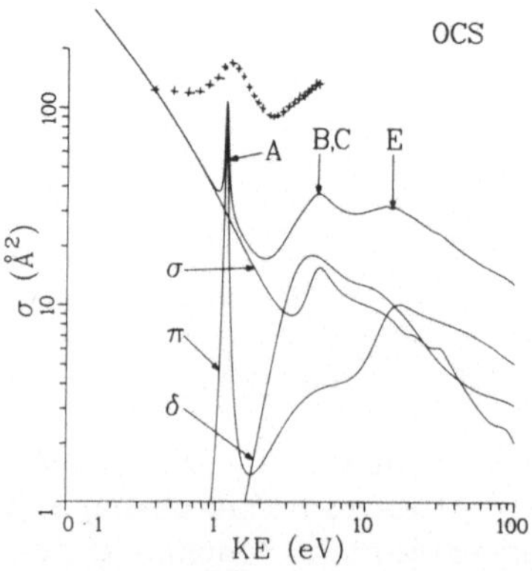

Fig. 13. Total and partial elastic e^--OCS scattering cross sections from 0-100 eV. Experimental data taken from ref. 48.

D. e^--SF_6 Scattering

Calculation of e^--SF_6 scattering represents a major jump in the complexity of the target molecule - to 70 electrons. As in the above cases, only the major results are summarized here, and the reader is referred to the original article[31] for numerical details. Four channels exhibit distinct resonance behavior - a_{1g}, t_{1u}, t_{2g}, and e_g. As expected, the integrated elastic cross section in Figure 14 shows resonant enhancement at the energy locations of each step in the eigenphase sums - at 2.1 eV (a_{1g}), 7.2 eV (t_{1u}), 12.7 eV (t_{2g}), and 27.0 eV (e_g). In addition, the a_{1g} cross section increases sharply near zero energy due to the departure of its eigenphase sum from zero phase shift near zero energy. This feature is considered tentative, however, as we normally do not attach too much credence to the results of the present model below a couple of tenths of an eV.

Although e^--SF_6 scattering has been measured by several authors, we will focus on recent measurements of total e^--SF_6 scattering by Kennerly, et al.,[52] since they are absolute, on a fine mesh, and are dominated by elastic scattering. In Figure 14, the (dashed) experimental spectrum exhibits three distinct peaks at 2.6 eV, 7.2 eV, and 11.8 eV, in good agreement with the calculated resonance positions. Moreover, the average magnitude of the cross section is in good agreement above ~5 eV, and the sharp rise at thermal energies is tentatively accounted for by the rise in the a_{1g} cross section. The

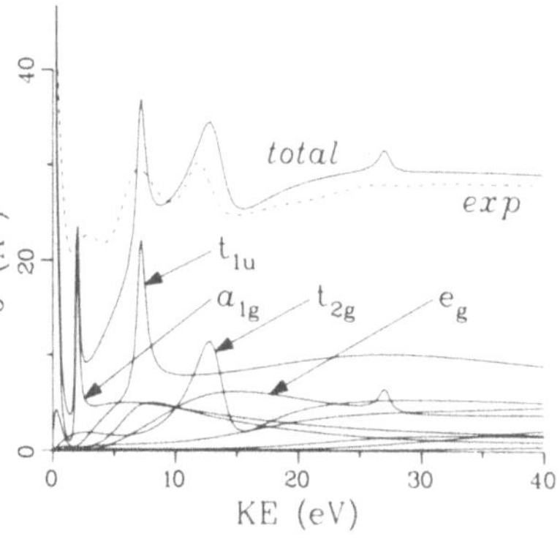

Fig. 14. Comparison of e^--SF_6 scattering cross sections. Solid curves are theoretical elastic cross sections. The dashed curve is the absolute total cross section measured by Kennerly et al. (ref. 52). Partial cross sections for resonant channels are labeled. To permit identification of other channels, we note that, in order of increasing cross section at 40 eV, the channel symmetries are a_{2g}, a_{2u}, e_u, a_{1g}, t_{1g}, e_g, t_{2u}, t_{2g}, and t_{1u}, respectively.

agreement is not exact, however, as the a_{1g} and t_{1u} resonances are narrower and stronger than experiment, the e_g resonance is not clearly observed, and the deep minimum in the nonresonant cross section near 2.5 eV is not present in the measurement. A final comparison must await consideration of effects of nuclear motion as this is likely to alter quantitative features of the cross section, especially at low energy, and to significantly lower and broaden resonance structures calculated at R_e.

Finally, we emphasize that localized states also dominate the K- and L-shell photoionization spectra[53] of SF_6, only these states are shifted to lower electron kinetic energy (photoelectron energy, in this case), owing to the decrease in the number of electrons in the system. In fact, the a_{1g} and t_{1u} resonances in e^--SF_6 become highly localized discrete states in SF_6.

After completion of this manuscript, an article[54] appeared which reported a calculation of the total e^--SF_6 scattering cross sections between 10-60 eV. That work is in qualitative but not quantitative agreement with the present work (e.g., the t_{1u}, t_{2g}, and e_g resonances are identified but lie at 11, 17, and 24 eV), due to their use of the Slater exchange approximation, a non-SCF charge distribution for SF_6, and lack of convergence.

IV. MOLECULAR PHOTOIONIZATION RESULTS

Before presenting specific results on the application of the CMSM to molecular photoionization, we offer a brief overview of this area. The original CMSM study[12,13] on K-shell photoionization of N_2 had several significant effects. It established the interpretation of the prominent π_g and σ_u features (see next section) as localized states, associated with high-ℓ components of the final-state wavefunction. The prototype study made it apparent that high-ℓ shape resonances and related localized discrete states would be very common in molecular spectra and offered a concrete explanation of the earlier interpretations[53,55,56] of striking non-atomic effects in inner-shell molecular spectra. The study went on to cover the manifestation of molecular shape resonances in random molecule[13] and fixed molecule[18,19,21,57] photoelectron angular distributions and EXAFS[13] in the K-shell spectra of N_2 and CO. Shortly after the original report on N_2, Davenport[20,58] began reporting CMSM results on photoionization from molecular valence shells where they expectedly found the same final state effects. Davenport's calculations on oriented-molecule photoelectron angular distributions in the vicinity of shape resonances in CO were utilized[59] to determine the orientation of molecules adsorbed on surfaces. This work has subsequently led to studies[60,61] on partial photoionization of molecules in the VUV range. At this writing, our group is completing a series of studies on the partial photoionization cross section and photoelectron angular distributions of the valence and inner shells of several molecules. This work is also testing alternative local exchange approximations and studying the effects of nuclear motion in photoionization spectra. We cannot review this large body of work here, but will briefly go over the N_2 K-shell calculation (including recent work on the effects of hole localization) because it exemplifies the use of the CMSM to gain physical insight into molecular photoionization dynamics. We will also include previously unpublished electron-optical results for the K shells of N_2 and CO in order to illustrate this important aspect of inner-shell molecular photoionization.

A. K-Shell Photoionization of N_2

Inner-shell photoabsorption in molecules takes place deep inside the core of one of the constituent atoms, where the local field is dominated by the "atomic" field of the nearby nucleus. Nevertheless, molecular inner-shell spectra are known[53,55,56] to depart marketly from the corresponding atomic spectra. Here we review the first detailed accounting[12,13] of the nonatomic effects in a prototype molecular inner-shell spectrum - the K-shell spectrum of N_2. The "psuedo-photon" absorption spectrum[62] shown in Figure 15 best illustrates the deviations of this spectrum from the K-shell spectrum of atomic nitrogen, which would exhibit a normal $1s \rightarrow np$ Rydberg series converging to a monotonically decreasing $1s \rightarrow \varepsilon p$ continuum. Specifically, the N_2 spectrum below the I.P. (409.9 eV) is dominated by a single extremely intense peak (A) at ~401 eV with some weak, unresolved Rydberg structure (B-E) between ~406 to 410 eV. Above the K edge there is a band (G) of enhanced absorption in place of the smooth monotonic decay one might expect. The side band F is attributable[62] to double excitation, and is not accounted for in this work. The novel features A and G must result from the interaction between the "atomic" electron escaping from the K shell and the anisotropic molecular field. In fact, we show that these features are centrifugal barrier effects which manifest themselves as shape resonances in high-ℓ components of the final-state wavefunction. The molecular field is crucial in this phenomenon acting both to cause the resonance behavior in these high-ℓ components and to couple these with the p wave produced by photoabsorption of a K electron in an essentially atomic field. By carrying the calculation to 100 Ry, we account for ~99.9% of the oscillator strength in the two N_2 K shells and bring out the EXAFS structure which results from the diffraction of the outgoing electron from the two atomic cores. We note that recent L^2 calculations[15] have been successful in achieving better quantitative agreement with experiment, while using the qualitative interpretation developed by this model calculation. Other examples of such L^2 approaches to molecular photoionization are presented elsewhere in this volume. Finally, it is worth nothing at

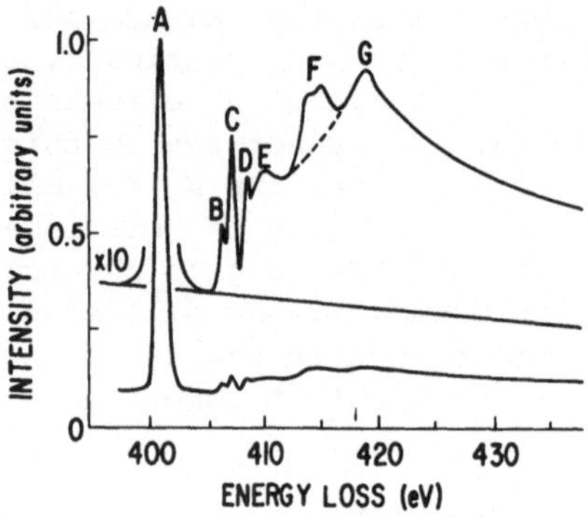

Fig. 15. "Pseudo-photon" absorption spectrum of N_2 in the vicinity of the K-shell edge (409.9 eV) by Wight et al. (ref. 62).

the outset that the K-shell spectra[62,63] of CO and NO are qualitatively similar to that of N_2, and that all of the major points made below carry over to these molecules.

Figure 16 contains the partial cross sections for the four-dipole-allowed channels for K-shell photoionization in the vicinity of the K edge. In this figure, the energy scale is referenced to the K edge ($h\nu$ = 409.9 eV) and is expanded by a factor of two below threshold. The numerical conditions, partial wave expansions, etc., are described in the original article.[13] Also note that we neglect, at this stage, the effects of hole localization, and thus retain the inversion symmetry quantum numbers for the excited electron and the N_2^+ core separately. As discussed in a later section, hole localization has been studied and found to affect total energies and photoelectron angular distributions, but not the general features of present discussion.

The results in Figure 16 show that the dominant peak in the experimental spectrum corresponds to the first member of the π_g sequence and dominates every other feature in the theoretical spectrum by a factor of ~30. (Note the first π_g peak has been reduced by a factor of 10 to fit in the frame.) The concentration of oscillator strength in this peak is a centrifugal barrier effect in the d-wave component of the π_g wavefunction. The final state in this transition is a highly localized state, about the size of the molecular core, and is the counterpart of the π_g shape resonance in e^--N_2 scattering at 2.4 eV. For the latter case, Krauss and Mies[32] demonstrated that the effective potential for the π_g elastic channel in e^--N_2 scattering exhibits a potential barrier due to the centrifugal repulsion acting on the dominant $\ell = 2$ lead term in the partial-wave expansion of π_g wavefunction. In the case of N_2 photoionization, there is one less electron in the molecular field to

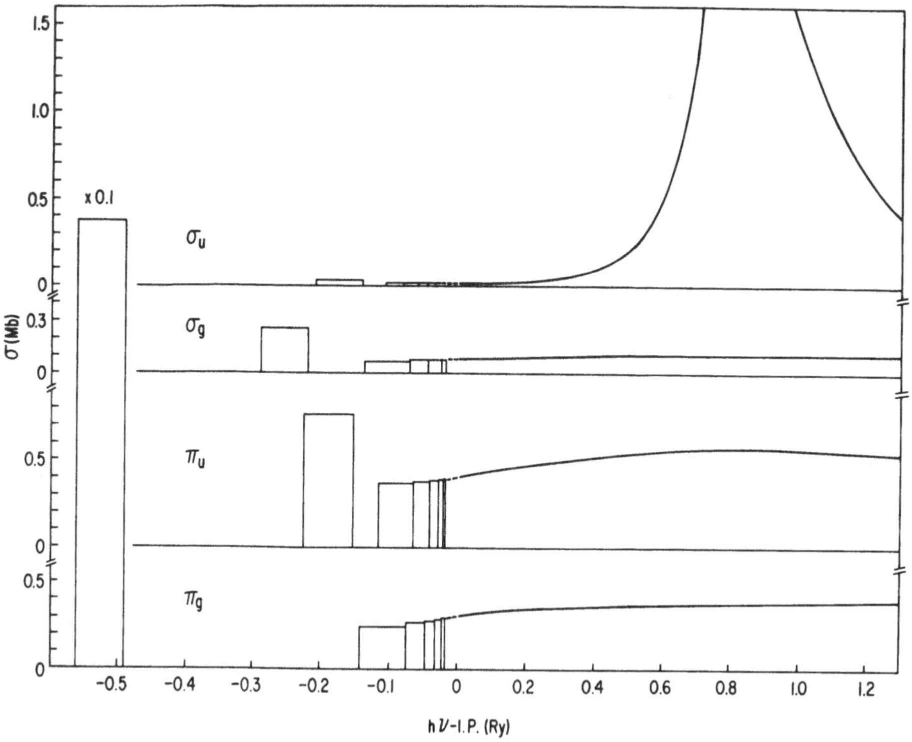

Fig. 16. Partial photoionization cross sections for the four dipole-allowed channels for K-shell photoionization of N_2. Note the energy scale is referenced to the K-shell I.P. (409.9 eV) and that the energy scale is expanded twofold in the discrete.

screen the nuclear charge so that this resonance features is shifted to lower energy and appears in the discrete region. It is in this sense that we refer to the peak A in Figure 16 as a "discrete" shape resonance. Therefore, it is clear that even for a first-row diatomic, molecular photoionization can exhibit prominent high-ℓ centrifugal barrier effects.

Within 4 eV below threshold, there appear four distinct features in Figure 15. Using Figure 16, we can assign peaks B and C to the first excited states of σ_g and π_u symmetry. Note that the relative intensity agrees very well. Peaks D and E are more difficult to assign on a one-to-one basis, but are formed from overlapping peaks of symmetries, σ_g, π_u, and π_g. The intensity of the σ_u channel is depleted in this part of the spectrum to balance the concentration of oscillator strength in the resonance above threshold.

The essence of the continuum phenomena can be described in mechanistic terms as follows. The electric dipole interaction, localized within the atomic K shell, produces a photoelectron with angular momentum $\ell=1$. As this p-wave electron escapes to infinity, the anisotropic molecular field can scatter it into the entire range of angular momentum states contributing to the allowed σ- and π-ionization channels ($\Delta\lambda=0, 1$). In addition, the spatial extent of the molecular field of the two atoms, separated by 1.1Å, enables the $\ell=3$ component of the σ continuum wavefunction to overcome its centrifugal barrier and penetrate into the molecular core at a kinetic energy of ~0.8 Ry. This penetration is rapid, a phase-shift of ~π occurring over a range of ~0.3 Ry. These two circumstances combine to produce a dramatic enhancement of photoelectron current at ~0.8 kinetic energy, with predominantly f-wave character. A clear demonstration[19] of the f-wave character of the resonance is given in Figure 17, which shows the angular distribution of photoelectrons ejected from molecules with fixed orientation relative to the polarization of the light ($m_p = 0$ for parallel and $m_p = 1$ for perpendicular orientation). The $m_p = 0$ cross section is much stronger than the $m_p = 1$ as only the former couples to the σ shape resonance. Moreover, the shape of the strong resonance peak is clearly f-like in character. The K shells of CO are also shown, and show the distortion of the basic f resonance due to coupling with $\ell=0,2$ caused by the loss of inversion symmetry.

The specifically molecular character of this phenomenon is emphasized by comparison with K-shell photoionization in atomic nitrogen and in the united-atom case, silicon. In contrast to N_2, there is no mechanism for the essential p-f coupling, and neither atomic field is strong enough to support resonant penetration of high-ℓ partial waves through their centrifugal barriers. With substitution of "d" for "f", this argument applied equally well to the d-type resonance in the discrete part of the spectrum. These results have profound implications for molecular photoionization in general. The effect of the shape resonance on photoelectron angular distribution for randomly oriented molecules is shown in Fig.18.

B. Electron-Optical Aspects of Inner-Shell Photoionization

We will now illustrate the point of view described in Sec. II.G, which relates the inner-shell photoionization cross section to the electron-optical parameter $|A^-_{LL'}|^2$. Recall that this is the square of the incoming-wave-normalized, dipole-allowed final state component(s) near the nucleus of the atomic core from which the electron is being ejected. For K-shell photoionization the cross section is given by Eq. (30) in terms of $|A^-_{L1}|^2$, the square of the p-wave component. This, then, relates relative probability for the p-wave to escape through the anisotropic molecular field to infinity in the asymptotic channel L, hence the concept of "electron optics" of the molecular field. Figure 19 shows the factor $|A^-_{L1}|^2$ for K-shell photoionization from N_2 and CO. The surfaces indicate the breakdown

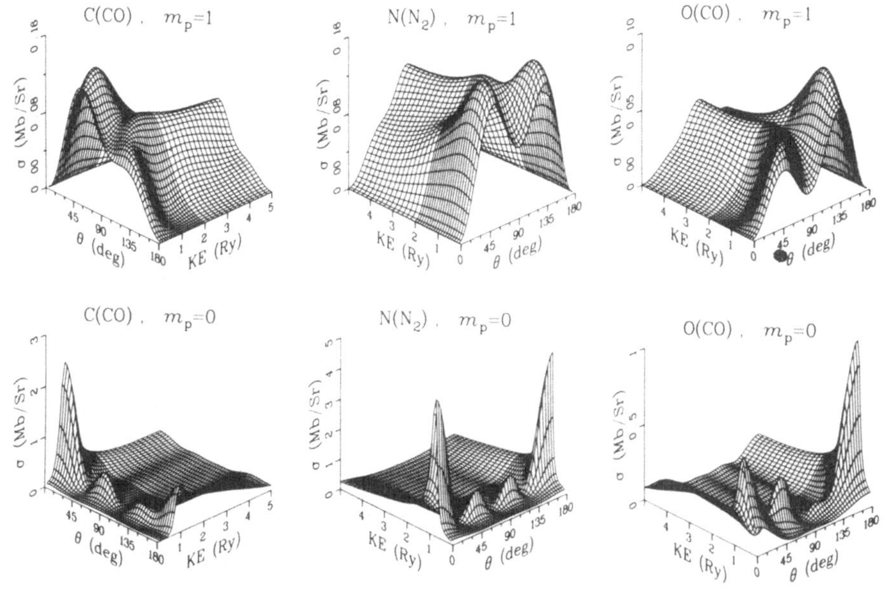

Fig. 17. Fixed-molecule photoelectron angular distributions for the K shells of N_2 and CO.

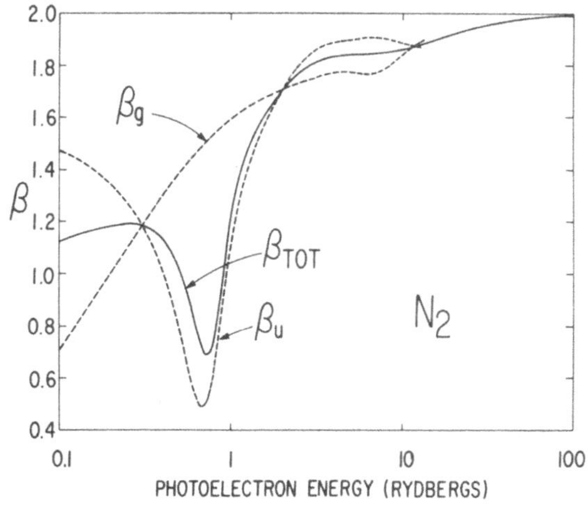

Fig. 18. Asymmetry parameter for photoelectron angular distributions from K-shell photoionization of randomly oriented N_2. Dashed curves show the decomposition into even (g) and odd (u) ionization channels.

into alternative asymptotic ℓ's and m's ($m=\lambda=0,1$ indicate σ and π ionization channels). At this stage we neglect hole localization in order to introduce the electron-optical concept in the context of the calculations reported in the last section. Hence, for N_2, the residual ion and photoelectron exhibit inversion symmetry separately, resulting in no coupling between even and odd asymptotic ℓ's.

In Figure 19, the surface for σ ionization in N_2 conveys at a glance the major features of the ionization dynamics given in the last section. The dominant peak in the $\ell=3$ channel at the resonance energy indicates that a p wave created in the core of a nitrogen atom with ~0.8 Ry excess energy is strongly coupled to and hence will escape to infinity in the f-wave component of the asymptotic wavefunction. This electron-optical enhancement is, of course, due to the $\ell=3$ shape resonance producing a large amplitude (also called enhancement factor or density of states) in the molecular core. Note that the escape of this p wave into other σ and π asymptotic channels is effectively blocked by the unfavorable transmission properties of the molecular field. The factor $|A_{31}^{-}|^2$ therefore accounts for the shape resonance in Fig. 16 as the other factors in Eq. (30) are only slowly varying functions of energy.

The CO surfaces in Figure 19 are likewise dominated by a large f-wave component in the σ channel. However, owing to the lack of inversion symmetry in this molecule, the $\ell=3$ resonance couples with the $\ell=0,2$ components, thereby inducing "sympathetic" resonances in those channels. The secondary resonances have the same profiles as the $\ell=3$ component so that the integrated cross sections for the CO K shells are nearly identical in shape to that for N_2, only the carbon cross section is larger than the oxygen cross section, reflecting both that the resonance is concentrated more heavily on the carbon end of CO and the larger size of the carbon K shell.

C. Effect of Hole Localization on Inner-Shell Photoionization

Ionization of molecular core states which are distributed on two or more equivalent atomic sites can lead to states of the residual ion in which the vacancy is effectively localized on a single equivalent site prior to becoming filled by Auger or radiative decay. We have recently shown[64] that the lowering of molecular symmetry accompanying such hole localization during the photoionization process (dynamic symmetry breaking) has a dramatic effect on the spectral variation of molecular photoelectron angular distributions. This effect can be traced to additional couplings in the final state, induced by the lowered symmetry of the molecular core. Hence, photoelectron angular distributions probe the effects of hole localization on the ejected electron's wavefunction, and thus complement the evidence based on the total energy of the ionic hole state first described by Bagus and Schaefer.[65]

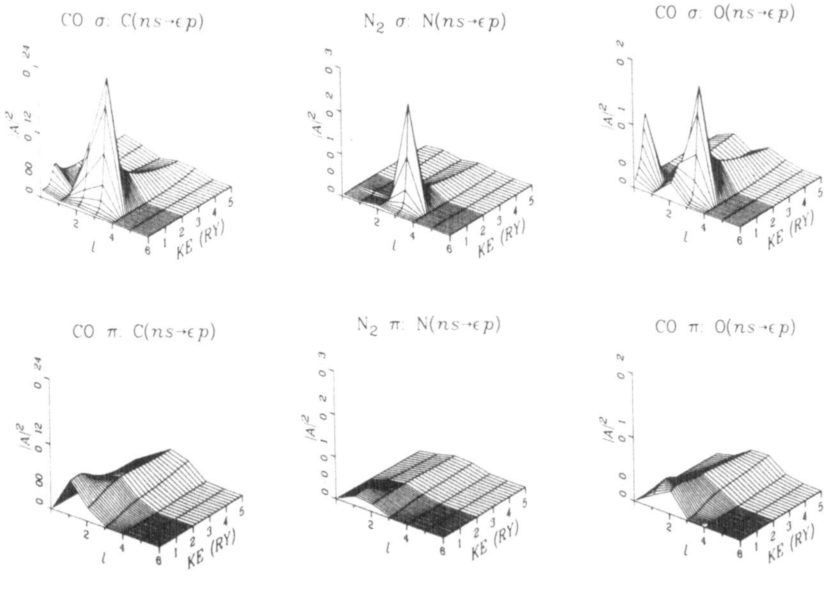

Fig. 19. $|A^{-}_{L1}|^2$ (in Ry^{-1}) for K-shell photoionization in N_2 and CO.

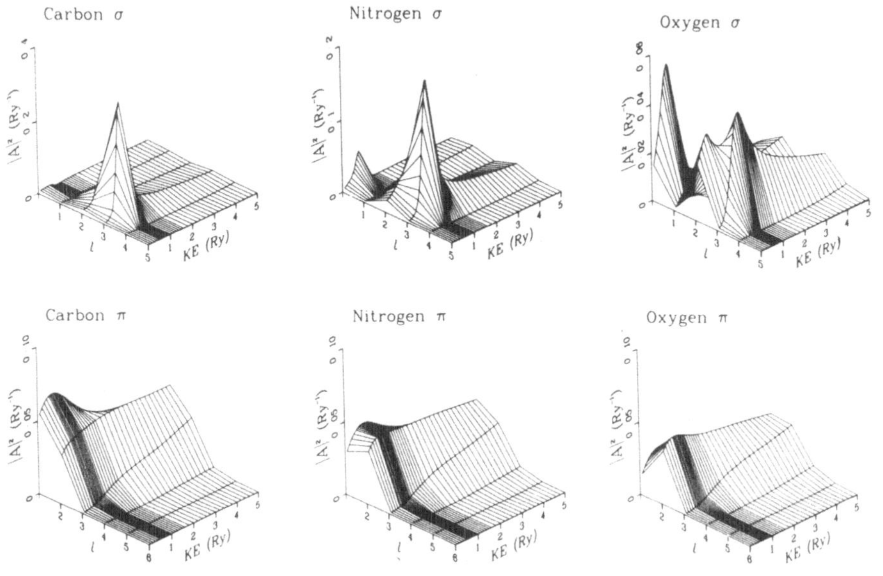

Fig. 20. $|A^{-}_{L1}|^2$ (in Ry^{-1}) for K-shell photoionization in N_2 and CO with hole localization in the final state.

To see this effect, we present in Figure 20 the $|A_{L1}^{-}|^2$ coefficients for N_2 and CO computed with the K-shell hole localized in the final state. For N_2, we see that the effect of dynamic symmetry breaking is to couple the resonant $\ell=3$ component to the $\ell=0,2$ components, as in the case of CO. In the CO surfaces, we see that the f- to s,d-wave coupling has been reduced in the carbon core and increased in the oxygen core. These observations follow from the fact that a localized K-shell hole increases the effective charge on the nucleus at that site.

Hole localization was found to have negligible effect in the integrated cross sections since the $\ell=3$ and induced $\ell=0,2$ resonances have nearly the same profile and are summed incoherently. However, the redistribution of photocurrent among different ionization channels has a dramatic effect on the photoelectron angular distributions, as emphasized in ref. 64. This is illustrated in Figure 21, where we show hole-localized calculations of the asymmetry parameter for K-shell photoionization in N_2 and CO. This figure has three aspects. First, the effect of hole localization is minor for CO, where the symmetry is not altered in the process. The β curves in localized and delocalized treatments of CO are very similar, and hence both are well represented by Figure 21. Second, dynamic symmetry breaking has a significant effect on N_2, as can be seen by comparing Figures 18 and 21. Third, the net result is to make the hole-localized N_2 result strongly resemble that for the oxygen rather than the carbon end of CO. This can be attributed to the effective increase in charge of the ionized nitrogen atom which would cause it to resemble more closely the higher atomic number atom in CO. Hence, independently of the accuracy of the present calculation, we predict that dynamic symmetry breaking can be experimentally observed by verifying that the nitrogen and oxygen

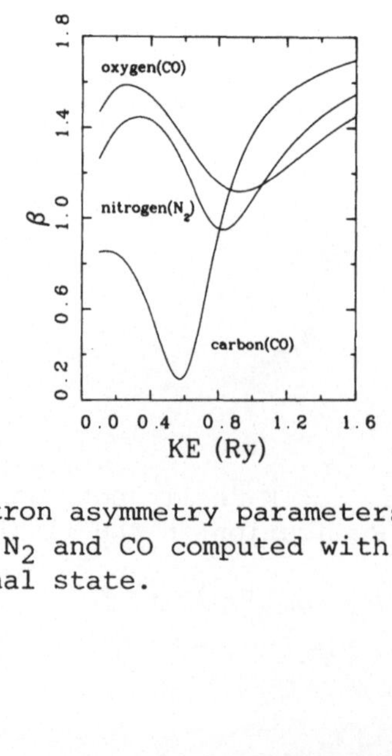

Fig. 21. Photoelectron asymmetry parameters for K-shell photoionization in N_2 and CO computed with the K shells localized in the final state.

K-shell β's are similar and substantially different that that for carbon.

V. CONCLUSION

The aim of this report was twofold: First, we presented in Sec. II a comprehensive theoretical framework for the continuum multiple scattering method itself, as well as its applications to electron-molecule scattering and molecular photoionization. Second, we reviewed highlights of recent applications in these two areas to illustrate the breadth of this technique. Specifically, the CMSM has circumvented several long-standing bottlenecks in the study of the molecular continuum, including (i) large energy ranges, (ii) complex molecules, and (iii) a large variety of physical processes. The major impact of the resulting studies over the last few years has been to establish the pervasive importance of shape resonances in electron collisions and photoionization of practically all (non-hydride) molecules. The consequences of these prominent features were found in many different physical observables.

We must emphasize that the CMSM in its present form is designed as a semiquantitative survey tool, and is no substitute for more rigorous approaches, some of which are described in this volume. In years to come, the complementarity of these different approaches must be exploited so that as more exact theoretical tools become available, the large scope of the CMSM survey calculations will point to those problems in molecular continuum physics in greatest need of deeper study.

Finally, we acknowledge the support received from the U. S. Department of Energy, the National Science Foundation (Grant CHE 78-08707) and the Petroleum Research Fund, administered by the American Chemical Society.

REFERENCES

1. J. L. Dehmer and D. Dill, Argonne National Laboratory Radiological and Environmental Research Division Annual Report, July 1972-June 1973, ANL-8060, Part I, pp. 119-157.
2. D. Dill and J. L. Dehmer, J. Chem. Phys. 61, 692 (1974).
3. K. H. Johnson, in Advances in Quantum Chemistry, P.-O. Löwdin, Ed. (Academic Press, New York, 1973), Vol. 7, p. 143.
4. J. Siegel, D. Dill, and J. L. Dehmer, J. Chem. Phys. 64, 3204 (1976)
5. J. Siegel, Ph.D. Thesis, Department of Chemistry, Boston University, 1978.
6. J. C. Slater, Quantum Theory of Atomic Structure (McGraw-Hill, N. Y., 1960) Appendix 22; and Quantum Theory of Molecules and Solids, Vol. IV (McGraw-Hill, N. Y., 1974).
7. S. Hara, J. Phys. Soc. Japan 22, 710 (1967).
8. J. B. Furness and I. E. McCarthy, J. Phys. B 6, 2280 (1973).

9. M. E. Riley and D. G. Truhlar, J. Chem. Phys. 63, 2182 (1975).
10. U. Fano and D. Dill, Phys. Rev. A 6, 185 (1972)
11. J. Siegel, D. Dill, and J. L. Dehmer, Phys. Rev. A 17, 2106 (1978).
12. J. L. Dehmer and Dan Dill, Phys. Rev. Letters 35, 213 (1975).
13. J. L. Dehmer and Dan Dill, J. Chem. Phys. 65, 5327 (1976).
14. R. B. Kay, Ph. E. Van der Leeuw, and M. J. Van der Wiel, J. Phys. B 10, 2513 (1977).
15. T. N. Rescigno and P. W. Langhoff, Chem. Phys. Lett. 51, 65 (1977).
16. D. Dill and U. Fano, Phys. Rev. Lett. 29, 1203 (1972).
17. E. S. Chang, J. Phys. B 11, L293 (1978).
18. D. Dill, J. Chem. Phys. 65, 1130 (1976).
19. D. Dill, J. Siegel, and J. L. Dehmer, J. Chem. Phys. 65, 3158 (1976).
20. J. W. Davenport, Phys. Rev. Lett. 36, 945 (1976).
21. S. Wallace, D. Dill, and J. L. Dehmer, Phys. Rev. B 17, 2004 (1978).
22. J. W. Davenport, W. Ho, and J. R. Schrieffer, Phys. Rev. B 17, 3115 (1978).
23. D. M. Chase, Phys. Rev. 104, 838 (1956).
24. N. Chandra and A. Temkin, Phys. Rev. A 13, 188 (1976).
25. R. K. Nesbet, Phys. Rev. A. (to be published).
26. D. Dill and J. L. Dehmer, Phys. Rev. A 16, 1423 (1977).
27. J. Siegel, J. L. Dehmer, and D. Dill, Phys. Rev. A, to be published.
28. J. L. Dehmer, J. Siegel, J. Welch, and D. Dill, Phys. Rev. A, to be published.
29. M. G. Lynch, D. Dill, J. Siegel, and J. L. Dehmer, J. Chem. Phys., to be published.
30. J. Welch, D. Dill, J. Siegel, and J. L. Dehmer, to be published.
31. J. L. Dehmer, J. Siegel, and Dan Dill, J. Chem. Phys. 69, 5205 (1978).
32. See, e.g., (a) M. Krauss and F. H. Mies, Phys. Rev. A 1, 1592 (1970); (b) D. T. Birtwistle and A. Herzenberg, J. Phys. B 4, 53 (1971); (c) P. G. Burke and N. Chandra, J. Phys. B 5, 1696 (1972); (d) B. D. Buckley and P. G. Burke, J. Phys. B 10, 725 (1977); (e) M. A. Morrison and B. I. Schneider, Phys. Rev. A 16, 1003 (1977); (f) M. A. Morrison and L. A. Collins, Phys. Rev. A 17, 918 (1978); (g) A. W. Fliflet, D. A. Levin, M. Ma, and V. McKoy, Phys. Rev. A. 17, 160 (1978).
33. R. E. Kennerly, Phys. Rev. A, to be published.
34. D. E. Golden, Phys. Rev. Lett 17, 847 (1966).
35. J. P. Bromberg, J. Chem. Phys. 52, 1243 (1970).
36. R. D. DuBois and M. E. Rudd, J. Phys. B 9, 2657 (1976).
37. S. K. Srivastava, A. Chutjian, and S. Trajmar, J. Chem. Phys. 64, 1340 (1976).
38. D. Hermann, K. Jost, and J. Kessler, J. Chem. Phys. 64, 1 (1976).

39. D. C. Cartwright, A. Chutjian, S. Trajmar, and W. Williams, Phys. Rev. A 16 1013 (1977).
40. H. Ehrhardt and K. Willmann, Z. Phys. 204, 462 (1967).
41. T. W. Shyn, R. S. Stolarski, and G. R. Carignan, Phys. Rev. A 6, 1003 (1972).
42. T. G. Finn and J. P. Doering, J. Chem. Phys. 63, 4399 (1975).
43. Z. Pavlovic, M. J. Boness, A. Herzenberg, and G. J. Schulz, Phys. Rev. A 6, 676 (1972)
44. M. Morrison, N. Lane, and L. Collins, Phys. Rev. A 15, 2186 (1977).
45. C. Ramsauer, Ann. Phys. (Leipzig) 83, 1129 (1927).
46. E. Bruche, Ann. Phys. (Leipz) 83, 1065 (1927); see also R. B. Brode, Rev. Mod. Phys. 5, 257 (1933).
47. T. W. Shyn, W. E. Sharp, and G. R. Carignan, Phys. Rev. A 17, 1855 (1978).
48. C. Szmytkowski and M. Zubek, Chem. Phys. Lett. 57, 105 (1978).
49. M. J. W. Boness and G. J. Schulz, Phys. Rev. A 9, 1969 (1974).
50. C. R. Claydon, G. A. Segal, and H. S. Taylor, J. Chem. Phys. 52, 3387 (1970).
51. P. D. Burrow, private communication.
52. R. E. Kennerly, R. A. Bonham, and M. McMillan, J. Chem. Phys., 70, 2039 (1979).
53. J. L. Dehmer, J. Chem. Phys. 56, 4496 (1972).
54. M. G. Benedict and I. Gyemant, Int. J. Quantum Chem. 13, 597 (1978).
55. J. L. Dehmer, Physica Fennica 9(S1), 60 (1974).
56. J. L. Dehmer and D. Dill, in Proceedings of the Second International Conference on Inner-Shell Ionization Phenomena - Invited Papers, Freiburg, Germany, 29 March-2 April 1976, Ed., W. Mehlhorn and R. Brenn (Fakultät für Physik, Universität, Freiburg, 1976), p. 221.
57. S. Wallace and D. Dill, Phys. Rev. B 17, 1692 (1978).
58. J. W. Davenport. Ph.D. Thesis, University of Pennsylvania (1976).
59. C. L. Allyn, T. Gustafsson, and E. W. Plummer, Chem. Phys. Lett. 47, 127 (1977).
60. E. W. Plummer, T. Gustafsson, W. Gudat, and D. E. Eastman, Phys. Rev. A 15, 2339 (1977).
61. H. Levinson, T. Gustafsson, and P. Soven, Phys. Rev. A, to be published.
62. G. R. Wight, C. E. Brion, and M. J. Van der Wiel, J. Electron Spectrosc. 1, 457 (1972/1973). A more quantitative measurement has been published recently in ref. 14.
63. G. R. Wight and C. E. Brion, J. Electron Spectrosc. 4, 313 (1974).
64. D. Dill, S. Wallace, J. Siegel, and J. L. Dehmer, Phys. Rev. Lett. 41, 1230 (1978); 42, 411 (1979).
65. P. Bagus and H. Schaeffer, J. Chem. Phys. 56, 224 (1972).

DISCUSSION

Rumble: I would like some clarification about the form of the potential you use in this model. What form does the potential take in the interstitial region - what you call Region II?

Dehmer: The potential is assumed to be constant in that region.

Rumble: How do you choose that constant?

Dehmer: It is the volume average of the potential taken over the surface of the spheres adjoining Regions I and II.

Rumble: Is it joined smoothly ....?

Dehmer: No, there is a discontinuity in the potential and if you think about it geometrically you find that must be the case.

Truhlar: What's the biggest value of the discontinuity at any point, just to get some idea of the magnitudes involved.

Dehmer: It's probably about one to two electron volts.

Truhlar: How do you decide how big to make the spheres? Is the calculation roughly independent of that?

Dehmer: The calculation is roughly insensitive to that. We nestle the spheres so as to minimize Region II and we usually apportion the sizes using covalent bond lengths or something like that.

Nesbet: What is the form of the polarization potential you use?

Dehmer: We use a simple adiabatic polarization potential $^{\alpha}/r^{4}$, where α is the spherical average of the dipole polarizability of the molecule.

Nesbet: How do you cut this potential off?

Dehmer: We only use a polarization potential in Region III.

Nesbet: Then the outer sphere boundary has the same status as the parameter used in cutting off the polarization potential and we know the results are very sensitive to that.

Dehmer: Our results are not.

Morrison: That's a point of some confusion.

Dehmer: The thought has never crossed our minds - to choose the geometry to make something fit. The geometry is always chosen to be close-packing spheres.

MOLECULAR RESONANCE PHENOMENA

J. N. Bardsley

Physics & Astronomy Department
University of Pittsburgh
Pittsburgh, Pennsylvania 15260

I. INTRODUCTION

Resonant scattering theory has been developed over more than fifty years. Its application in molecular physics began in earnest in 1962[1] although there were earlier treatments of predissociation,[2] dissociative recombination[3] and electron attachment[4] that contained some elements of the theory. There are two features of the theory as applied in atomic and molecular physics that are not so important in other fields. Firstly, since the non-relativistic Hamiltonian is known, one should be able to calculate the resonance parameters, such as the position and width, from first principles and not treat them solely as empirical parameters. Secondly, in studies of molecular resonant states one would like to incorporate the Born-Oppenheimer separation of electronic and nuclear motion into the definition of the resonances and the computational algorithms.

A brief summary of the formalism required for resonant electron-molecule collisions will be presented in Section II. It will be seen that the Born-Oppenheimer separation cannot be incorporated in a uniform manner in all applications. The techniques available for the computation of resonance parameters will be described briefly in Section III. Brevity is appropriate since few complete calculations have been performed and two of the most promising techniques will be described in accompanying papers by Hazi and McCurdy. Finally Section IV will contain a status report on work on electron collisions with some specific diatomic molecules.

Because of the limitations of time, the scope of this talk will be limited. There will be no discussion of polyatomic molecules and few comments on electron collisions with ions. The important role of quasi-bound states of molecules in heavy particle collisions will be ignored. The primary goals will be to explore the frontiers of our knowledge and to point out what is not understood, rather than to catalogue the successful applications of the theory. A comprehensive review of theoretical and experimental work before 1973 has been given by Schulz.[5]

II. RESONANT SCATTERING THEORY FOR ELECTRON-MOLECULE COLLISIONS

The traditional approaches in resonant theories of electron-molecule scattering have been through the Kapur-Peierls and Siegert theories[6] and the Fesbach formalism.[7,8] The Kapur-Peierls formalism is a version of R-matrix theory using complex boundary conditions. The Siegert theory, based on earlier ideas of Gamow, allows one to obtain results that are independent of the size of the box that defines the internal region. We will see below that by appropriate techniques of analytic continuation the box can be allowed to become infinite in size, provided that one is concerned only with the properties of isolated resonances. The Feshbach approach has been useful formally but its implementation is difficult in molecular problems.

This presentation will involve a generalization of Fano's theory of configuration interaction.[9,10] In application to electron-atom scattering the resonant width Γ is expressed by the golden-rule formula,

$$\Gamma = 2\pi \; |V(E)|^2 \tag{1}$$

where V(E) is a matrix element between a discrete representation of the quasi-bound state and a continuum function ψ_E that describes the non-resonant scattering,

$$V(E) = \langle\phi_d|H-E|\psi_E\rangle \; . \tag{2}$$

The continuum functions ψ_E are normalized according to

$$\langle\psi_E|\psi_{E'}\rangle = \delta(E-E') \; , \tag{3}$$

which implies an asymptotic form

$$\psi_E \approx \frac{C}{k^{\frac{1}{2}}} \sin\,(kr + \eta) \; . \tag{4}$$

The scattering wave function Φ_E is expanded as

$$\Phi_E = a\phi_d + \int dE' \; b(E') \; \psi_E \; . \tag{5}$$

Diagonalization of the Hamiltonian within the space spanned by ϕ_d and the $\psi_{E'}$ leads to linear equations for a and b(E') which have a unique solution consistent with the appropriate boundary conditions.

To apply this theory to electron-molecule collisions let us replace ϕ_d by an electronic wave function $\phi_d(q,R)$ which describes the resonant state within the Born-Oppenheimer framework. The symbol q is used to denote all of the electron coordinates, while R is the internuclear separation. The continuum function ψ_E is replaced by a set of functions $\overline{\psi}_{vE}(R)$ that describe the non-resonant scattering of an electron by a molecule in the vibrational state v. The expansion of the complete scattering wave function $\Phi_{vE}(q,R)$ then takes the form

$$\Phi_{vE}(q,R) = \phi_d(q,R)\ \xi_d(R) + \sum_{v'} \int dE'\ b_{v'}(E')\ \overline{\psi}_{v'E'}(q,R) \tag{6}$$

The constant a has now become a function $\xi_d(R)$ which describes the motion of the nuclei during the lifetime of the resonance. The aim of the subsequent development will be to obtain a differential equation for $\xi_d(R)$.

The most immediate generalization of Eq. (2) is obtained by defining matrix elements of the electronic Hamiltonian $H_{el}(q,R)$ by

$$\chi_{vE}(R) = \int dq\ \phi_d^*(q,R)(H_{el}-E)\ \overline{\psi}_{vE}(q,R) \quad . \tag{7}$$

Note that $\chi_{vE}(R)$ is a matrix element between an electronic function $\phi_d(q,R)$ and a scattering function which allows for nuclear motion. Diagonalization of the full Hamiltonian, with imposition of the boundary condition that there are incoming waves only corresponding to the initial vibrational state v, then gives the equation

$$\left\{-\frac{\hbar^2}{2M}\nabla_R^2 + E_d(R) - E\right\}\xi_d(R) = \chi_{vE}(R)$$

$$+ \sum_{v'} \int dE'\ \frac{1}{E-E'+i\varepsilon}\ \chi_{v'E'}(R) \int dR'\ \chi^*_{v'E'}(R')\ \xi_d(R') \tag{8}$$

in which M is the reduced mass of the nuclei and $E_d(R)$ is the expectation value of the electronic Hamiltonian in the state $\phi_d(q,R)$.

The terms on the R.H.S. of eq. (8), which are not present in the corresponding nuclear wave equation for bound molecular states, represent the formation of the resonance through electron capture and its decay through electron emission. The decay term has the form of a complex non-local potential.

Eq. (8) incorporates one desired feature of the Born-Oppenheimer concept in that the electronic energy $E_d(R)$ acts as a potential for the nuclear motion. However in many applications it is reasonable to assume that the probability of electron capture and the rate of electron emission, although functions of R, are independent of the motion of the nuclei. Some simplifying assumptions are needed for this to be apparent in Eq. (8).

Let us introduce a local wave number k(R) at which the resonance would occur if electrons were scattered by molecules with the nuclei fixed at separation R, by setting

$$\frac{\hbar^2}{2M_e} k^2(R) = E_d(R) - E_o(R) \tag{9}$$

where $E_o(R)$ is the electronic energy of the target molecule. For simplicity we will assume that the energy is below the threshold for electronic excitation of the target, so that we need define only one such wave number. Let us now assume that we can write

$$\frac{1}{f(k_v)} \chi_{vE}(R) = \frac{V(R)}{f(k(R))} \zeta_v(R) \tag{10}$$

where $\zeta_v(R)$ is the wave function describing the nuclear motion in the target molecule and V(R) is a purely electronic matrix element analogous to that used for atomic resonances; f(k) is some simple function that will be discussed below.

In this approximation we are assuming that the matrix element $\chi_{vE}(R)$, which describes the coupling between the quasi-bound state and one of the degenerate non-resonant continua, can be expressed as a product of an electronic matrix element and the target nuclear functions. The presence of the wave numbers k_v and k(R) in Eq. (10) attests to the importance in the decay rate of the energy available to the escaping electron. With this approximation the decay term can be simplified, using the completeness of the functions $\zeta_v(R)$, to give a complex local potential which can be transferred to the L.H.S. of the equation, to give

$$\left\{-\frac{\hbar^2}{2M}\frac{d^2}{dR^2} + \tilde{E}_d(R) - \frac{i}{2}\Gamma(R) - E\right\} \xi_d(R) = \frac{f(k_v)}{f(k(R))} V(R)\, \zeta_v(R) \tag{11a}$$

where

$$\Gamma(R) = 2\pi |V(R)|^2 \tag{11b}$$

and where $\tilde{E}_d(R)$ contains a small shift given by the usual principal part integral.

Two of the major applications of resonant scattering theory are in the calculation of the cross sections for dissociative attachment and vibrational excitation. The dissociative attachment cross section can be expressed as

$$\sigma_{DA} = \frac{\pi^2}{k_v^2} g \frac{K}{M} \lim_{R\to\infty} |\xi_d(R)|^2 \tag{12}$$

in which K is the wave number corresponding to the motion of the dissociation fragments in the center-of-mass frame and g is the ratio of the statistical weights of the resonant state and the target electronic state. The vibrational excitation cross section is best described in terms of the Green function G(R,R') appropirate to the operator on the LHS of Eq. (11a). It is

$$\sigma_{v\to v'} = \frac{2\pi^3}{k_v^2} g\, f(k_v)\, f(k_{v'})$$

$$\times \left| \int\int dR dR' \, \zeta^*_{v'}(R) \frac{V^*(R)}{f^*(k(R))} G(R,R') \frac{V(R')}{f(k(R'))} \zeta_v(R') \right|^2 . \tag{13}$$

From Eq. (13) it is clear that detailed balance is guaranteed by the resonance formalism. However to ensure the correct behavior of the vibrational excitation cross section near threshold we must specify the function f(k). Let us suppose that the lowest partial wave that contributes to resonance formation has angular momentum ℓ. Then Wigner's threshold laws for electron collisions with neutral molecules can be satisfied by taking f(k) to be $k^{\ell+\frac{1}{2}}$ or a barrier penetration factor. However for resonances with very large widths calculations suggest that if this is done then the elastic scattering cross sections may violate the unitarity restrictions.

If one wishes to determine the absolute magnitude of the resonant scattering cross sections and to consider elastic scattering, and is not concerned about cross sections near threshold, then one should take f(k) to be identically one, as is done by Dube and Herzerberg.[11] For calculations on vibrational excitation in which the threshold region is important I would recommend that f(k) be taken as $k^{\frac{1}{2}}$, to give zero threshold values for the cross sections while minimizing the likelihood of unitarity violation. This choice was made in most previous applications.[6,12] However in most applications the cross sections away from threshold should not be very sensitive to the choice of f(k). For electron-ion collisions f(k) should always be taken to be one.

For shape resonances Eq. (13) can be used with V(R) calculated from the width Γ(R), using Eq. (11b). The width is determined by a barrier penetration factor and so may depend strongly on the energy available to the escaping electron. Several authors have suggested that in determining the partial decay rates to each final vibrational state one should explicitly allow for the changes in the barrier penetration probabilities that arise from the different electron wave numbers in each channel. This can be accomplished within this formulation by taking f(k) to be the barrier penetration factor and setting

$$V(R) = \sqrt{\frac{\Gamma(R)}{2\pi}} = Cf(k(R)) \qquad (14)$$

The formalism described above can be applied to all resonances in which the potential curve for the negative ion lies well above that of the neutral molecule near to its equilibrium separation R_e. However the ground states of NO^- and O_2^- lead to shape resonances at small R but are electronically stable at large R, with crossing points R_c that are close to R_e. In studies of vibrational excitation of NO and O_2 by Koike[13] and by Fiquet-Fayard and collaborators,[14,15] the energy dependence of the partial decay rates was treated in the manner just described. However no account was taken of the fact that within the Born-Oppenheimer picture electron emission can occur only for $R < R_c$.

When the potential curve for the quasi-stationary state lies below that of its parent for all R, the decay can clearly not be described in terms of the Born-Oppenheimer picture. So to describe the autoionization of vibrationally or rotationally excited Rydberg states one must compute the rate of transfer of energy from nuclear to electronic motion. Significant progress has been made in this area in recent years[16-18] and it has been shown that the autoionization rates can be expressed simply in terms of the quantum defects of the Rydberg states. The rate of decay due to transfer of vibrational energy depends on the R-derivative of the quantum defects whereas that for rotational energy transfer is determined by the difference in the quantum defects for Rydberg states with different angular momentum projections. Thus even in cases where the decay is due to Born-Oppenheimer breakdown the decay rates can be deduced from calculations made within the B-O framework.

For further discussion of the relative merits of local, non-local and R-independent decay rates the reader should consult the review by Fiquet-Fayard.[19]

III. The Computation of Resonance Parameters

Potential curves for resonant states of molecules can be calculated, with a reasonably high confidence level, by the stabilization method[20] or, in a limited number of cases, by the use of projection operators based on symmetry operators or on a one-electron model. The computation of resonance widths is much more difficult and in this section we will review the available techniques that give the width or, preferably, the energy and the width simultaneously. These can be loosely classified as

A. Golden rule formula
B. Variational calculations of complex energies
C. Fits to the results of complete scattering calculations.

In the methods of class A, Eq. (11b) is used with states that are defined within the Fano or Feshbach formalisms. Examples of this approach are the calculations by Bottcher and Docken[21] on doubly excited states of H_2 and by Pearson and Lefebvre-Brion[22] for the 10.04 eV resonance in e-CO scattering.

The major difficulty in performing such calculations arises in the generation of the continuum wave functions. An exciting new development in this regard involves the use of Stieltjes-imaging techniques to replace the continuum by a set of discrete functions.[23] This approach, which is described by Hazi in an accompanying paper, permits the use of the powerful integral packages that have been developed for bound state problems.

The earliest example of the methods of group B was the Siegert state calculation on H_2^- by Bardsley, Herzenberg and Mandl[6] in which a complex energy was obtained by imposing outgoing boundary conditions on a finite surface. This technique lay dormant for many years, partly because of computational difficulties and partly due to protestations that molecules do not live in boxes - R matrix devotees take note. For finite range interactions the Siegert eigenvalues should be independent of the size of the internal region provided that it extends beyond the range of the interaction. The boundary can then be allowed to approach infinity without serious problems. However the existence of long range forces leads to divergent integrals if this procedure is followed for most atomic and molecular problems.

The divergent behavior of Siegert state wave functions for atoms can be handled through the use of coordinate rotation,[24] in which all electronic coordinates r_i are replaced by $r_i e^{i\alpha}$. However in molecular problems computational difficulties arise from the Born-Oppenheimer separation, since one would rather not be forced into using complex values for the internuclear separation. This means that one cannot easily apply a uniform rotation to each electron-nucleus coordinate vector.

Recent progress using these techniques will be described by McCurdy.

When close-coupling, R-matrix or algebraic techniques for calculating the scattering matrices appropriate to electron-molecule collisions are fully developed, it will be possible to derive resonance parameters. This approach should soon lead to useful results for e-H_2^+ collisions, but improved treatment of electronic distortion in the target will be required before reliable results can be obtained for other molecules.

IV. Examples From Electron Collisions With Neutral Molecules

A. $\underline{H_2}$:

The ground $^2\Sigma_u$ state of H_2^- is unstable against electron emission for $R \lesssim 3$ a.u. and leads to a shape resonance that is so wide that its qualifications as a resonance have been questioned. However ab initio calculations have confirmed that vibrational and rotational excitation arise predominantly from the Σ_u portion of the scattering wave function. Although these excitation processes can be described by non-resonant theories, no alternative theory of dissociative attachment has yet been applied.

The value of resonant scattering theory for such a short-lived state is being tested through a study of the dependence of the dissociative attachment cross section near threshold upon the initial rotational or vibrational state of the H_2 molecule. This investigation was prompted by the observation by Nicolopoulou et al[25] of an anomalously large number of H^- ions in a low pressure hydrogen plasma. Experiments by Allen and Wong[26] and calculations by Wadehra and Bardsley[27] have shown that the threshold cross section for dissociative attachment is enhanced by a factor of the order of 30 for H_2 molecules with v = 1 compared with that of ground state molecules. For D_2 the ratio is higher (~40) even though the excitation energy is less. Significant enhancement is also found from rotational excitation, as was originally suggested by Chen and Peacher.

The first excited state (B $^2\Sigma_g$) of H_2^- has a repulsive potential curve and appears to lie above the repulsive b $^3\Sigma_u$ state of H_2 for a broad range of R-values ($1.4 < R < 5.0$). Its effect can be seen in dissociative attachment near 10 eV and in the excitation of high vibrational levels in H_2 and D_2.[28] However even at 10 eV vibrational excitation to the v=1 level is dominated by the lower X $^2\Sigma_u$ resonance. Wadehra and Bardsley have attempted to analyze all of the available scattering data in terms of the contributions from these two resonances. They have not yet been successful in finding a single theoretical model that describes all of the data. A comprehensive investigation of these phenomena, involving both theorists and experimenters, would be worthwhile at this time.

Further theoretical information [29,30] on the resonance between 11 and 12 eV has become available and it appears that the most prominent features arise from a predissociating state A $^2\Sigma_g$ and the C $^2\Pi_u$ state. It has been suggested that the structure at 11.19 eV that has been attributed to the Π_u resonance may instead be a threshold feature associated with the B $^1\Sigma_u$ state of H_2. Further study of the predissociation of the A $^2\Sigma_g$ state is warranted.

One unresolved question regarding H_2^- concerns the possible existence of a state that is stable against electron emission within the non-relativisitic theory. Although there is some experimental evidence in support of such a state[31] recent theoretical calculations have not produced one. The most promising candidate appears to be the $^4\Sigma_g^-(1\sigma_g)(1\pi_u)^2$ state, which would have a long lifetime if it were below the c $^3\Pi_u$ state of H_2.

B. N_2:

Although resonant scattering of electrons by N_2 at energies close to 2 eV has been studied for many years it remains of interest due to the attempts that are underway to calculate the resonance parameters in ab initio calculations for comparison with the values that have been deduced from semi-empirical analyses. These attempts have so far met with little success. Furthermore, due to the theoretical work of Birtwistle, Dube and Herzerberg[11,12] and the excellent experimental work performed by several groups, one can compare theory and experiment at a much deeper level for N_2 than for other molecules, and thus develop more stringent tests of the theory. Two particular areas of current concern are the absolute values of the total cross section around 2 eV, and the values of the vibrational excitation cross sections between 2.5 eV and 3.5 eV. In regard to the total cross section one needs reliable estimates of the non-resonant contributions to the elastic cross section. A careful comparison between theory and experiment on inelastic scattering at energies near the top of the resonant region, particularly with respect to the excitation of very high vibrational levels, might provide information regarding the relative merits of the different resonant models described above.

Mazeau et al[32] have studied the differential cross sections for excitation of the A $^3\Sigma_u^+$ and B $^3\Pi_g$ states of N_2 with electron energies between 7.5 eV and 13 eV. They suggest that six resonances are required to explain their results. In order to help experimentalists to interpret this type of data a set of angular distribution formulae appropirate to the most common inelastic electron transitions would be useful. I do not believe that this information has been collected together anywhere, although the techniques needed for calculations of the distributions have long been available.

Ab initio calculations by Thulstrup and Andersen[33] suggest the possible presence of $^2\Pi_u$, $^2\Delta_g$, $^2\Sigma_u^+$ and $^2\Sigma_g^+$ resonances in this area.

Although there is evidence of a $^2\Delta_g$ resonance at 9.8 eV in both excitation channels and of a $^2\Pi_u$ resonance in the A channel at 8.6 eV, the identification of these resonances is far from complete.

C. NO:

Vibrational excitation of NO at energies below 2 eV has been analyzed by Koike[13] and by Teillet-Billy and Fiquet-Fayard.[15] They have both derived information on the potential energy curve and partial decay rates for the ground state of NO^- from the observed excitation cross sections. Their models differ mainly in the amount of p and d wave mixing in the resonant wave function. Teillet-Billy and Fiquet-Fayard[15] deduce a value for R_e in NO^- that is slightly higher than that derived by Siegel et al.[34] from photodetachment data. They deduced the widths by fitting to the absolute values of the measured cross sections, obtaining values that are significantly smaller than the apparent widths of the experimental peaks. On the other hand Koike assumed R_e for NO^- to have the value deduced by Siegel et al. and fitted the width to the breadth of the observed peaks.

Although some of the differences between the two models could be tested, in principle, by examination of the angular distributions of the scattered electrons, the experimental results of Tronc et al.[35] appear at first sight to agree with neither model. This experiment supports the existence of three resonant states of NO^- below 2 eV. Clearly more work is needed.

D. O_2^-:

The low energy vibrational excitation has been analysed by Koike[13] and by Parlant and Fiquet-Fayard.[14] Both analyses suggest that R_e for O_2^- is greater than the value deduced from photodetachment data by Celotta et al.[36] However the discrepancy is only 10% of the difference in R_e between O_2 and O_2^-. The spin-orbit splitting within the ground $^2\Pi_g$ state of O_2^- is about 20 meV. The mean energy of the (v=4, J=0) levels is approximately 80 meV about the ground state of O_2, and the lifetime is deduced to be $\sim 8 \times 10^{-11}$ secs ($\Gamma \approx 8\ \mu eV$). The lifetime agrees well with the value of $(1.0 \pm 0.3) \times 10^{-10}$s derived by Shimamori and Hatano[37] from measurements of the pressure at which the three-body attachment of electrons to O_2 molecules exhibits saturation. This agreement confirms that the lowest resonance level plays a dominant role in three-body attachment.

A remarkable experiment concerning e-O_2 interactions at thermal energies was performed by Bartels[38] who measured the attachment frequency of electrons at 77°K in mixtures of oxygen and helium for helium densities up to 7×10^{21} cm^{-3}. A plot of attachment rate versus number density shows a strong sharp peak with $N_{He} \approx 3 \times 10^{21}$ cm^3. This phenomenon can be explained as being due to a pressure shift. However if the energy of electrons moving freely through the helium gas and that of the O_2^- ions are shifted by equal amounts there will be no change in the effective position of the resonance.

Let us assume that the two important effects are the polarization of the dense He gas and the Fermi interaction[39] arising from close encounters of the electrons with the He atoms. The He atoms will clearly be polarized both by free electrons and by O_2^- molecules. For low energy continuum electrons the Fermi interaction leads to a positive energy shift which is proportional to the e-He scattering length. However this interaction is momentum-dependent and is unlikely to be so strong for electrons which are bound in negative ions. If one assumes that there is no Fermi shift for the O_2^- ions one finds that at densities of 3×10^{21} cm^{-3} the v=4 level of O_2^- becomes accessible as a resonance to "zero energy" continuum electrons. At higher densities it becomes stable whereas at lower densities the resonance is too high to be formed by electrons at 77^0K.

In e-O_2 scattering at around 7 eV there appears to be a $^2\Pi_g$ and a $^4\Sigma_u^-$ resonance.[40] There is a large body of experimental data in this energy range on vibrational excitation,[41] electronic excitation[41,42] and dissociative attachment to the ground state[43] and excited[44] O_2 molecules. However there has been little theoretical analysis of this data since O'Malley's excellent work on the temperature dependence of dissociative attachment.[45]

E. HCl:

The discovery of large threshold structures[46] in the vibrational excitation cross sections for HF, HCl and HBr has attracted considerable theoretical attention. Dube and Herzenberg[47] have constructed an R-matrix model to explore the effects of the dipole field upon vibrational excitation. The short range interaction is described through the imposition of an R-dependent logarithmic boundary condition on the surface of the inner region. In the outer region the electron moves in a dipole potential with a strength that varies with R. Dube and Herzenberg were able to reproduce the essential features of the threshold peaks using reasonable values of the adjustable parameters in the logarithmic boundary condition. They show that the scattering wave function has maximum density on the hydrogen side of the molecule and has a large s-wave component, as is indicated by the observed differential cross sections.

Goldstein, Segal and Whetmore[48] have calculated potential curves for HCl^- using the stabilization method. They find six states that have been related to the experimental scattering data by Taylor et al.[49] The most controversial aspect of their calculation is the prediction of two $^2\Sigma^+$ states with potential curves that are parallel to, and just above, the ground state of HCl near its equilibrium separation. It seems clear that there is considerable localization in the wave functions for low energy e-HCl scattering that leads to time delay in the collisions. However it is not obvious that one can pick specific energies from the continuum at which this localization is maximum and so construct meaningful negative ion potential curves. Nesbet[50] has suggested that the threshold phenomena may be caused by a virtual state. This would imply that the localization and time delay are greatest at threshold and that the corresponding HCl^- potential curve should be below that of HCl, but on an unphysical energy sheet.

Another interesting feature of these calcultions is the avoided crossing between the 4 $^2\Sigma^+$ and 5 $^2\Sigma^+$ states which could explain the anomalous effective threshold for H^- production at 6.9 eV. Further investigation is warranted to see whether the inner minimum in these curves should lead to any observalbe structure in the scattering cross sections.

F. F_2, Cl_2, and I_2:

The development of rare-gas halide lasers has lead to renewed interest in inelastic electron collisions with halogen molecules. There has been a need for direct measurements and calculations of collision cross sections and for study of the negative ion potential curves through photodetachment experiments and ab-initio or pseudo-potential calculations. Semi-empirical analyses of the experimental data on dissociative attachment have been carried out by Shipsey[51] and Hall.[52] More work is required as more accurate data becomes available, both to check the consistency of the experimental results and to provide information on cross sections that have not yet been measured. Since many of the experiments involve swarms of electrons, analyses of the electronic speed distributions would be helpful in the interpretation of the measured reaction rates.

V. Summary and Conclusions

Most of the success of resonant scattering theory has been through semi-empirical analyses in which the resonance parameters are chosen to fit the most accurate experimental information and used to predict cross sections for which the measurements are less reliable or have not been performed. The above examples were chosen to show that there is much work still to be done in this area, even for diatomic molecules. Some successful analyses have been carried out for polyatomic molecules that are not described above. In this context the collaboration of physicists and chemists is most desirable.

The remainder of this workshop should indicate how close we are to being able to perform useful ab initio calculations of resonance parameters. My own belief is that we will see significant progress in the next year or so on a few diatomic molecules, but that it will be many years before the major value of the theorist in this field comes from his ability to do ab initio calculations.

In concluding, I would like to acknowledge the support of the Advanced Research Projects Agency (Contract No. NOOO-14-76-C-0098) and the National Science Foundation, Theoretical Physics Section (PHYS 76-21456).

REFERENCES

1. A. Herzenberg and F. Mandl, Proc. Roy. Soc. A270, 48 (1962).
2. O. K. Rice, J. Chem. Phys. 1, 375 (1933).
3. D. R. Bates, Phys. Rev. 78, 492 (1950).
4. T. Holstein, Phys. Rev. 84, 1073 (1951): F. Bloch and N. Bradbury, Phys. Rev. 48, 689 (1935).
5. G. J. Schulz, Rev. Mod. Phys. 45, 423 (1973).
6. J. N. Bardsley, F. Mandl and A. Herzenberg, Proc. Phys. Soc. 89, 305, 321 (1966).
7. J. C. Y. Chen, Phys. Rev. 148, 66 (1966); 156, 12 (1967).
8. T. F. O'Malley, Phys. Rev. 150, 14 (1966); 156, 230 (1967).
9. U. Fano, Phys. Rev. 124, 1866 (1961).
10. J. N. Bardsley, J. Phys. B 1, 349 (1968).
11. L. Dube and A. Herzenberg, Phys. Rev. A (1979) (accepted).
12. D. T. Birtwistle and A. Herzenberg, J. Phys. B 4, 53 (1970).
13. F. Koike, J. Phys. Soc. Japan 39, 1590 (1975).
14. G. Parlant and F. Fiquet-Fayard, J. Phys. B 9, 1617 (1976).
15. D. Teillet-Billy and F. Fiquet-Fayard, J. Phys. B 10, L111 (1977).
16. G. Herzberg and C. Jungen, J. Mol. Spectrosc. 41, 425 (1972).
17. O. Atabek, D. Dill and C. Jungen, Phys. Rev. Lett. 33, 123 (1974): O. Atabek and C. Jungen, in Electron and Photon Interactions with Atoms, eds., H. Kleinpoppen and M.R.C. McDowell (Plenum, New York, 1975).
18. P. M. Dehmer and W. A. Chupka, J. Chem. Phys. 65, 2243 (1976).
19. F. Fiquet-Fayard, Vacuum 24, 533 (1974).
20. H. S. Taylor and A. U. Hazi, Phys. Rev. A 14, 2071 (1976).
21. C. Bottcher and K. Docken, J. Phys. B 7, L5 (1974).
22. P. K. Pearson and H. Lefebvre-Brion, Phys. Rev. A 13, 2106 (1976).
23. A. U. Hazi, J. Phys. B 11, L259 (1978).
24. J. N. Bardsley, Int. J. Quant. Chem. 14, 343 (1978), and following papers.
25. E. Nicolopoulou, M. Bacal and H. J. Doucet, Journal de Physique 38, 1399 (1977).
26. M. Allan and S. F. Wong, Phys. Rev. Lett. 41, 1791 (1979).
27. J. M. Wadehra and J. N. Bardsley, Phys. Rev. Lett. 41, 1795 (1979).
28. R. Hall, in Invited Papers and Progress Reports, ICPEAC X (North Holland, Amsterdam, 1978), p. 25.
29. B. D. Buckley and C. Bottcher, J. Phys. B 10, L635 (1977).
30. J. N. Bardsley and J. S. Cohen, J. Phys. B 11, 3645 (1978).
31. W. Aberth, R. Schnitzer and M. Anbar, Phys. Rev. Lett. 34, 1600 (1975).

32. J. Mazeau, F. Gresteau, R. I. Hall, G. Joyez and J. Reinhardt, J. Phys. B 6, 862 (1973).
33. A. Thulstrup and A. Andersen, J. Phys. B 8, 965 (1975).
34. M. W. Siegel, R. J. Celotta, J. L. Hall, J. Levine and R. A. Bennett, Phys. Rev. A 6, 607 (1972).
35. M. Tronc, A. Heutz, M. Landau, F. Pichou and J. Reinhardt, J. Phys. B 8, 1160 (1975).
36. R. J. Celotta, R. A. Bennett, J. L. Hall, M. W. Siegel and J. Levine, Phys. Rev. A 6, 631 (1972).
37. H. Shimamori and Y. Hatano, Chem. Phys. 21, 187 (1977).
38. A. Bartels, Phys. Lett. 45A, 491 (1973).
39. E. Fermi, Nuovo Cim. 11, 157 (1934); H. Margenau and W. W. Watson, Rev. Mod. Phys. 8, 22 (1936).
40. M. Krauss, D. Neumann, A. C. Wahl, G. Das and W. Zemke, Phys. Rev. A 7, 69 (1973): G. Das, A. C. Wahl, W. T. Zemke and W. C. Stwalley, J. Chem. Phys. 68, 4252 (1978).
41. S. F. Wong, M.J.W. Boness and G. J. Schulz, Phys. Rev. Lett. 31, 969 (1973).
42. S. Trajmar, D. C. Cartwright and W. Williams, Phys. Rev. A4, 1482 (1971).
43. W. R. Henderson, W. L. Fite and R. T. Brackmann, Phys. Rev. 183, 157 (1969): D. R. Spence and G. J. Schulz, Phys. Rev. 188, 280 (1969): P. J. Chantry and G.J. Schulz, Phys. Rev. 156, 134 (1967): R. J. Van Brunt and L. J. Kieffer, Phys. Rev. A 2, 1899 (1970).
44. P. D. Burrow, J. Chem. Phys. 51, 4922 (1973).
45. T. F. O'Malley, Phys. Rev. 155, 59 (1967).
46. K. Rohr and F. Linder, J. Phys. B 8, L200 (1975); J. Phys. B 9, 2521 (1976); K. Rohr, J. Phys. B 10, L399 (1977).
47. L. Dube and A. Herzenberg, Phys. Rev. Lett. 38, 820 (1977).
48. E. Goldstein, G. A. Segal and R. W. Wetmore, J. Chem. Phys. 68, 271 (1978).
49. H. S. Taylor, E. Goldstein and G. A. Segal, J. Phys. B 10, 2253 (1977).
50. R. K. Nesbet, J. Phys. B 10, L739 (1977).
51. E. J. Shipsey, J. Chem. Phys. 52, 2274 (1970).
52. R. J. Hall, J. Chem. Phys. 68, 1803 (1978).

STIELTJES-MOMENT-THEORY TECHNIQUE FOR CALCULATING RESONANCE WIDTHS

A. U. Hazi

Theoretical Atomic and Molecular Physics Group
Lawrence Livermore Laboratory
University of California, Livermore, CA 94550

I. INTRODUCTION

Resonant autoionizing states of molecules and autodetaching states of molecular negative ions play important roles in many collision phenomena involving low energy electrons, e.g. dissociative attachment, associative and Penning ionization, dissociative recombination, etc.[1,2] Although the formal theory[3] of resonant scattering of electrons from molecules has been developed during the past 15 years, there have been very few quantitive studies of resonant phenomena even in diatomic molecules. The main reason for this has been the lack of practical methods for calculating the width of molecular resonances within the Born-Oppenheimer approximation.[4] This situation has prevailed for quite some time, even though methods for calculating the resonance parameters of atoms[5] and the potential energy curves of molecular resonances[6] have been available.

In this talk, I will review a recently developed method[7] for calculating the widths of atomic and molecular resonances. The method is based on the golden-rule definition of the resonance width, $\Gamma(E)$.[8] The method uses only square-integrable, L^2, basis functions to describe both the resonant and the non-resonant parts of the scattering wavefunction. It employs Stieltjes-moment-theory techniques[9] to extract a continuous approximation for the width $\Gamma(E)$ from a discrete representation of the background continuum. Since the method utilizes L^2 basis functions exclusively, it has several advantages. Its implementation requires only existing atomic and molecular structure codes. Many electron effects, such as correlation and polarization, are easily incorporated into the calculation of the width via configuration interaction techniques.

Once the width, $\Gamma(E)$, has been determined, the energy shift[8] can be computed by a straightforward evaluation of required principal value integrals. The main disadvantage of the method is that it provides only the total width of a resonance which decays into more than one channel in a multi-channel problem.

The remainder of the talk will be divided into two parts: a review of the various aspects of the theory (Section II) and a discussion of representative results which have been obtained with this method for several atomic and molecular resonances (Section III). Finally, I will give a brief summary (Section IV).

II. THEORY

A. General Remarks

In Fesbach's theory of resonances,[8] the width of an isolated resonance which decays into a single open chnnel is defined according to the "golden-rule" formula:

$$\Gamma(E) = 2\pi \left| \langle \phi_r (H - E) \psi_E^+ \rangle \right|^2 \tag{1}$$

Here ϕ_r is a localized, L^2 function describing the resonance state. The function $\psi_E^\pm$ is an energy-normalized scattering function which represents the non-resonant, background continuum at energy E. The operator H is the full, many-electron Hamiltonian. For narrow resonances, the physical width is $\Gamma(E_r)$, where E_r is the resonance energy.

For molecular resonances, the resonance wavefunction ϕ_r can be computed using well-known techniques, e.g., the stabilization method,[6] which require only standard electronic structure codes. The calculation of the non-resonant wavefunction $\psi_E^\pm$, however, is very difficult for molecules because of the non-spherical and non-local nature of the electron-molecule interaction potential. As a result, previous ab initio calculations of molecular resonance widths utilized very simple approximations for $\psi_E^\pm$, e.g., uncoupled, undistorted Coulomb waves in the case of autoionizing states of neutral molecules like H_2[10] and HeH.[11]

One of the major advantages of the present method is that it allows for a more accurate description of the background continuum without sacrificing the computational simplicity inherent in the exclusive use of electronic structure codes. The method accomplishes this through the following essential features.

(i) The scattering wavefunction is divided into resonant and non-resonant parts using appropriately defined projection operators Q_o and P_o.

(ii) The resonance state is obtained as a discrete eigenstate of Q_oHQ_o in the usual way.

(iii) The non-resonant scattering functions are expanded in terms of L^2, many-electron basis functions.

(iv) The width matrix-elements are calculated from the "golden-rule" formula.

(v) Stieltjes-moment-theory techniques[9] are employed to extract correctly normalized widths from the discrete representation of the background continuum.

B. Calculation of the Resonant and Non-resonant Wave Functions

For a given problem, one starts by selecting a suitable, orthonormal set of one electron basis functions. The choice of the one-electron orbitals is governed by the electronic structure of the target, the expected nature of the resonance under consideration, and the symmetry of the open-channel into which the resonance decays. (The optimization of the basis in stabilization calculations has been discussed previously.[6]) From the one-electron basis, one constructs all the many-electron configurations which will be eventually included in the total wavefunction. This set defines the total space $P_o + Q_o$, which, of course, must be incomplete in practice. Hopefully, the space $P_o + Q_o$ is adequate to describe all the essential physics: the structure of the resonance, its energy and decay lifetime.

The projection operator P_o is constructed from those configurations which provide a reasonably good approximation for the decay channel, i.e., the target + scattered electron. All the other configurations are included in the subspace Q_o. This procedure is illustrated with a specific example in Table I, which shows the choice of configurations, and of the subspaces P_o and Q_o, for the $(1s\ 2s^2)\,^2S$ resonance of He^-. This resonance decays to the ground state of He, i.e., $He^-\,(^2S) \rightarrow He(^1S) + e\ (ks)$, and a large set of s-type functions must be used in order to accurately represent the scattered, s-wave electron. In this particular case, the SCF function $(1s^2)\,^1S$ provides a sufficiently reasonable approximation for the target state so that only the "static-exchange-like" configurations $(1s^2 s_n)$, $n = 1, 2, \ldots$ needed to be included in subspace P_o. The resonance wavefunction ϕ_r, and the corresponding (unshifted) energy ε_r, are obtained by diagonalizing Q_oHQ_o. For the choice of subspaces P_o and Q_o given in Table I, the $(1s\ 2s^2)$ He^- resonance corresponds to the lowest eigenvalue of Q_oHQ_o, and the function ϕ_r includes electron correlation effects in the resonance state. However, it should be recognized that in other cases it may be necessary to include configurations constructed from more than one target configuration in the subspace P_o.

Table I

Orbital Basis and Partitioning of Configurations for $He^-(1s\ 2s^2)\ ^2S$

Orbital Basis: 1s, 2s, s_1, s_2, ... p_1, p_2, ... d_1, d_2, ...

1s: SCF orbital for He $(1s^2)\ ^1S$

2s: first approximation to resonance orbit

s_n, n = 1, 2, ...:	adequate to describe both relaxation of 1s, 2s and the continuum orbital ks.
p_n, d_n, n = 1, 2, ...:	adequate to describe correlation in target and in the resonance state.

Configurations:

P_o:	1s 1s 2s	1s 1s s_n	n = 1, 2, ...	
Q_o:	1s 2s 2s	s_n 2s 2s	1s 2s s_n	1s s_n s_n' etc.
	1s p_n p_n'	2s p_n p_n'	s_n p_n' p_n''	
	1s d_n d_n'	2s d_n d_n'	s_n d_n' d_n''	
	p_n p_n' d_n''			

In order to obtain an accurate representation of the non-resonant continuum in many-electron targets, two new projection operators P and Q are defined as

$$Q = \sum_i \phi_{ri} >< \phi_{ri}$$

and

$$P = 1 - Q = P_o + (Q_o - Q)$$

The index i runs over all the resonances of a given symmetry under consideration, e.g., 2s2p, 2s3p, 3s2p $^1P^o$ in e + He^+, or $(1s2s^2)\,^2S$ in He^-, etc. It is important to note that P space contains not only those configurations which approximate the decay channel (originally in space P_o) but also the higher, non-resonant solutions of Q_oHQ_o. I have found that this repartitioning of the spaces is essential to incorporate, as fully as possible, many electron correlation and polarization effects in the description of the non-resonant continuum. In the case of many-electron targets, such effects must be accounted for if accurate widths are to be obtained. For example, in the case of the ^2S resonance of He^-, if one used only the original subspace P_o (see Table I) to approximate the background continuum, one would neglect correlation in the $(1s^2)\,^1S$ target state and would treat the ejected electron only within the static-exchange approximation. In practice, such a procedure yields a width of about 14 meV compared to a more accurate value of 11.5 - 12 meV.[7,12]

Once the projection operator P is defined, the discrete representation of the non-resonant continuum is obtained by diagonalizing PHP in the basis of all L^2 configurations which make-up the total space $P_o + Q_o$. The resulting eigenfunctions satisfy

$$< \chi_n | PHP | \chi_m > = \delta_{nm}\, \varepsilon_n \tag{2}$$

and

$$< \chi_n | \chi_m > = \delta_{nm} \tag{3}$$

All the solutions with non-zero eigenvalues are orthogonal to the resonance functions $\{\phi_{ri}\}$ by construction.

C. Calculation of the Resonance Widths

It is now recognized that the eigenvalues and the eigenfunctions which result from diagonalizing the Hamiltonian for a scattering problem in an L^2 basis will form a discrete representation of the scattering continuum. For sufficiently large basis sets, each solution χ_n approximates ψ_E^+ with $E = \varepsilon_n$ in a region near the nuclei, except for an overall normalization factor. Since the "golden-rule" formula in Equation (1) contains the bound function ϕ_r localized near the nuclei, the matrix element has no significant contribution from the asymptotic region where χ_n fails to approximate $\psi_E^{\pm}$.[13] Consequently, it is possible[14] to use $\{\chi_n\}$ to calculate the width matrix-elements:

$$\gamma_n = 2\pi |< \phi_r \; H \; \chi_n >|^2 \tag{4}$$

However, χ_n cannot be used directly to approximate $\Gamma(\varepsilon_n)$ because χ_n is unit-normalized whereas $\psi_{\varepsilon_n}^+$ is energy normalized. Some way must be found to determine the normalization constant relating χ_n and $\psi_{\varepsilon n}^+$.

The same problem occurs when one employs L^2 basis functions to calculate continuum oscillator strengths or photoionization cross sections which describe transitions from bound initial states to continuum final states. In that case the work of Langhoff and co-workers have shown that accurate photoionization cross sections can be computed from variationally constructed, discrete pseudo-spectra employing Stieltjes moment theory.[9] In the present context, the same technique is used to extract appropriately normalized resonance widths from the pseudospectrum $\{\gamma_n \varepsilon_n\}$ defined in Equations (2) - (4).

It is useful to consider the so-called cumulative function F(E) defined according to the equations:

$$F(E) = \int dE' \; \Gamma(E') \tag{5a}$$

$$\Gamma(E) = \frac{dF}{dE} \tag{5b}$$

The pseudospectrum $\{\gamma_n \varepsilon_n\}$ associated with the non-resonant solutions of PHP defines a histogram approximation to F(E):

$$\tilde{F}(E) = 0, \qquad 0 \leqslant E \leqslant \varepsilon_1 \tag{6a}$$

$$\tilde{F}(E) = \sum_{n=1}^{k} \gamma_n = \sum_{n=1}^{k} 2\pi |< \phi_r \; H\chi_n >|^2, \qquad \varepsilon_k < E < \varepsilon_{k+1} \tag{6b}$$

At the rise points, $E = \varepsilon_k$, the cumulative function is approximated by

$$\tilde{F}(E) = \frac{1}{2} [\tilde{F}(\varepsilon_k -) + F(\varepsilon_k +)]$$

$$= \sum_{n=1}^{k-1} \gamma_n + \frac{1}{2} \gamma_k' \qquad E = \varepsilon_k \tag{7}$$

It is interesting to compare the expression in Equation (6b) to the result that one would obtain by evaluating the integral in Equation (5a) using numerical quadrature, i.e.,

$$F(E) \approx 2\pi \sum_n \omega_n |< \phi_r \, H \, \psi^+_{E_n} >|^2 , \tag{8}$$

where E_n and ω_n are the quadrature points and weights. It is clear that in the Stieltjes development the non-resonant eigenvalues $\{\varepsilon_n\}$ are used as the points associated with an equivalent quadrature.[15] Furthermore, the normalization constants relating χ_n and $\psi^+_{\varepsilon_n}$ n = 1, 2, ... determine implicitly the quadrature weights, ω_n.

The Stieltjes derivative of the histogram approximation to F(E) is obtained as the slope of a straight line connecting the values of F at two neighboring rise points [Equation (7)]. The resulting histogram approximation for $\Gamma(E)$ has the form

$$\tilde{\Gamma}(E_k) = \frac{\gamma_k + \gamma_{k+1}}{2(\varepsilon_{k+1} - \varepsilon_k)} \qquad k = 1, 2, \ldots \tag{9}$$

where E_k are the half-way points $E_k = \frac{1}{2}(\varepsilon_k + \varepsilon_{k+1})$. The result in Equation (9) shows that the "correct" normalization constants associated with the discrete, non-resonant eigenfunctions χ_n are determined by the density of eigenvalues representing the continuous spectrum of PHP.

In actual computations employing Stieltjes imaging, one does not work directly with the pseudospectrum

$$\{\gamma_n, \varepsilon_n, n = 1 \cdots\} \; .$$

Instead, one performs a moment analysis to obtain a "smoothed" spectrum

$$\{\bar{\gamma}_n^{(M)}, \bar{\varepsilon}_n^{(M)}, n = 1 \cdots M\}$$

whose elements are uniquely determined by the first 2M inverse power moments of the original pseudospectrum provided $M \leqslant N$. Usually, the calculations must be repeated for several values of M, ($M \ll N$) until mutually consistent results are obtained. The details of the moment analysis have already been discussed at this workshop by Peter Langhoff, so I will not discuss them further.

It should be noted, however, that to calculate the physical width of the resonance, $\Gamma(E_r)$, some interpolation of the Stieltjes values given in Equation (9) seems to be required. In the present work the approximate cumulative function given by Equation (7) was fitted to an appropriate analytic function of E which could be differentiated at any energy to yield $\Gamma(E)$.

In closing this discussion of the Stieltjes technique, I wish to point out that Hickman, Isaacson and Miller[14] were the first to calculate resonance widths using L^2 approximations for the non-resonant wavefunction in the "golden-rule" formula. However, in their work, Hickman et al had to rediagonalize PHP in different basis sets until one of the background eigenvalues ε_n came close to the resonance energy E_r. In addition, they used a "box-normalization" idea[16] to normalize correctly the L^2 approximations to ψ_E^+. In contrast, the Stieltjes-moment theory procedure provides both the required renormalization of the discrete eigenfunctions and the <u>entire</u> non-resonant continuum in a <u>single</u> diagonalization of PHP. The latter feature makes the method practical even when very large CI matrices must be diagonalized to obtain accurate results for molecular resonances.

D. Calculation of Resonance Shifts

In Feshbach's theory of resonances,[8] the energy shift, $\Delta(E)$ of an isolated resonance is given by the principal value integral:

$$\Delta(E) = (2\pi)^{-1} P\int \frac{\Gamma(E')dE'}{E - E'} \tag{10}$$

For narrow resonances, the resonance energy E_r satisfies the equation:

$$E_r = \varepsilon_r + \Delta(E_r)$$

where ε_r is the eigenvalue of $Q_o H Q_o$.

The Stieltjes-moment-theory technique allows for calculating the shift, $\Delta(E)$, since it provides an approximation for not only the physical width, but also the width as a function of energy, i.e., $\Gamma(E)$. To obtain the results which I will show shortly for the $^2\Pi_g$ shape-resonance of N_2^-, $\Delta(E)$ was evaluated by using the histogram representation of $\Gamma(E)$ to approximate the integrand in Equation (10). In terms of the pseudospectrum $\{\gamma_n \varepsilon_n\}$ representing the non-resonant continuum, this approximation yields the result:[17]

$$\tilde{\Delta}(E) = (2\pi)^{-1} \sum_{k=1}^{M} \tilde{\Gamma}(E_k) \ln\left|\frac{\varepsilon_k - E}{\varepsilon_{k+1} - E}\right| \tag{11}$$

where $\tilde{\Gamma}(E_k)$, k = 1, 2, ... are the Stieltjes values given by Equation (9).

III. RESULTS

A. Atomic Resonances

The method which I have just described has been applied to several, well-known resonances in He, He^- and Mg^- for which previous calculations are available for comparison. Table II summarizes[7] the computed resonance energies and widths. In these cases, moderately sized basis sets of 130-260 configurations were used to define the total space $P_o + Q_o$. As the results for He and He^- show, the method is quite successful for core-excited, or Feshbach-type resonances. Accurate widths are obtained even for cases which involve many-electron targets e.g., He^- ^2S, because polarization and correlation effects are included in the description of the non-resonant, background continuum. The method is applicable to higher resonances of a given symmetry; for example, accurate resonance parameters were obtained for the third lowest ^1P resonance below the n = 3 threshold in e + He^+. The method can also be used for shape-resonances, although the definition of the projection operators P_o and Q_o are not as straightforward as in the case of core-excited resonances. The width calculated for the ^2P shape-resonance of Mg^- is quite encouraging since this is a very broad resonance close to threshold.

In closing, I wish to emphasize that the Stieltjes-moment-theory procedure is not designed to provide very precise widths for atomic resonances but it is intended for molecular problems where the implementation of other methods becomes prohibitively difficult.

B. $^2\Pi_g$ Shape-Resonance of N_2^-

The first molecular application of the method that I wish to discuss is the well-known, low-energy shape-resonance of N_2^-. This resonance has been studied extensively, and there are several recent static-exchange calculations[22-25] from which the resonance parameters have been determined. The objective of the present calculation was not to obtain definitive results for N_2^-, but to test the Stieltjes-imaging procedure on a well-defined model for which several different methods have been used to determine Γ.

The present calculations were carried out in the static-exchange model, where the bound orbitals were taken from an SCF calculation of the $^1\Sigma_g^+$ ground state of N_2. The one-electron, GTO basis set was augmented with up to 20 π_g orbitals to describe both the resonance and the non-resonant background. The calculations were repeated for several internuclear distances between 1.744 and 2.3916 bohr.

The computed resonance energies and widths are plotted as a function of internuclear distance in Figure 1. As I have mentioned before, in this case the energy shift, $\Delta(E)$, was computed utilizing

Table II

Comparison of Computed Resonance Parameters for Atoms

Resonance	Present Method E_{res}(eV)	Present Method Γ(meV)	Previous Calculations E_{res}(eV)	Previous Calculations Γ(meV)
He $(2s2p)\,^3P$	58.31	8.3	58.30	8.4[a]
			58.31	10 [b]
He $(2s2p)\,^1P$	60.19	36	60.15	38 [c]
			60.13	37 [b]
He $(2s3p)\,^1P$	63.67	8.3	63.68	8.0[a]
$He^-(1s2s^2)\,^2S$	19.40	11.5	19.40	12.1[a]
			19.39	12.1[e]
$Mg^-(3s^23p)\,^2P$	--	200	0.15	230 [f]

[a] Bhatia and Temkin, Ref. 18

[b] Drake and Dalgarno, Ref. 19

[c] Bhatia et al, Ref. 20

[d] Bain et al, Ref. 21

[e] Junker and Huang, Ref. 12

[f] Robb (unpublished)

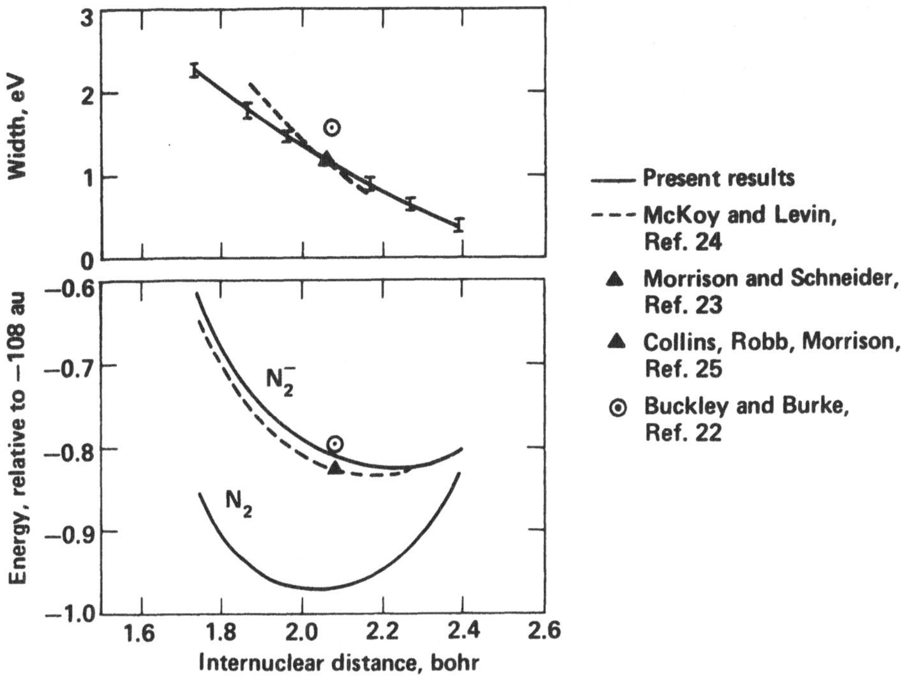

Fig. 1. Energy and Width of N_2^- $^2\Pi_g$ in the Static-Exchange Approximation.

the histogram approximation for $\Gamma(E)$ to evaluate the required principal value integral. The shift varied between -0.25 and -0.40 eV for the internuclear distances considered. The present resonance energies are somewhat higher than those obtained by Levin and McKoy[24] from the T-matrix calculations, except at R = 2.3916 bohr. At the equilibrium geometry, (R = 2.068 bohr) all the calculated widths, ≈1.2 eV, agree very well except the static-exchange result of Buckley and Burke.[22] This discrepancy has been attributed[25] to the neglect of exchange with the $1\sigma_g$ and $1\sigma_u$ occupied orbitals in their calculations. The width obtained with the Stieltjes-imaging procedure increases smoothly with decreasing internuclear distance. However, significant discrepancies between the present widths and those Levin and McKoy[24] are observed for internuclear distances less than 2.068 bohr. The source of this discrepancy is not understood at this time. It should be recognized, however, that in the static-exchange model, the $^2\Pi_g$ resonance of N_2^- is very broad and rather large uncertainties may be associated with the computed resonance parameters. In addition, for such broad resonances it is not trivial to extract unique E_r and Γ from the computed energy dependence of the eigenphase sums, as required in scattering calculations.

C. Core Excited Resonances of H_2

The Stieltjes technique for calculating resonance widths has been also applied to the lowest $^1\Pi_u$ and $^1\Sigma_u^+$ doubly-excited states of H_2 which are core-excited resonances associated with the repulsive, $^2\Sigma_u^+$ state of H_2^+.

The $(1\sigma_u 1\pi_g)\,^1\Pi_u$ resonance decays to the $^2\Sigma_g^+$ ground state of H_2^+ and an ejected $k\pi_u$ electron. In this case, the total space $(P_o + Q_o)$ included all 2-electron configurations constructed from a basis of $5\sigma_g$, $5\sigma_u$, $8\pi_g$ and $22\pi_u$ orbitals. The very large π_u basis set was required to describe accurately the non-resonant $1\sigma_g\ k\pi_u$ continuum. Since this $^1\Pi_u$ resonance decays into both the $p\pi_u(\ell = 1)$ and $f\pi_u(\ell = 3)$ components of the open channel, the present calculations were performed with two different π_u basis sets. In one case, the $f\pi_u$ molecular orbitals were constructed as linear combinations of $d\pi$-GTO's centered on the two nuclei (basis 2c-$d\pi$); in the other case, they were chosen as $f\pi$-GTO's on the center of the molecule (basis 1c-$f\pi$). The results of these calculations for R = 2.0 bohr are shown in Table III. The computed resonance energy agrees well with that obtained by Robb (private communication) in a 2-state close-coupling calculation. The present widths, 11-12 meV, are somewhat higher than his value of 8.0 meV. In order to test whether any of this difference is due to truncating his expansion to the lowest two states of H_2^+ ($1\sigma_g$ and $1\sigma_u$), we have repeated our calculations with only these sigma orbitals. As Table III shows the computed width decreased to 10.5 meV.

Table III

Width of $H_2\ (1\sigma_u 1\pi_g)\,^1\Pi_u$ State R = 2.0 bohr

Calculation	E_{res} (eV)	Γ (meV)	Γ (meV) "decoupled"
$5\sigma_g$, $5\sigma_u$, $8\pi_g$, $22\pi_u$ (2c - $d\pi$)	25.62	12.2 ± 0.5	10.1 2.3
$5\sigma_g$, $5\sigma_u$, $8\pi_g$, $22\pi_u$ (1c - $f\pi$)	25.62	11.1 ± 0.5	8.2 2.8
$1\sigma_g$, $1\sigma_u$, $8\pi_g$, $22\pi_u$ (2c - $d\pi$)	25.64	10.6 ± 0.5	7.7 3.0
$1\sigma_g$, $1\sigma_u$, $8\pi_g$, $22\pi_u$ (1c - $f\pi$)	25.64	10.5 ± 0.5	7.4 3.1
Robb's 2-state close-coupling	25.62	8.0 ± 0.1	6.9 1.1

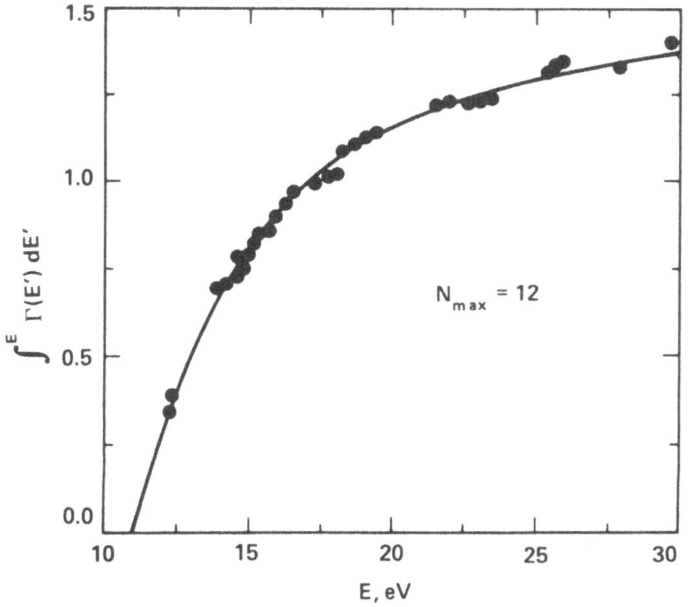

Fig. 2. Cumulative Function From Orders 3-N_{max} for H_2 $(1\sigma_u 1\pi_g)^1\Pi_u$ R = 2.0 bohr.

The current results for this particular resonance of H_2 shed some light on channel-coupling as it affects Stieltjes calculations of molecular continuum processes - an interesting question raised earlier by Hugh Kelly. As I already mentioned, the $(1\sigma_u\ 1\pi_g)^1\Pi_u$ resonance decays into both the $p\pi_u$ and $f\pi_u$ components of the open $(1\sigma_g\ k\pi_u)$ channel but with most of the flux going into the latter. Since our π_u basis contained both components, the resulting pseudospectra $\{\gamma_n \varepsilon_n\}$ consisted of successive pairs of one large and one small width-matrix elements, γ_n, in this case. The resulting cumulative function for orders 3-12 and the best analytic fit are shown in Fig. 2. The data points, which correspond to Equation (7), clearly show oscillations which are due to alternating $p\pi$ and $f\pi$ non-resonant eigenfunctions in the spectrum of PHP. In such a case, one may use two procedures to obtain meaningful results. First, one can use only those Stieltjes values which are derived by solving the inverse moment problem for low orders. This procedure tends to smooth out the oscillations in the cumulative function, as illustrated in Fig. 3, where only points from orders 3-7 are plotted. The widths given in the third column of Table III were obtained in this way. A second procedure is possible when each non-resonant eigenfunction can be uniquely associated with each of the channel components. In the present case, we divided the pseudospectrum into $p\pi$ and $f\pi$ components by inspecting each of the eigenfunctions, χ_n.

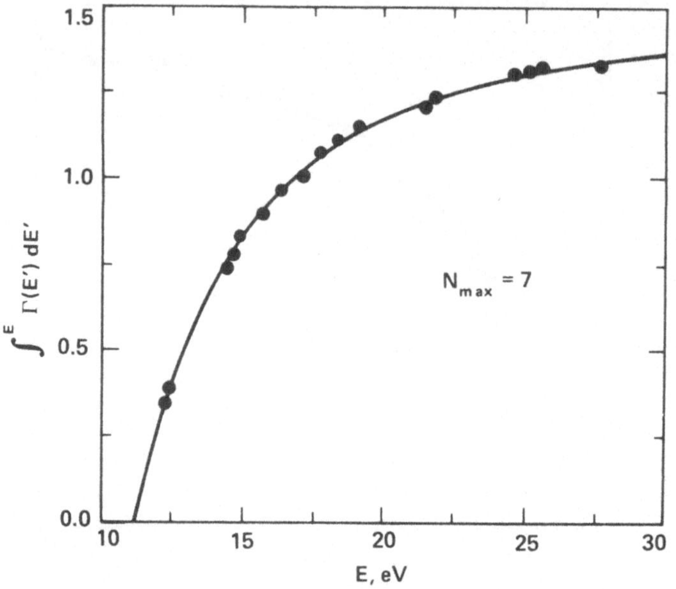

Fig. 3. Cumulative Function From Orders 3-N_{max} for H_2 $(1\sigma_u 1\pi_g)\,^1\Pi_u$ R = 2.0 bohr

Then, the two subspectra were imaged separately to yield the "partial" widths shown in the last column of Table III. The fact that the width calculated with the first procedure (column 3) is consistent with the sum of the two "partial" widths (column 4) for each calculation is very encouraging and gives us confidence in the results.

As the last application, I wish to discuss briefly the widths calculated for the $(1\sigma_u\ 2\sigma_g)\,^1\Sigma_u^+$ resonance of H_2. The dependence of the width on internuclear distance is uncertain for this resonance since two previous calculations,[26,27] both using undistorted Coulomb waves for ψ_E^+ predicted totally different behavior. The Stieltjes calculations utilized all $^1\Sigma_u^+$ configurations constructed from a basis of $5\sigma_g$, $18\sigma_u$, $4\pi_u$ and $4\pi_g$ orbitals. Table IV shows the calculated resonance parameters for internuclear distances between 1.0 and 3.0 bohr. The present results confirm the conclusion of Kirby et al[27] that the width of this resonance is an increasing function of internuclear separation. However, the present widths, which were obtained by including both distortion and polarization effects in the description of the non-resonant continuum, are consistantly higher than those calculated by Kirby et al using one- and two-center Coulomb waves (columns labelled "a" and "b," respectively) in the "golden-rule" formula. A recent, independent study of autoionizing states in e + CH^+[28] has shown that using undistorted,

Table IV

Comparison of Calculated Widths for $^1\Sigma_u^+$ Autoionizing State of H_2

		Γ(eV)		
R(au)	E_r (eV)	Present	Previous Results	
1.0	39.46	0.22	0.17[a]	0.17[b]
1.4	30.71	0.40	0.29[a]	0.29[b]
1.8	25.54	0.65	0.38[a]	0.37[b]
2.0	23.71	0.74	0.43[a]	0.40[b]
2.5	20.55	0.88	0.56[a]	0.48[b]
3.0	18.67	1.05	0.70[a]	0.54[b]

[a,b] K. Kirby, S. Guberman and A. Dalgarno, J. Chem. Phys., to be published.

uncoupled Coulomb waves to describe the background continuum seriously underestimtes the widths, at least, in that system.

IV. SUMMARY

In this talk I have attempted to show that Stieltjes-moment-theory provides a practical and a reasonably accurate method for calculating the widths of atomic and molecular resonances. The method seems to possess a number of advantages for molecular applications, since it avoids the explicit construction of continuum wavefunctions. It is very simple to implement the technique for molecular resonances, because it requires only existing bound-state structure codes. Through the use of configuration interaction techniques, many electron correlation and polarization effects can be included in the description of <u>both</u> the resonance and the non-resonant background continuum. However, it is also clear that additional work is required to refine the method and to gain computational experience. In application to shape-resonances, the definition of the projection operators P_o and Q_o is somewhat ambiguous. The question of channel coupling and its effect on the computed widths need to be studied further. The criteria for choosing augmented GTO basis sets which can give uniformly accurate widths at several internuclear distances have not yet been established. Finally, the applicability of the method to resonances occuring in electron-polar molecule collisions needs to be investigated.

This work was performed under the auspices of the U. S. Department of Energy by the Lawrence Livermore Laboratory under contract number W-7405-ENG-48.

REFERENCES

1. G. J. Schulz, Rev. Mod. Phys. 45, 423 (1973).
2. J. N. Bardsley, Rev. Mod. Phys. to be published. (1979).
3. J. N. Bardsley and F. Mandl, Rep. Prog. Phys. 31, 472 (1968).
4. F. Fiquet-Fayard, Vacuum 24, 533 (1974).
5. One example is the quasi-projection-operator technique: A. Temkin, A. K. Bhatia and J. N. Bardsley, Phys. Rev. A5, 1663 (1972).
6. One example is the stabilization method: H. S. Taylor, Adv. Chem. Phys. 18, 91 (1970); H. S. Taylor and A. U. Hazi, Phys. Rev. A14, 2071 (1976).
7. A. U. Hazi, J. Phys. B11, L259 (1978).
8. H. Feshbach, Ann. Phys. (New York) 19, 287 (1962).
9. P. W. Langhoff, Int. J. Quant. Chem. Symp. 8, 347 (1974).
10. C. Bottcher and K. Docken, J. Phys. B 7, L5 (1974).
11. W. H. Miller, C. A. Slocomb, and M. F. Schaefer, J. Chem. Phys. 56, 1347 (1972), A. P. Hickman, A. D. Isaacson, and W. H. Miller, J. Chem. Phys. 66, 1483 (1977).
12. B. R. Junker and C. L. Huang, Phys. Rev. A18, 313 (1978).
13. W. H. Miller, Chem. Phys. Lett. 4, 627 (1970).
14. A. P. Hickman, A. D. Isaacson, and W. H. Miller, Chem. Phys. Lett. 37, 63 (1976).
15. E. J. Heller, W. P. Reinhardt and H. A. Yamain, J. Comput. Phys. 13, 536 (1973).
16. A. U. Hazi and H. S. Taylor, Phys. Rev. A1, 1109 (1970).
17. P. W. Langhoff and W. P. Reinhardt, Chem. Phys. Lett. 24, 495 (1974).
18. A. K. Bhatia, and A. Temkin, Phys. Rev. 182, 15 (1969).
19. G. W. F. Drake and A. Dalgarno, Proc. Roy. Soc. A 320, 549 (1971).
20. A. K. Bhatia, P. G. Burke and A. Temkin, Phys. Rev. A8, 21 (1973).
21. R. A. Bain, J. N. Bardsley, B. R. Junker and C. V. Sukumar, J. Phys. B 7, 2189 (1974).
22. B. D. Buckley and P. G. Burke, J. Phys. B 10, 725 (1977).
23. M. A. Morrison and B. I. Schneider, Phys. Rev. A 16, 1003 (1977).
24. D. A. Levin and V. McKoy, Phys. Rev. A to be published.
25. L. A. Collins, W. D. Robb and M. A. Morrison, J. Phys. B 11 L777 (1978).
26. C. Bottcher, J. Phys. B 7, L352 (1974), and unpublished results.
27. K. Kirby, S. Guberman and A. Dalgarno, J. Chem. Phys. to be published.
28. G. Raseev, A. Giusti-Suzor, H. Lefebvie-Brion, J. Phys. B 11, 2735 (1978).

DISCUSSION

Temkin: Concerning the choice of projection operators in the case of He^-, why did you include the configurations of the type $2p^2ns$ in Q_o?

Hazi: It is true that the ground state wavefunction of He has the form $(1s^2) + \lambda(2p^2) + \lambda'(3d^2)$... However, in this case λ, λ', etc. are sufficiently small so that projecting out only the configurations $(1s^2ns)$ will ensure that the $(1s2s^2)$ ^2S resonance is the lowest eigenvalue of Q_oHQ_o.

Rescigno: What about cases where the correlating configurations, e.g., $(2p^2)$, are more important?

Hazi: In that case, those configurations must also be included in subspace P_o. In general, you include enough configurations in P_o so that the resonance is the lowest solution of Q_oHQ_o but leave Q_o large enough to give an accurate resonance function ϕ_r.

Bardsley: Concerning N_2^-, there may be a problem with fitting the energy dependent eigenphases to extract the resonance parameters of broad resonances.

Hazi: That may be a possible explanation of the discrepancy between the Stieltjes and the T-matrix results for the width. Debbie Levin estimated an uncertainty of 20-30% due to fitting to Breit-Wigner or effective-range formulas. Vince McKoy may want to comment on that further.

McKoy: Yes. Another possible explanation is that the present T-matrix results are not variationally corrected.

Rescigno: The T-matrix calculations on N_2^- have been underway for some time. The basis sets have been optimized extensively at R = 2.068 bohr. However, this optimization has not been done at other internuclear separations.

Nesbet: For broad resonances it is possible to correct the Feshbach resonance parameters. Since you know the energy dependence of the width from your calculations, perhaps you can do that.

Dehmer: In the calculations of Buckley and Burke, does the $^2\Pi_g$ resonance of N_2^- fall at the right energy?

Hazi: All the results which are compared in Fig. 1 were obtained with the static-exchange approximation, i.e., no polarization was included. The resonance energies, all about 4 eV at R_e, agree reasonably well, however, they are all too high relative to the real resonance position.

Langhoff: Do the oscillations in the width matrix elements for the $^1\Pi_u$ resonance of H_2 appear in the raw spectrum?

Hazi: Yes, but they are still present in the spectrum after the moment analysis, as Fig. 2 shows.

Rescigno: Are the oscillations more pronounced in the cumulative function derived directly from the raw spectrum?

Hazi: I do not know, since in practice we never work with the raw spectrum directly.

McKoy: You showed that the sum of the "partial" widths obtained by imaging the $p\pi_u$ and $f\pi_u$ spectra separately agreed with the value obtained by imaging the total spectrum. Does this mean the coupling between the components is weak?

Hazi: Not necessarily. It shows that the results are internally consistent. However, if the coupling between $p\pi_u$ and $f\pi_u$ were strong, I would not expect the matrix elements to oscillate; the raw spectrum would be smoother. In general, my prescription for treating channel-coupling is the following. If the channels are strongly coupled, image the total spectrum. If the channels are weakly coupled and the components can be identified, separate the subspectra, image them separately and then add the resulting widths. But I would not attach physical significance to the "partial" widths.

Rescigno: There is a way to test whether the coupling is dynamic. One can diagonalize PHP in the $p\pi_u$ and $f\pi_u$ subspaces separately, say two 10 x 10 matrices instead of one 20 x 20. This procedure ensures that no dynamic coupling is allowed in the background solutions. Then, if the sum of the widths obtained this way is roughly the same as before, then dynamic coupling is indeed small.

Reinhardt: Our experience with atomic photodetachment, e.g., in H^-, has shown that, in the case of a structured spectrum, if you use only the low order moments in the Stieltjes calculation you will average out the structure beautifully, i.e., you will do well in the average sense.

PROGRESS TOWARD THE APPLICATION OF COMPLEX COORDINATE AND COMPLEX BASIS FUNCTION TECHNIQUES TO MOLECULAR RESONANCE CALCULATIONS

C. William McCurdy

Department of Chemistry
The Ohio State University
Columbus, Ohio 43210

I. INTRODUCTION

In this talk I will discuss some recent developments in complex coordinate and basis function techniques which are beginning to make possible the application of these techniques to the treatment of molecular resonance phenomena. I will avoid any detailed discussion of the history of successful applications to atomic resonances and atomic photoionization. That history is in itself sufficient to furnish material for a review.[1] Instead I will concentrate mainly on the discussion of two problems which have, until recently, slowed the development of generalizations of complex coordinate and basis function approaches suitable for molecular resonance calculations.

The first of those problems is the fact that calculations using the simple rotated coordinate approach on atomic systems with more than two electrons seem to require extremely large basis sets. Even for resonances of He^- this difficulty has hindered the straightforward application of rotated coordinates. I will discuss both the origin of this problem and its solution in later sections.

The second problem appears, on the surface at least, to be more serious. For reasons discussed below it is imperative that we treat the molecular resonance problem within the framework of the Born-Oppenheimer approximation. The theorems of Balslev and Combes,[2] on which the complex coordinate approach for atoms is based, were derived for a class of Hamiltonians (those which are dilatation analytic) which does not include the Born-Oppenheimer electronic Hamiltonian. This fact certainly poses at least a formal problem. In our calculations[3] it also posed an apparent computational difficulty. It has been argued both that this problem is a serious one,[3] and also that it is not a problem at all for practical resonance calculations.[4] In either case it is an unresolved question which must be answered before we can proceed with more involved applications of complex coordinates, particularly calculations of molecular photoionization cross sections from resolvent matrix elements, such as have been performed for atomic photoionization cross sections.[5]

In Section II I will give a brief discussion of the essentials of the complex coordinate technique, and in Section III I will discuss the two problems mentioned above as well as proposed solutions of them which are commonly being referred to as complex basis function techniques. In Section IV I make some concluding remarks.

II. COMPLEX SCALING FOR DILATATION ANALYTIC HAMILTONIANS

The complex coordinate technique for treating atomic resonances is a literal application of the theorems of Balslev and Combes[2] on the spectrum of dilatation analytic Hamiltonians under a complex scaling transformation of all radial coordinates. Briefly, the portion of the Balslev and Combes theorems relevant to our discussion states that: The spectrum of the analytically continued center of mass Hamiltonian, $H(\theta)$, in which all radial coordinates have been subjected to the transformation

$$r_i \to e^{i\theta} r_i$$

consists of:

1) A system of parallel half lines in the complex plane in the direction $e^{-2i\theta}$ originating at the elastic and all inelastic thresholds.

2) A discrete set of <u>real</u> eigenvalues corresponding to the bound state energies of the untransformed Hamiltonian.

3) A discrete set of complex eigenvalues, lying anywhere in the segment between the line originating at the elastic threshold and the real axis, corresponding to the resonance poles of the S matrix.

This spectrum can be understood in a mathematically pedestrian way which suggests that generalizations to Hamiltonians not covered by the original theorem should be possible. For simplicity, consider a one particle system for which the Schrödinger equation is

$$H(r)\ \Psi_E(r) = E\ \Psi_E(r) \tag{1}$$

The eigenvalues, E, determined by the usual scattering and bound state boundary conditions consists of a set of isolated, real bound state eigenvalues together with the continuous (real) scattering eigenvalues. Making transformation $r \rightarrow re^{i\theta}$ on this equation, once it is solved subject to the above boundary conditions on Ψ_E of real argument, is simply a change of variables

$$H(re^{i\theta})\ \Psi_E(re^{i\theta}) = E\ \Psi_E(re^{i\theta}) \tag{2}$$

and obviously does not change the spectrum. In order to obtain the spectrum addressed by the Balslev and Combes theorems we solve

$$H(re^{i\theta})\ \chi_{E_\theta}(r) = E_\theta\ \chi_{E_\theta}(r) \tag{3}$$

subject to the boundary conditions: (1) for bound states $\chi_{E_\theta}(r)$ is square-integrable (L^2) and (2) for continuum states χ_{E_θ} satisfies

$$\chi(r) \xrightarrow[r \to \infty]{} \sin\ (k'r + \delta) \tag{4}$$

where k' is real. These are not the boundary conditions satisfied by $\Psi_E(re^{i\theta})$ in equation (2) and hence the eigenvalues E_θ of equation (3) can differ from those of equations (1) and (2). The original bound states of equation (1) ($E < 0$) satisfy

$$\psi_E(r) \xrightarrow[r \to \infty]{} e^{-\sqrt{-2E}\quad r} \tag{5}$$

and remain L^2 under the change of variable $r \rightarrow re^{i\theta}$

$$\psi_E(re)^{i\theta}) \xrightarrow[r \to \infty]{} e^{-\sqrt{-2E}\ r\ \cos\theta}\ e^{-i\sqrt{-2E}\ r\ \sin\theta} \tag{6}$$

So for bound states

$$\chi_{E_\theta}(r) = \Psi_E(re^{i\theta}) \tag{7}$$

$$E_\theta = E$$

and this portion of the spectrum remains unchanged.

However, the original continuum solutions under the same change of variable (s waves for simplicity) satisfying

$$\Psi_E(re^{i\theta}) \xrightarrow[r\to\infty]{} \sin(kre^{i\theta}\alpha + \delta) \qquad (8)$$

are not acceptable solutions for $\chi_{E\theta}(r)$ unless $k \to ke^{-i\theta}$ because otherwise they do not satisfy the boundary condition of equation (4). Thus the continuous spectrum is rotated

$$E_\theta = \left(\frac{ke^{-i\theta}}{2}\right)^2 = Ee^{-2i\theta} \qquad (9)$$

and the continuum solutions of equation (3) are not obtained by a simple change of variable in those of equation (1).

Resonance states appear as additional solutions of equation (3) satisfying the boundary condition of square-integrability. They can be thought of as solutions of the original differential equation (eg. (1)) under a change of variable

$$\Psi_R(re^{i\theta}) \xrightarrow[r\to\infty]{} e^{ik_R re^{i\theta}} \qquad (10)$$

which are L^2 (for a range of θ) because they correspond to complex values of k for which continuum solutions of equation (1) contain only outgoing waves.

The point to be made by this simple discussion is that it is the boundary conditions and hence the asymptotic behavior of the solutions of equation (3) which determine its eigenvalues. Recent work which generalizes the complex coordinate approach to the Stark Hamiltonian,[6] which is not a dilatation analytic operator, can be understood in terms of a simple argument like the above in which the continuum boundary conditions and the complex coordinate transformation are replaced by those appropriate to the Stark problem[7] (in which translational analyticity is exploited). Also, it should be noted that the very recent work by Simon[8] on the complex coordinate transformation in the electronic Born-Oppenheimer Hamiltonian avoids the problems arising from the fact that the nuclear attraction potential lacks dilatation analyticity while preserving the nature of the simple $re^{i\theta}$ scaling asymptotically.

III. DIFFICULTIES IN THE MOLECULAR PROBLEM AND SOME OF THE PROPOSED SOLUTIONS

A. The Many-Electron Problem

Until recently the application of complex coordinate methods to atoms with more than two electrons has been hindered by the apparent need for very large basis sets in those calculations - far larger than those of the two electron calculations of reference

1. To see the origin of this difficulty clearly, consider elastic scattering from an N electron system. (See for example the discussion in reference 9.) The elastic scattering wavefunction for this N + 1 electron system is antisymmetric under interchange of the coordinates of any pair of electrons and has the asymptotic form as any one coordinate becomes large (suppressing dependence on angular coordinates, purely for notational convenience),

$$\psi_k(r_1,\ldots,r_{N+1}) \rightarrow \psi_o(\ldots r_{i-1},r_{i+1}\ldots)$$

$$\times \left[A(k)e^{-ikr_i} - A(-k)\ e^{ikr_i} \right]. \qquad (11)$$

At the complex value of k corresponding to a resonance, A(k) vanishes. The resonance wavefunction evaluated with all coordinates scaled as $re^{i\theta}$ is square integrable when $\theta > |\arg (k_R)|$ just as in the potential scattering case. This wavefunction is an eigenfunction of the analytically continued Hamiltonian, $H(\theta) = H(re^{i\theta}, \ldots, r_{N+1}e^{i\theta})$, with the boundary condition of square integrability. The important point to notice is that the asymptotic form of the resonance eigenfunctions of $H(\theta)$ is

$$\psi_\theta \xrightarrow[r_i \rightarrow \infty]{} \psi_o(\ldots r_{i-i}e^{i\theta},\ r_{i+i}e^{i\theta}\ \ldots)\ e^{ik_R r_i e^{i\theta}} \qquad (12)$$

and particularly that it involves a bound, N-electron target wavefunction evaluated at $re^{i\theta}$.

If, for example, the target ground state wavefunction were well approximated by a single Slater determinant

$$\psi_o\ (re^{i\theta},\ldots,r_N e^{i\theta}) \simeq A\ (\varphi\ (re^{i\theta})\ \ldots\ \varphi_N\ (r_N e^{i\theta})) \qquad (13)$$

the orbitals appearing in the determinant would be decaying but oscillatory functions of r, since they would have the asymptotic form (atomic units)

$$\varphi_i(re^{i\theta}) \sim e^{-\sqrt{2|E_i|}\ re^{i\theta}} \qquad (14)$$

where E_i is the orbital binding energy.

Now the difficulty in the complex coordinate method for many electron systems becomes apparent. If we diagonalize $H(\theta)$ in a set of (N + 1) electron configurations made up from determinants of <u>real</u> orbitals, we are trying to expand the now oscillatory complex bound orbitals $\varphi_i(re^{i\theta})$ in terms of the <u>real</u> basis orbitals. Whereas a single real determinant might have sufficed to approximate the target state for the real Hamiltonian, we now need many such determinants just to approximate the target state for the complex Hamiltonian. Moreover the problem is worse for the inner electrons because their greater binding energy causes faster oscillation in equation (14).

The solution of this problem hinges on the fact that we can see from the asymptotic form given in equation (12) that it is necessary only to make the coordinate dependence of a single electron complex in order to render the wavefunction square integrable, provided we preserve the overall antisymmetry of the wavefunction in so doing. One can show[10] that matrix elements of the rotated Hamiltonian, $H(re^{i\theta}, \ldots, r_{N+1} e^{i\theta})$ between functions of the form $\chi(r,\ldots r_{N+1})$ are the same (except for an overall scale factor) as matrix elements of the real Hamiltonian, $H(r_1, \ldots, r_{N+1})$ between functions $\chi(r_1 e^{-i\theta}, \ldots, r_{N+1} e^{-i\theta})$, whose coordinates are scaled by $e^{-i\theta}$. This fact suggests using configurations of the form

$$\chi = A \left[\psi_o(r_1 \ldots r_N)\, \varphi_i\, (r_{N+1} e^{-i\theta}) \right],$$

where ψ_o is a one configuration or multiple configuration approximation to the target state wavefunction, to diagonalize the real Hamiltonian. Although this procedure is not equivalent to using the usual complex Hamiltonian, it does correspond to the analytic continuation of the effective one electron Hamiltonian obtained by integrating over the coordinates of the target electrons.

This prescription was proposed by Rescigno, McCurdy and Orel[9] and was verified computationally for a shape resonance in electron-beryllium atom scattering. This calculation also made use of the approximation that the real Hamiltonian could be written in the form

$$H \sum_{\alpha\beta=1}^{M} |\Phi_\alpha > H_{\alpha\beta} < \Phi_\beta| \tag{16}$$

where the Φ_α are real configurations, for the purpose of computing matrix elements between complex configurations of the form given in equation (15). This procesure greatly simplified the numerical aspects of the Be^- calculation, and allowed us to treat a much more complicated problem than was previously possible.

One way of looking at this calculation is that it makes use of the transformation $r \to re^{i\theta}$ to analytically continue the nonlocal, effective one-electron Hamiltonian. Recently, Yaris et al.[4] have argued that the complex scaling transformation can be used for computing resonance positions in many cases when the Hamiltonian has a nonlocal potential. The work of reference 9 is in agreement with that assertion.

Junker and Huang[11] have proposed a solution to the many electron difficulty along similar lines. They diagonalize the complex Hamiltonian in a basis set containing complex basis functions which are capable of providing the oscillatory behavior required to describe the "bound-like" part of the resonance wavefunction. The similarities between the approaches of references 9 and 11 become apparent after a few contour distortions in the Hamiltonian matrix elements.

B. Problems with the Born-Oppenheimer Hamiltonian

The Balslev and Combes theorems[2] apply to the center of mass Hamiltonian of a system of particles interacting via Coulomb potentials. A molecule is such a system, so one might argue that a reasonable way to proceed would be to abandon the usual Born-Oppenheimer approximation, scale both nuclear and electronic coordinates by $e^{i\theta}$, and attempt to find resonance eigenvalues of the complete Hamiltonian. This approach would require the use of a basis set in <u>both</u> the nuclear and electronic coordinates. Since obtaining convergence with respect to electronic configurations within current computer limitations is already a problem, even for atomic resonances, the far larger basis sets required to treat nuclear and electronic motion simultaneously would clearly be impractical. Aside from the practical drawbacks of this brute force procedure, it also suffers the defect that it would miss many interesting electronic resonances completely.

Consider the common case of a molecular autoionizing state or a resonance state in electron-molecule scattering which corresponds to a dissociative potential surface. Plotting the real part of the energy of this state as a function of internuclear distance, for a diatomic molecule for example, would result in a purely repulsive potential curve. Such a state is a continuum state for nuclear motion and is thus a continuum state for the entire system of electrons plus nuclei. It therefore does not correspond to an isolated resonance eigenvalue of the complete Hamiltonian. Nevertheless, this resonance has physical meaning in the sense that it may appear as a feature of the electron scattering cross section, or, in the case of an autoionizing state, may be responsible for Penning ionization. A good example is the autoionizing state responsible for Penning ionization in the collision of $He(2^1S)$ with ground state hydrogen atoms.[12,13] Although the real part of the energy of this state displays a very shallow well, the Penning ionization process obviously takes place in the continuum with respect to nuclear motion.

Thus for several reasons it is desirable to retain the Born-Oppenheimer approximation in the description of molecular resonances. The principal obstacle to the extension of the complex coordinate approach to molecular resonances is the fact that the Born-Oppenheimer electronic Hamiltonian is not dilatation analytic, and the Balslev and Combes theorems consequently are inapplicable. Let us explore this point further.

In the <u>atomic</u> center of mass Hamiltonian the transformation, $r \to \sigma r$, of all electronic coordinates leads to a transformed Hamiltonian, $H(\sigma)$, of the form

$$\hat{H}(\sigma) = \hat{T}(\sigma) + \hat{V}(\sigma)$$
$$= \sigma^{-2}\,\hat{T} + \sigma^{-1}\,\hat{V} \qquad (17)$$

where $\hat{T}$ and $\hat{V}$ are the untransformed kinetic and potential energy operators. In this case the apparent analyticity of the transformed potential as a function of σ (except at $\sigma = 0$) is sufficient to establish the property of dilatation analyticity for the atomic Hamiltonian. A case where the analyticity of the potential as a function of σ is not sufficient to establish the dilatation analyticity of the Hamiltonian is the case of the Stark Hamiltonian,[6] because the Stark potential, $eEr\cos\theta$, where E is the field strength, is a pathological perturbation (not "relatively compact"). However, this is not the type of complication which concerns us in the case of the Born-Oppenheimer Hamiltonian.

The nuclear attraction potential in the Born-Oppenheimer Hamiltonian is

$$V = \sum_{i,j} \frac{-Z_j\, e^2}{|\vec{r}_i - \vec{R}_j|} \qquad (18)$$

where r_i and R_j denote electronic and nuclear coordinates respectively, and if we interpret $|\vec{r}_i - \vec{R}_j|$ as $[(\vec{r}_i - \vec{R}_j)^2]^{\frac{1}{2}}$ this potential has a branch point where $\vec{r}_i = \vec{R}_j$. To see how this branch point leads to severe problems consider the scaling properties of the analogous one dimensional Hamiltonian

$$H = \frac{d^2}{dr^2} + \frac{1}{(r - 1)^{\frac{1}{2}}} + \frac{1}{(r + 1)^{\frac{1}{2}}} \qquad (19)$$

The transformation $r \to \sigma r$ yields the transformed Hamiltonian

$$H(\sigma) = \sigma^{-2}\frac{d^2}{dr^2} + \frac{1}{(\sigma r - 1)^{\frac{1}{2}}} + \frac{1}{(\sigma r + 1)^{\frac{1}{2}}} \qquad (20)$$

As a function of σ the transformed potential has a square root branch point at $\sigma = 1/r$. If r can have any value greater than zero then the potential has a continuous line of branch points in σ

along the real σ axis. This Hamiltonian is thus not dilatation analytic as a result of the non-analyticity of the potential in the scaling parameter σ. The square root branch point of the nuclear attraction potential in the Born-Oppenheimer Hamiltonian causes that Hamiltonian to lack dilatation analyticity in a similar way.

McCurdy and Rescigno[3] proposed a solution to this problem which exploits the fact that it is the asymptotic form of the eigenfunctions of the Hamiltonian which gives rise to the modification of the spectrum under the complex scaling transformation. The McCurdy and Rescigno[3] procedure is to choose basis functions which are equivalent to scaling r by $e^{i\theta}$ only asymptotically while avoiding the non-analyticity problem at the nuclear centers. This is accomplished by first noting the useful identity that scaling the coordinates in each basis function by $e^{-i\theta}$, so that the basis functions and all of the form $\varphi_i(re^{-i\theta})$, and using these basis functions to diagonalize the real Hamiltonian, is equivalent to diagonalizing the complex Hamiltonian, $H(\theta)$, in real basis functions.[10] The basis functions corresponding to asymptotic scaling which they used are Gaussians with complex exponents, $\alpha e^{-2i\theta}$

$$\varphi_{\ell mn}(\alpha e^{-2i\theta},\vec{r},\vec{A}) = N_{\ell mn}(\alpha e^{-2i\theta})\ (x-A_x)^{\ell}\ (y-A_y)^{m}\ (z-A_z)^{n} \times e^{-\alpha e^{-2i\theta}(\vec{r}-\vec{A})^2} \tag{21}$$

Consider the matrix elements with respect to these complex basis functions of a nonspherical Hamiltonian $H(\vec{r})$, which is, however, spherical at large $|\vec{r}|$

$$H_{AB} = \int \varphi_{\ell mn}\ (\alpha e^{-2i\theta},\ \vec{r},\ \vec{A})\ H(\vec{r})\ \varphi_{ijk}\ (\beta e^{-2i\theta},\ \vec{r},\ \vec{B})d\,\vec{r} \tag{22}$$

The first point to note is that the normalization constant in equation (21) is such that[14]

$$N(\alpha e^{-2i\theta}) = e^{-3i\theta/2}N(\alpha)e^{-(\ell+m+n)\ i\theta} \tag{23}$$

It then becomes clear, from examining the integrand of equation (22), that as the orbital exponents, α and β, becomes small, one obtains the same numerical value for the integral whether one associates a phase of $e^{-2i\theta}$ with the orbital exponents or a factor $e^{-i\theta}$ with the electronic coordinates $\vec{r}$. The latter is true because the matrix element becomes independent of the orbital center in the limit that both exponents become small. But it is precisely these diffuse basis functions which determine the asymptotic form of the wavefunction. So McCurdy and Rescigno[3] proposed that using basis functions of the form prescribed above to diagonalize the real Hamiltonian is equivalent to using rotated coordinates in the Hamiltonian asymptotically.

This procedure was tested on H_2^+ and on a shape resonance of a model nonspherical Hamiltonian, and it appears to produce a spectrum which contains, at least, the expected bound state and resonance eigenvalues. Of course, the important question at this point is "Precisely how does this prescription avoid the branch point problem in the Born-Oppenheimer Hamiltonian?" In section IV I will discuss some recent mathematical developments which may bear on this question, but at the moment I hazard no detailed speculations on this very important point.

Another solution to the problem of the lack of dilatation analyticity of the Born-Oppenheimer Hamiltonian has been suggested by Isaacson, McCurdy and Miller[15] which is based on the complex basis function method of Bardsley and Junker[16] for atoms. (See also Bain et al. in ref. 1.) Since a nonspherical potential couples partial waves, the question which arises immediately in the attempt to generalize the method of Bardsley and Junker[16] is what angular function to associate with the outgoing wave. For an atomic problem one can construct eigenfunctions of the total angular momentum, and in each configuration which contains the outgoing wave basis function,

$$\varphi(\vec{r}) = Y_{\ell m}(\theta,\varphi)\; g(r)\; \frac{e^{ikr}}{r} \tag{24}$$

the correct $Y_{\ell m}$ is determined by the overall symmetry of the resonance wavefunction being computed. We can see what happens in a molecular case by considering the asymptotic form of the scattering solutions of a nonspherical potential scattering problem (symmetry corresponding to a diatomic molecule in this example), which are of coupled channel form in the angular momentum representation

$$\psi_{\ell}(\vec{r}) \sim \sum_{\ell} Y_{\ell m}(\hat{r})\,(\delta_{\ell\ell'}\, e^{-ikr} - S^{m}_{\ell\ell'}(k)\, e^{ikr}) \tag{25}$$

Every nonvanishing S matrix element $S_{\ell\ell'}(k)$, has a pole at the complex resonance momentum k_R. From this fact it can be shown that one can find the resonance energy using the method of Bardsley and Junker by including an outgoing wave basis function of the form given in equation (24) for each value of ℓ and m which contribute to the molecular symmetry in question.[15] In practice it appears this infinite set can be truncated to fewer than five partial waves. Isaacson, McCurdy and Miller[15] have investigated the practicality of this method using atomic examples to demonstrate its convergence properties. The method as just described for molecules has been successfully applied to He(2^1S, 2^3S) + H by Isaacson and Miller.[13] The difficulty here is to construct a convenient way to evaluate the required integrals. In the calculations of Isaacson and Miller[13] the molecular nuclear attraction integrals were evaluated numerically, and this drawback must clearly be remedied before the method is a feasible alternative for general molecular calculations.

This concludes the discussion of recent computational experiments in using complex coordinate and related techniques to treat the molecular resonance problem. In the following section I will discuss some recent mathematical developments and make some concluding remarks.

IV. CONCLUSION: RECENT MATHEMATICAL DEVELOPMENTS AND SOME UNSOLVED PROBLEMS

Recently, some rigorous extensions of the complex scaling procedure have been developed which overcome, for particular cases, the limitation to dilatation analytic potentials of the original Balslev and Combes[2] work. The first advances which appeared dealt with the Stark Hamiltonian. I will not discuss this problem in any detail here because it is not part of the molecular resonance story. However, the nature of the generalization is well worth noting. Avron and Herbst[17] have used complex coordinate translations

$$x \to x + iq \tag{26}$$

as opposed to simple scaling, to discuss the existence of wave operators for scattering in the presence of a dc field. Coordinate translation leads to a complex translation of the continuous spectrum of the Stark Hamiltonian and a consequent uncovering of resonance eigenvalues. A discussion of numerical applications of this approach to the hydrogen atom in dc and ac fields is given in a recent article by Cerjan et al.[6], who also discuss the consequences of the application of complex scaling to the Stark Hamiltonian. Herbst and Simon[18] have also recently discussed some additional mathematical points concerning the Stark problem.

More directly relevant to the problem of how to generalize the complex scaling procedure to the case of the Born-Oppenheimer Hamiltonian is the recent work of Simon.[8] Simon proposes a procedure he calls the "Method of Exterior Complex Scaling" and has shown that the spectrum of the molecular Born-Oppenheimer Hamiltonian under this coordinate transformation is transformed in the same manner as that of the atomic Hamiltonian under the dilatation transformation. Briefly Simon's proposed method is as follows. The electronic coordinates are scaled only outside a sphere of radius R_o which contains the nuclear centers, that is, we define the transformation R(r) by

$$R(r) = r \qquad 0 < r \leq R_o$$
$$= R_o + e^{i\varphi}(r - R_o) \qquad R_o \leq r \quad . \tag{27}$$

The branch point difficulty is avoided by keeping r real for $r \leq R_o$. The transformed Hamiltonian is given by

$$H_{R_o}(\theta) = U_{R_o}(\theta)\, H\, U_{R_o}^{-1}(\theta) \tag{28}$$

where $U_{R_o}(\theta)$ is an operator which, when operating on a function of the electronic coordinates, modifies each electronic coordinate according to equation (27) and multiplies the function by an overall Jacobian factor.

This theory solves the formal problems in the molecular problem, but as yet it appears to be quite difficult to implement. The electron repulsion integrals are particularly messy after the potential is transformed according to equations (26) and (27). However work is currently underway to develop variants of this theory which are easier to apply, if not as rigorous as the original.[19] It is worth noting that an intuitive understanding of Simon's proposed method[8] is given by the simple discussion of Section II.

A good question at this point is why bother to pursue a procedure which appears to be as awkward as Simon's does at first glance. This brings us to the consideration of some of the unsolved problems in this area. One of the most intriguing of those unsolved problems is how to compute matrix elements of the resolvent of the Born-Oppenheimer Hamiltonian. Complex coordinates have been applied to the calculation of the photoionization cross sections of atoms by using the fact that the photoionization cross section can be expressed in terms of a matrix element, $(\varphi, (z - H)^{-1} \chi)$, of the resolvent between square integrable functions.[5] The essential point here is that the integrations in that matrix element are evaluated along rotated contours. To apply this method to molecules we require a generalization which specifies the contours appropriate to that problem. Moreover we need to be able to evaluate the matrix element in some practical way once the correct contours are known. Simon's proposal is a promising move in this direction, although it may not be a practical solution. Unfortunately, it is not yet clear how the method of McCurdy and Rescigno[3] bears on this problem. The list of other questions about which little is known includes the following.[20]

To solve the many electron difficulty it appears we want to make complex transformations on the coordinates of only one electron.[9] Can a rigorous theory be developed for scaling the coordinates of only one electron? Certainly, we should currently require any method proposed for molecules to allow for treating systems of several electrons in some manner similar to that discussed in Section III.

Also, as Simon has asked,[20] can error bounds be found for the resonance positions determined by complex scaling methods? The advantages are obvious, but the difficulties are substantial.

Finally, are there generalizations of complex scaling which can treat the molecular problem in a both rigorous and practical manner? For example, can the McCurdy and Rescigno proposal be put on a mathematically sound footing? Also, can the generalization of the Bardsley and Junker[16] method which Isaacson and Miller[13] have applied to HeH be further simplified computationally, so as to make it practical for more general systems? Precisely how is it related, if at all, to complex scaling?

As the mathematical subleties become more fully understood (Simon's work is particularly promising in this regard) solutions to the practical problems will be easier to find. If recent history is any indication, we should see advances in all these areas between the time of this writing and the appearance of the proceedings of the workshop.

REFERENCES

1. I give here only a very abbreviated bibliography of papers dealing with the atomic applications: J. Nuttall and H. L. Cohen, Phys. Rev. 188, 1542 (1969); G. Doolen, J. Nuttall and R. Stagat, Phys. Rev. A10, 1612 (1974); G. D. Doolen, J. Phys. B 8, 525 (1975); R. A. Bain, J. N. Bardsley, B. R. Junker and C. V. Sukumar, J. Phys. B 7, 2189 (1974), T. Rescigno and V. McKoy, Phys. Rev. A12, 552 (1975); A. P. Hickman, A. D. Isaacson, and W. H. Miller, Chem. Phys. Lett. 37, 63 (1976). Also see: International Journal of Quantum Chemistry 14 number 4 (October 1978) which is devoted entirely to complex scaling and contains articles on computational aspects of the problem as well as more rigorous mathematical discussions.
2. E. Balslev and J. M. Combes, Commun. Math. Phys. 22, 280 (1971)
3. C. W. McCurdy and T. N. Rescigno, Phys. Rev. Letts. 41, 1364 (1978).
4. R. Yaris, John Bendler, Ronald A. Lovett, Carl M. Bender and Peter A. Fedders, Phys. Rev. A18, 1816 (1978).

5. T. N. Rescigno and V. McKoy, Phys. Rev. A12, 552 (1975); T. N. Rescigno, C. W. McCurdy and V. McKoy, J. Chem. Phys. 64, 422 (1976).
6. J. E. Avron and I. Herbst. Commun. Math. Phys. 52, 239 (1977); C. Cerjan, R. Hedges, C. Holt, W. P. Reinhardt, K. Scheibner and J. J. Wendoloski, Int. J. Quant. Chem. 14, 393 (1978) and references therein; I. W. Herbst and B. Simon, Phys. Rev. Letts. 41, 67 (1978).
7. I am indebted to W. P. Reinhardt for pointing this fact out to me.
8. B. Simon, Phys. Letters, accepted for publication.
9. T. N. Rescigno, C. W. McCurdy and A. E. Orel, Phys. Rev. A17, 1931 (1978).
10. T. N. Rescigno and W. P. Reinhardt, Phys. Rev. A8, 2828 (1973).
11. B. R. Junker and C. L. Huang, Phys. Rev. A18, 313 (1978).
12. W. H. Miller, C. A. Slocomb and H. F. Schaefer, J. Chem. Phys. 56, 1347 (1972).
13. A. D. Isaacson and W. H. Miller, Chem. Phys. Letts., to appear.
14. I. Shavitt, in Methods in Computational Physics, edited by B. Alder, S. Fernbach, and M. Rotenberg (Academic, New York 1963), Vol. 2, p. 1.
15. A. D. Isaacson, C. W. McCurdy and W. H. Miller, Chemical Physics 34, 311 (1978).
16. N. Bardsley and B. Junker, J. Phys. B 5, L 178 (1972).
17. J. E. Avron and I. Herbst, Commun. Math. Phys. 52, 239, (1977).
18. I. W. Herbst and B. Simon, Phys. Rev. Lett. 41, 67 (1978).
19. C. W. McCurdy, work in progress. It may be possible to combine complex scaling in the Hamiltonian with contour rotations in all matrix elements to produce a practical version of "exterior complex scaling."
20. Simon has discussed some unsolved mathematical problems in a rigorous overview of complex scaling. See B. Simon, Int. J. Quant. Chem. 14, 529 (1978).

DISCUSSION

Taylor: You mentioned that Miller, Isaacson and yourself have reinvestigated the Bardsley, Junker method for atoms. What did you find?

McCurdy: We looked at a potential scattering problem and examined the convergence to the true resonance values in the limit of a large number of expansion functions. We used up to 100 Laguerre functions and found that the method converges very nicely. Now recall that the Siegert function $e^{ikr/r}$ diverges in the limit $r \to \infty$ for k in the lower half plane and that Siegert originally applied his boundary condition at finite r. We are in a sense applying the boundary condition at larger and larger values of r by letting the number of basis functions increase and you might suppose that this would ultimately fail. But it seems to work.

Rescigno: In this connection, I believe that some people have implied that applying this technique was like calculating terms in an asymptotic series and that if you pushed it too hard it would ultimately diverge. Your results seem to refute this notion.

Taylor: When you apply this technique to molecular problems, where do you place the Siegert functions? Is it on a nuclear center or at some arbitrary point?

McCurdy: Presumably, the results are independent of the center chosen but there has not been much checking of this yet. In the He^* + H applications, the Siegert functions were centered on the helium atom. Of course, in that case, the Siegert function goes as an outgoing Whittiker function and not a simple plane wave.

Davidson: How were the matrix elements done?

McCurdy: They were numerically evaluated along rotated contours. You can do the integrals that way for a linear problem, but it becomes impractical for non-linear polyatomic systems.

Rescigno: The practicality of implementing these methods for realistic problems is an important question. In the Be^- work that Bill and I did, we first projected the Hamiltonian onto a set of real configurations so that when we then came to the evaluation of matrix elements with complex functions, we only needed to calculate overlap matrix lements. Now I'm not saying that this is the best way of proceeding, but I do think we have to give a lot of serious thought to implementing some of these ideas without completely redoing quantum chemistry.

DISCUSSION ON PHOTOIONIZATION AND MOLECULAR RESONANCES

Chairman: Robert Nesbet

Davidson: I want to make some remarks concerning an alternative scheme for carrying out a moment analysis in connection with photoionization (see article by Davidson following this discussion).

Hazi: What systems have you applied this technique to?

Davidson: We have looked at the K-shell ionization in atomic neon. We haven't looked at molecules yet, but we are building up to that. We have found that even-tempered Gaussian functions provide the best way to build a non-singular overlap matrix.

Hazi: What did your calculations show?

Davidson: We found that once you extend the recurrence coefficients, the α's and β's, using the known asymptotic limits, the Tchebyscheff procedure gives perfectly smooth results, but only if you carry it to high order.

Langhoff: Let me try to figure out what you're doing. If I take a continued fraction and truncate it at some point by assuming all the α's and β's subsequent to that point have the same numerical value, then I simply get a quadratic equation for the value of the continued fraction and if I solve it of course, I get a square root branch cut. Is that your branch cut?

Davidson: It's a square root branch cut.

Langhoff: But you don't want that.

Davidson: But we do this as a function of order.

Langhoff: Yes, the order is the last value of α you use before you assume they have become constant.

Davidson: Again, our procedure still oscillates as a function of order, so we then take a Cesaro mean.

Langhoff: But that differs a little bit from what the correct answer is. I'd like to point out that your procedure is similar to one used by Roy Gordon in connection with imaging the spectral densities of normal mode oscillations. The problem of taking a set of variationally determined α's and β's and then a specific extension or extrapolation-- that problem can be solved exactly for the associated density. It's the so-called reference density approximation.

Davidson: We should also point out that the time involved in any of these schemes is negligible compared to the time spent in getting the oscillator strengths to begin with. So it doesn't really make much difference how you do it, so long as you choose a stable algorithm that gives you the result you're looking for. My essential point is that if you are going to use the Tchebyscheff procedure, you must use extended coefficients and be careful not to truncate the procedure at too low an order.

Nesbet: It sounds like the various procedures, if done properly, will give numbers that are essentially the same. One very important thing that everyone should realize, however, is that it is well known in the applied mathematics literature that the algorithms which were put forward for going from the moments to the recursion coefficients (α,β) which are needed for the polynomials in the general theory of quadrature, which then go on to give the points and weights that become the principal representation, are numerically very unsound. You cannot really go in this direction. The only way you can go through that path is when you have the moments as rational numbers with an infinite number of significant figures. When I came across this point I realized rather soon that this wasn't ever going to work and so I found a way of going directly from the transition amplitudes to the recursion coefficients (α) and (β). This procedure is extremely stable and always works beautifully. You can always express it as adding sums of positive numbers. But everyone should realize how bad the first procedure is because there's a lot of earlier literature on the quotient-difference algorithm which you really can't use in practice.

The second point is that this technique among other things is very useful in describing the oscillator strength distribution, which is something Dalgarno and others have exploited very much in the past. Turning the thing around, it turns out that one can actually use this scheme of the cumulative oscillator strength to develop out of the computation something like the whole oscillator strength distribution. And then you can go back and fit this to Rydberg series and you can get, I suppose, reasonable estimates of entire Rydberg series of oscillator strengths by fitting under these curves and turning the imaging around. So we use the physically known positions of the Rydberg states which you of course can get in terms of the simple quantum defect analysis.

Another thing that you get out of it, again going back to the older Dalgarno work, is that you have closed form formulas for quantities such as the Van der Waals coefficients, the two-body and three-body dipole constants. In the trivial calculation I did for helium, the C_6 coefficient is the same as the best available experimental number. So this appears also to be a very good way to calculate these kinds of things.

Rescigno: If you have a simple Rydberg series, after you get beyond $n = 4$ or 5, the oscillator strengths just begin to fall off like $1/n^3$ anyway. What about a case where you have strong perturbations in the Rydberg series where you would need multi-channel quantum defect theory? Does it work then?

Nesbet: Then your oscillator strength distribution will have an irregular curve of some sort and you may have to separate the series. Now, that's another issue which we'll talk about later on, i.e., how to separate several excitation series superposed on each other. It's not entirely clear how to do that. Now these issues are dealt with in my Phys. Rev. paper [P.R. A14 , 1065 (1976)] which I thought it would be nice to add to Peter Langhoff's list of references and also this more recent paper by Barshun and myself on CH, which was published earlier this year in the Journal of Chemical Physics [JCP 68, 2783 (1978)]. Now we're seeing this method being applied to physical problems where we don't know the results experimentally and it is difficult to obtain results by any other method. We're seeing now, in the applications that have come up in the last year, that this really is becoming a quite powerful method.

Taylor: I have a question on Stieltjes imaging. Suppose you have some testing function, ϕ, and some continuum, ψ_ε. Stieltjes imaging generally obtains $|\langle\phi|\psi_\varepsilon\rangle|^2$. If you're doing photoionization ϕ is $r\psi_0$. What are the limitations on ϕ? I've seen two problems that are very tantalizing, because they're not photoionization but do use Stieltjes imaging. In one case, in a one-electron problem, Walter Kohn took ϕ to be $\delta(r - r')$. Then of course the quantity here becomes the one-electron density as a function of ε. He used that in the Kohn-Hohenberg Model and here he's using a delta function to be the testing function. He thought that he had discovered Stieltjes imaging and I had to point out that he hadn't and that this was just a special case of this testing function. I think this is very relevant.

The next question is this: Bill Reinhardt has recently done a problem where he calculated a T-matrix. He noticed that the imaginary part of the T-matrix is of the previous form and once he gets this form, he does as Hazi did for resonance widths. In the T-matrix case, the testing function turned out to be nothing more than the potential times a scattering function. Certainly the scattering function doesn't go to zero but maybe the potential at large distance, which would be $\alpha/_{r}4$, goes to zero. Does the testing function have to be square integrable? Can it be a delta-function? What kind of problems can you apply this to?

Langhoff: We've looked at a problem where the moments don't exist formally.

Taylor: The positive moments?

Langhoff: Yes. And that works. That is to say, you can justify this procedure mathematically using what methematicians call regularization which to us means that if you have some integrals that diverge, you put in something to make them converge and then go through the moment theory carrying it around.

Taylor: Does that mean you can let ϕ be anything you want?

Langhoff: I'm not saying that. I'm saying you can make the moment theory work in a number of ways even when you're dealing with moments that are not formally defined. Another regularization is provided simply by the fact that when you calculate moments in a square-integrable basis set, you just have a finite sum and because there's always a largest ε, you always get a finite number in a square-integrable basis. So you can say that a finite Hilbert space representation of the spectrum itself provides a regularization. We used this procedure on H^- and hydrogen, we used positive power moments in that case, and successfully imaged the spectra.

Nesbet: But in that case, the moments could not be bounded?

Langhoff: No, as you let n get large they diverge.

Taylor: But do you then use the first trick?

Langhoff: No, we're saying that the basis set itself provides the regularization.

Taylor: This is turning out to be a very important question because the same thing is coming up in complex rotation.

Rescigno: No, there you have a different problem. In Langhoff's case you have a representation where the basis set itself provides the regularization, because we use a finite number of energies. In some of the complex coordinate extensions you have in mind, if you look at this problem as one of trying to calculate matrix elements of some resolvent, with respect to some testing function, if the testing function is not square integrable then the matrix elements you need in the complex coordinate procedure may not exist.

Taylor: Yes, but again, if you expand the plane waves out to some large distance where the potential has become small, you get an answer.

McCurdy: In the complex coordinate system, the scattering functions, the $j_\ell(kr)$, are blowing up exponentially and if the potential doesn't decrease exponentially along the ray that you've rotated the coordinates, then the whole thing is diverging.

Langhoff: One more point about positive moments. I'll direct this to you, Bob, because the method is very important in connection with the way you formulate the procedure for calculating the recurrence coefficients. You talk about multiplying the inverse of the Hamiltonian matrix into a vector.

Nesbet: There's no doubt it's easier to multiply by a matrix than by its inverse.

Langhoff: Yes, that's the essential point. When you have positive moments, you deal with the Hamiltonian itself.

Nesbet: Would it be possible to put in these exponential factors and wouldn't that make it better behaved.

Langhoff: Yes, we have done it both ways in test cases on H and H^-, so we have looked at this in detail.

Kelly: A question that has been coming up several times is the question of the interweaving oscillator strengths in a Stieltjes imaging problem. It was a problem for us in this HF calculation where we noted that in fact any molecular oscillator strength, in some sense, is an infinite number of partial cross sections. And so, I'm wondering if people have seen this problem. Have you seen it, Tom?

Rescigno: Yes, we've seen that in our calculations. But what kind of a basis set did you use?

Kelly: It's similar to what Hazi already discussed in his talk.

Rescigno: Hazi only had two channels to worry about. You

have an infinite number of channels.

Kelly: This is in HF and so we're using a basis set of s, p,d and f orbitals and all of them can couple and I forget how many s's and p's we used.

Nesbet: Do you actually compute the principal moments, that's the most important question?

Kelly: Yes, and we wanted to break them up of course into series. Now I don't mean to say it's impossible to solve this, because we didn't work very long at it. But we decided that maybe if we did a careful population analysis, that might help to see if this series has more d-character and this one has more p-character, but we couldn't do it by just eyeballing quickly.

Nesbet: We do need good working criteria to separate series that are superimposed like this so we won't get into false structure or oversimplify the problem by just drawing smooth curves where there might be some structure. That's the difficulty.

Taylor: I wanted to mention an observation that Bob Yaris and I made when we were looking at the work of McCurdy, Rescigno and Orel and also of Bobby Junker. This concerns complex rotation and resonances. As McCurdy said this afternoon, the original prescription was to rotate the Hamiltonian, and use real basis sets. People didn't get convergence and they didn't understand why. They were supposed to get an answer that was hopefully independent of θ, but they got ones that were terribly dependent on θ. There was a long period where a lot of groups, including Bob Yaris, the Wisconsin group, the Swedish group, started looking for theorems that would tell you the best θ to use, even though the answer was supposed to be independent of θ. There is and will be a lot of literature coming out, based on virial theorems which try to pin down the best θ, which I feel is quite irrelevant.

Now these individuals pointed out that, if you use the rotated Hamiltonian, $H(re^{i\theta})$, what you really want to do - and let's talk with reference to a specific resonance, say $1s\ 2s^2$, which is represented by Q - is to rotate this set completely and the outer part which we'll call $1s^2\ k\ell$ - this is the P- space - only has $k\ell$ unrotated. Notice all the bound-like electrons are rotated. Then you preserve the Hartree-Fock model. You preserve the independent particle model. So the real problem was that over the years people were using the wrong inner orbitals. You could be doing uranium but if the 1s orbital is off, no matter how many basis functions you use, if the energetically heavy 1s is wrong, it's going to pull everything in.

Now I'll make another statement and here I realize I'm going out on a limb. Whenever you have quantities like these matrix elements of a resolvent with respect to a response function, you can use either Stieltjes imaging or rotated coordinates. It seems the rotated coordinate technique has the advantage of being able to use two different testing functions, although I must say that since we haven't applied any of these things, we may get some rude surprises. It turns out that the photoionization problem was done by Rescigno and McKoy using complex coordinates, but they had the same problem that people had before they discovered the Rescigno, McCurdy, Orel and Junker trick.

Rescigno: One of the attractive features, in fact, the whole motivating feature behind some of the complex coordinate approaches for looking at molecular resonance problems is that we want to be able to take advantage of this wealth of computer codes that the quantum chemistry community has built up over the last 10 to 15 years. I think it's fairly obvious from what's been said here during the last two days that people like Bill McCurdy and I and others who use L^2-techniques for doing scattering calculations are very heavily indebted to the individuals who have developed those codes and in a certain sense we just take those codes off the shelf. Now whenever one thinks of a new idea, the reason we're able to do it in a reasonable amount of time is because we can to a certain extent just take the codes off the shelf with a minimum number of modifications. But if one comes up with an idea which involves substantial modifications of those codes, it's fairly clear that if we're talking about a code like ALCHEMY or SCREEPER or one of the other big CI codes that people use - those things took 10 years to develop, and if you're thinking of something that requires a radical change you're going to spend the next 10 years debugging it after you've made the changes. So with reference to the practicality of using rotated coordinate methods for molecular scattering calculations, I think this inner/outer business is pretty well resolved now. We know that the resonance effects are basically one-electron effects, so we don't want to rotate the inner electrons, we want to use a set of complex orbitals - and it doesn't make a difference if they're simply scaled or whether they're not - for the outer electron. But since not all the basis functions are complex, we've lost the essential simplicity of the original dilatation transformation. We can't simply compute real matrix elements and scale them by a complex phase factor. We have to contend with the fundamental problem that we have new types of integrals to do. We should give serious consideration to the fact that the basis functions that we decide we're going to use should be not only physically reasonable but we should actually be able to do something in a reasonable amount of time.

If one pursues the ideas that Bardsley and Junker have shown to be very successful in atomic problems, that is, the idea of using energy dependent Siegert-type functions, I think you're going to find out very rapidly that you're not going to be able to do molecular problems. You can do linear molecules that way and in fact the case that Bill McCurdy showed - the results he got with Bill Miller and Alan Isaacson were for a linear system. Those matrix elements had to be done numerically. But that was a linear system. I think it's fairly clear that no one is going to do that for a polyatomic because no one is going to do numerical complex quadrature in three dimensions.

Now the implementation that Bill and I were talking about where we simply use Gaussian functions whose orbital exponents are complex - at least we know that we can do the integrals. The analytic expressions we have are valid for complex orbital exponents, but I just want to suggest that perhaps there are alternatives that are even easier.

Barry Schneider has already talked about this idea in some of his R-matrix work. Let's suppose I want to calculate a matrix element of a many-electron Hamiltonian between two many-electron basis functions $\langle \psi_\alpha | H | \psi_\beta \rangle$. Say these are determinantal wave-functions so they're a product of orbitals and on each side maybe one of the orbitals is complex. Let's say we represent the operator on a real basis set as an N-term separable expansion

$$H \sim \sum_{ij} |\phi_i\rangle H_{ij} \langle \phi_j|$$

where the (ϕ_i) are many-electron configurations and everything is real. This is just a matrix representation of the operator. Now for the final step where we have to calculate complex matrix elements of this Hamiltonian, of course once we close on this separable representation of the operator, we get

$$\langle \Psi_\alpha | H | \Psi_\beta \rangle \sim \sum_{ij} \langle \Psi_\alpha | \phi_i \rangle H_{ij} \langle \phi_j | \Psi_\beta \rangle ,$$

that is, all we have to do is calculate complex overlap matrix elements. In fact, the point which Bill McCurdy didn't emphasize when he showed those results on Be^-, a five-electron problem, is that we did use this technique. That's why we were able to do the calculation so easily. Now I'm not saying that this is the only way to proceed. I'm saying this is a reasonable first step. We haven't tried this in a molecular calculation, but I think that worrying about making these kinds of calculations practical is important. If you're really serious about doing this you're going to have to think of how to implement some of these ideas without completely re-doing quantum chemistry.

Bardsley: I had a list of questions I wanted to ask people. I'd like to ask three questions and could I ask them first and then see whether people would like to respond? The first one concerns the locality of the width and how much we have to worry about the energy dependence of the width. Bob Nesbet brought that up last night. I think one good place to examine this is in N_2 since the most work has been done in N_2 and we can extend the resonance model to its greatest lengths there. There are two things that worry me about the N_2 results. They picked the experiment extremely well. You get a lot more information out of the model than you put in. But there are two problems. One is the absolute normalization that seems to be wrong by about 10% and there are three possible explanations. One is that Kennerly, Bonham and Golden are wrong. The second, as Arvid Herzenberg pointed out, is that the non-resonant contributions are wrong. Thirdly, is that the elastic scattering contribution from the resonance model might be wrong, as I pointed out this afternoon. Now that's the least likely explanation because the width is not really so large that you get into trouble. The other problem with N_2 that I see is that as you go to the high energy end of the structure, all theories, whether they're resonance models or some hybrid models, seem to underestimate the cross section. And I think that may be because in not building in the barrier penetration factor as a function of energy rather than R, you're underestimating the probability of capturing the high energy electron and re-emitting the high energy electron. And I was wondering whether either Arvid or Bob would have some comments on that.

A second question is directed at Barry. We have an experimentalist in disguise among us. One of the topics of high current interest is the dissociative attachment measurements of electrons on F_2. These are all measurements done essentially the same way using swarm methods, measuring the apparent cross section as a function of the mean energy in various different buffer gases.

Schneider: I think that Chantry's measurement is not a swarm experiment, it's a beam experiment.

Bardsley: What I'd like to ask Barry is if he could say how consistent these are. How much can we learn from this data in terms of getting attachment cross sections as a function of energy? Are we at the stage, for example, where we can start to model this process and learn vibrational excitation cross sections? There's one paper that's already been published in which people tried to analyze these results and get vibrational excitation cross sections [JCP 68, 1803 (1978)].

My third question concerns HCℓ, DCℓ and polar molecules in general. What I'd really like to know is, after a lot of theoretical work, what have we learned? What tests can we apply to find out whether we understand what's going on? Are there any predictions that we can make? Can we say something about the isotope effect in vibrational excitation? What's the link between the dipole strength and the vibrational excitation cross section? Are there any systematics in going from one molecule to another?

Herzenberg: I'll start at the end which is the one I still remember. I'm sure the isotope effect for DCℓ is quite straightforward to work out, but I can't do it in my head because we haven't done it. It should come out of the model with no problems. As far as understanding what's going on in HCℓ, I'm not quite sure if this was in your mind but remember this afternoon we were arguing what sort of state we were dealing with here. I suppose we could certainly establish, within the approach Dubé and I used, where the singularity in the scattering amplitude really is. You do a fixed-axis calculation and continue the scattering amplitude off the real axis. You can decide whether there is in fact a pole on the imaginary axis or whether it's somewhere else. If it's on the imaginary axis, then one would call it a virtual state.

Taylor: What stabilization can calculate is large amplitudes. And that's what we got, and we got two of them. There's no question ther are two diabatic states there and that they die out at large distance. I don't think the energy is going to be zero frankly. We tried very hard to make it zero, because the trick in stabilization is to make sure the electron will go away as you add to your basis set. But of course in any finite size calculation, you can make an error, especially now if you are only 0.3 eV from zero.

Bardsley: But what predictions can we make?

Taylor: That's always very tricky. With respect to these two "resonances," there's a third one certainly at high energy and it's of the same type. It's not a valence resonance. It's a big orbital. That was the one that gave us a 6.8 eV barrier. It gave us a potential curve with a 6.8 eV barrier which explains the disassociative attachment leading to H^- and why it has that onset.

Bardsley: But that wasn't a prediction, that was an explanation.

Taylor: I know. But, very honestly, we didn't think we were going to get it there. That same series of states, if you want to call it a series, the third one agreed beautifully with another experimental point. That's all I can say. The point is that each state can be correlated with two different experiments. I'm not saying it's a definitive calculation. Stabilization never is. Eventually you have to do a scattering calculation. But every one of those curves can be related to two separate experiments. You may think of other explanations for each one experiment, but when you put them all together, it's hard to get one set of curves that explain all the experiments and these do. I'm talking about dissociative attachment, HCℓ, the structure in it, H^-, the onsets, the whole thing.

Bardsley: But the explanation has not gained you anything.

Taylor: Sometimes we can make predictions. It happens that in $HCℓ^-$, we didn't make any predictions. There is something now in CO_2^- which seems a bit more definitive. That's the fact that these same types of states seem to show up, according to Peyerimhoff and to Ken Jordan, in CO_2^-. If you bend CO_2^- and do a calculation including diffuse orbitals - and by the way, bent CO_2^- has a 1.8 Debye dipole moment - you actually find quite a few dipole Rydberg states packed against the CO_2 curve. And as the state opens up, those states fail to parallel CO_2 by a very small amount. So what they do is to actually cross over - at 170° they are there, packed against the bottom with the same type of localization and this was found by Peyerimhoff. And this is very much the nature of the states I found in HCℓ. They are close enough so that they may be overlapping resonances.

Herzenberg: Since you were asking for predictions there's one I'll try with CO_2. If you agree that when CO_2 bends, it develops a dipole moment, and that therefore you get these bound dipole Rydberg states, then they are probably not bound when CO_2 is linear. Then I would predict a large temperature dependence in the low energy scattering cross section, simply from the bending vibrations. After all the bending vibrations, I think, are at 80 meV which is 800°.

Taylor: I want to think about that. I'm not arguing with you, but I'd like to reserve judgement.

Rescigno: I'll make a pitch for theory. You talked about some F_2 results. I think that with the stabilization calculations we did on F_2 and F_2^- two or three years ago, we suggested that the dissociative attachment cross section for electrons on F_2 would have a zero energy threshold.

Taylor: You mean zero kinetic energy at threshold?

Rescigno: Yes. Bender and I predicted the cross section would be largest at zero energy. And that prediction was made before any of the measurements were done and that, I think, is consistent with the data you talked about.

Bardsley: That seems to be consistent with all the swarm experiments. Is that consistent with Chantry's experiment?

Schneider: Yes, within his resolution. He does see this very large rise and then a tailing off. There are three experiments now that I think one could say are in reasonable agreement, as far as shape. The cross sections we got, Peter Chantry's cross sections and unnormalized cross sections of Wong at Yale all agree beautifully in shape. Now Peter says he's got an absolute cross section. We claim we have absolute cross sections. We differ by a factor of about 1.7. Chantry says his measurements could be in error by no more than 40%. So he says for one reason or another that ours are wrong. We looked at all the possibilities that could make our cross section too small and everything in our experiment says it should go the other way.

Herzenberg: Bob Hall fitted a swarm experiment for F_2 by a negative ion potential energy curve. Having done that, he then predicted the dependence of vibrational excitation and dissociative attachment on the initial vibrational state. Now there's a lot of predictions starting from one experiment there.

Schneider: I think the data Hall used was old data.

Bardsley. You say you did a calculation similar to Hall's?

Schneider: Yes, but the kind of calculations I was talking about were simple back-of-the-envelope things - trying to take the cross section information that was given and parametrize it in terms of a compound state model and play with the parameters. There were three sets of data at that time - this was about a year ago - and I tried to say whether one set of data looked more reasonable than another just because in trying to fit them, I got unreasonable physical parameters. And there was one set of data which was clearly wrong and at about the same time it was retracted in the literature. So I don't know if the calculation I did proves anything.

Herzenberg: But anyway, we do know that if we have one decent set of data on this dissociative attachment from F_2 at low energies, then we could get a pretty good compound state model from this and we could make a whole set of predictions from it.

Schneider: If you're talking about shapes and not absolute cross sections, I think any of the three experiments - ours, Wong's or Chantry's - are all in agreement.

Bardsley: But I think you need both. You need the absolute magnitude and the shape in order to get both the width and the potential curve.

Herzenberg: Yes, I'm afraid so. We have two parameters. Norm, you also had some questions about N_2.

Bardsley: I just wondered whether there was any missing physics in the underestimation of the high energy structure.

Herzenberg: I'd be surprised if it came from the non-locality of the width in this case.

Bottcher: Actually, I can answer that specifically. Sukamar and I did a model calculation. It was a more or less accurate calculation in which we really didn't know how the width depended on energy, but we made the usual extrapolation to zero energy. We used, actually, a discretization technique for the non-local problem and we compared it with the local approximation done both with discretization and the numerical differential equation technique. You get almost the same answers every way. If you look at it closely, it hinges on the open or accessible vibrational wavefunctions being a reasonably complete set. So whenever you've got a set of vibrationally bound intermediate states, and a set of bound final states, you do not expect non-locality to be very important. However, if you have a dissociative intermediate state, then it can become much more important. But in the N_2 problem it's a waste of time looking there. It just makes no difference to the numbers.

Nesbet: I wanted to add just one comment on the N_2 question. We're going to run the Dill and Dehmer calculation through my model adiabatic approximation and see what the vibrational structure is. We're going to do the same thing with the most recent calculation from Aaron Temkin.

Herzenberg: On the question of predictions on the N_2 problem, at the Seattle ICPEAC we actually produced a folder which had two calculations in it, one of which was measured afterwards and that fitted very well. Dubé actually worked out the vibrational excitation starting from a vibrationally excited state. And it fit like a glove. So this was one prediction that worked. There was a similar aspect which touched on the vibrationally elastic cross section. After Dubé finished all his fitting on v = 0 to 1 and v = 0 to 2, etc., one actually managed to do an experiment on the vibrationally elastic scattering by picking up the large changes in angular momenta. This gives you enough broadening in the energy of the outgoing electrons to know that he must have excited some rotational states. And Dubé did some fitting on that and picked out the $\Delta J = 4$ component which could only come from the resonance and that did pretty well. And for the $\Delta J = 4$, v = 0 to 0, the old resonance model made a prediction and that did pretty well. That was actually published by Dubé and Wong in the February issue of Physical Review.

Nesbet: I might just comment that the results I showed for the $0 \rightarrow 1$ transitions should essentially be identical to the

Herzenberg, Birtwistle results because I used their parameters. So in magnitude they should be the same and it's in apparent agreement with Wong's present statement about the normalization.

A MODIFICATION OF THE LANGHOFF IMAGING TECHNIQUE

Ernest R. Davidson
and Richard L. Martin
Chemistry Department BG - 10
University of Washington
Seattle, Washington 98195

P. W. Langhoff[1] has introduced Stieltjes and Tchebycheff imaging techniques for smoothing a discrete L^2 approximation to the continuous photoionization cross section. These techniques are discussed at length in Langhoff's accompanying presentation. In this technique the energy-weighted moments

$$S(-k) = \sum_{i=1}^{N} \varepsilon_i^{-k} f_i$$

of the oscillator strength from a L^2 psuedo-spectrum are matched to a continued fraction generating function,

$$\beta(z) = \sum_{k=0}^{\infty} S(-k) z^k$$

of the form

$$\beta(z) = \cfrac{b_1}{1-a_1 z-\cfrac{b_2 z^2}{1-\dots}}$$

The nth order truncation of the continued fraction gives a Padé approximant

$$\beta^n(z) = P_{n-1}(z)/Q_n(z)$$

The poles and residues of this yield the Stieltjes psuedospectrum $(\tilde{\varepsilon}_i^n, \tilde{f}_i^n)$. A formal series expansion of $\beta^n(z)$ agrees with the first 2n terms of $\beta(z)$ for n<N.

The photoionization cross section is given in the Stieltjes method by

$$\sigma(\omega) = 2\pi^2 c^{-1} g(\omega)$$

where g (ω) is the derivative of the cumulative oscillator strength

$$g = dF/d\varepsilon$$

$$F = \int^{\varepsilon} df$$

While this method gives very reliable results, it has the disadvantage that it gives estimates of σ only near the midpoints on the intervals $(\tilde{\varepsilon}_i^n, \tilde{\varepsilon}_{i+1}^n)$.

Langhoff et al. also showed that it is fairly easy to preassign one of the $\tilde{\varepsilon}$ and find a Padé approximant $\beta^n(z,\tilde{\varepsilon})$ which agrees with the first 2n-1 terms of the series for $\beta(z)$. From this the Tchebycheff approximation is produced by computing the cumulative oscillator strength and finding its derivative at the preassigned $\tilde{\varepsilon}$. Although this gives a continuous approximation to σ, this approach has not worked satisfactorily since neither high nor low order Tchebycheff results are reasonable when derived from psuedo-spectrum moments.

The higher moments S(-k) computed from the psuedo-spectrum tend to be less accurate than the lower ones. Consequently it is beneficial to replace the higher continued fraction coefficients (a_n, b_n) by an extended set which have the correct asymptotic behavior. Langhoff et al. have suggested using

$$a_n = (2\varepsilon_T)^{-1} + c_1 n^{-1} + c_2 n^{-2}$$

$$b_n = (4\varepsilon_T)^{-2} + d_1 n^{-1} + d_2 n^{-2}$$

where ε_T is the threshold energy for photoionization and (c_1, c_2, d_1, d_2) are found by fitting to the lower a_n, b_n. Use of these extended coefficients in very high order Tchebycheff calculations gives reasonable results.

Once this extension is made, however, an alternative scheme for determining σ becomes available. The continued fraction β now has a branch cut for real values of z greater than ε_T. Further, on this branch cut the continued fraction fails to coverge but the limit as z approaches the real axis is well defined and has a non-vanishing imaginary part. The cross section is easily re-expressed as

$$\sigma(\omega) = 2\pi\omega^{-2} \lim_{\eta \to 0^+} \mathrm{Im}\ \beta(\omega + i\eta)$$

Because the fraction does not converge for $\eta=0$ and any finite truncation gives zero for the imaginary part, this equation has not been used. Gordon[2] has, in fact, previously suggested using this equation with all (a_n, b_n) beyond a certain N replaced by their asymptotic values a_∞, b_∞. Langhoff et al. have argued, however, that this cannot work since it gives a square root branch cut which is known to be different in form from the branch cut in exactly soluble models such as the hydrogen atom.

We have found a method for evaluating σ from β which exactly reproduces the numerical values of σ for all cases considered by Langhoff and agrees with the infinite order Tchebycheff limit for all other examples tried to date. In this procedure σ^N is defined as the approximation to σ resulting from replacing all the (a_n, b_n) beyond n=N by their asymptotic limit. Since the continued fraction is not convergent, this results in a function which oscillates with N about the correct value (once N is sufficiently large). The Cesaro mean

$$\tilde{\sigma} = \lim_{K \to \infty} K^{-1} \sum_{p=L}^{K+L} \sigma^p$$

has been found to converge rapidly to the desired value. For the hydrogen atom a relative error of 10^{-4} was obtained with $L \simeq 15$ and $K \simeq 35$.

Explicit equations for evaluation σ^N can be found following Wall.[3] Suppose we define

$$x = 4\varepsilon_T z^{-1} - 2$$

$$A_n = -2 + 4\varepsilon_T a_n$$

$$B_n = 16\varepsilon_T^2 b_n$$

so

$$A_\infty = 0$$

$$B_\infty = 1$$

Then

$$\sigma = 8\pi\varepsilon_T\, b_1\, \omega^{-3}\, t$$

$$t = \lim_{\eta\to 0^+} \text{Im}\ t_1(\omega+i\eta)$$

$$t_1(z) = (x-A_1-t_2)^{-1}$$

$$t_n(z) = -B_n\ (x-A_n-t_{n+1})^{-1} \qquad n \geq 2$$

Wall has shown that, with $T = t_{N+1}$, t_1 can be written as

$$t_1 = (Y_{N-1}T+Y_N)\ (Z_{N-1}T+Z_N)^{-1}$$

where Y_n and Z_n satisfy the recursion relation for X_n,

$$X_n = (x-A_n)\ X_{n-1} - B_n\ X_{n-2}$$

with $Y_1 = Y_{-1} = Z_0 = 1$ and $Y_0 = Z_{-1} = 0$.

Since Y and Z are real for real z it is convenient to use the identity

$$Y_n Z_{n-1} - Y_{n-1} Z_n = \prod_{K=2}^{n} B_k \qquad n \geq 2$$

to write

$$t = (\prod_2^N B_k)\ |Z_{N-1}T_+ + Z_N|^{-2}\ \text{Im}(-T_+)$$

where

$$T_+ = \lim_{\eta\to 0^+} T(z).$$

For T we have used the value resulting from replacing all the remaining a_n, b_n by their asymptotic limit so that

$$T = -(x+T)^{-1}$$

or

$$T = -\tfrac{1}{2}\ (x \pm i\ \sqrt{4-x^2}) \qquad \text{for } |x|<2.$$

This gives

$$T_+ = -\tfrac{1}{2}\ (x + i\ \sqrt{4-x^2})$$

for real x ($\eta=0^+$). Thus we have finally,

$$t^N = \tfrac{1}{2} \left(\prod_{2}^{N} B_k\right) (4-x^2)^{\frac{1}{2}} \left[(Z_N - \tfrac{1}{2} x Z_{N-1})^2 + \tfrac{1}{4} (4-x^2) Z_{N-1}^2 \right]^{-1}$$

This equation, in connection with the Cesáro mean, has been found to work well in practice. Because of the simple relation between Z_n and Z_{n-1}, Z_{n-2} this equation is easily used to produce a sequence of t^n.

REFERENCES

1. P. W. Langhoff, C. T. Corcoran, J. S. Sims, F. Weinhold and R. M. Glover, Phys. Rev. A, 14, 1042 (1976) and references therein.
2. R. G. Gordon, Adv. Chem. Phys. 15, 79 (1968).
3. H. S. Wall, Analytic Theory of Continued Fractions, (Van Nostrand, New York, 1948).

WORKSHOP ON L^2 - METHODS

Chairman: Ernest R. Davidson

Schneider: I want to discuss some problems one faces computationally in using these L^2 - methods. Most of what we've already heard on the L^2 - methods, aside from the CI work on photoionization, has been limited to the static-exchange approximation or static-exchange plus some sort of model polarization or optical potential. But the general question I want to throw out for discussion is - when you are really faced with the full multi-channel scattering problem and you want to implement it, what kinds of difficulties are you faced with? I alluded to some of those problems in my talk concerning integral transformations. What seems to be common to all the L^2 - methods is that they all require fairly large basis sets. If we write down the scattering wavefunction as

$$\Psi(1,,N+1) = \sum_{i\alpha} C_{i\alpha} \mathcal{A}(\phi_i(1,,N)\ \phi_{i\alpha}(N+1)) + \sum_q C_q\ \Psi_q(1,,N+1)$$

where ϕ_i is a target state, or if you like, a psuedo-state, then the first terms represent the open channels and the other terms can be anything else that dies exponentially. This is a fairly general wavefunction, and we want to be able to set up all the integrals that are required to construct the Hamiltonian. In the T-matrix language, you would still be faced with very similar problems. You'd need a T-matrix which was labeled by channel indices and then in addition you would have a one-particle basis to represent the scattering orbital in each channel.

Rescigno: That's exactly what Joe Macek did.

Macek: It's certainly true that exactly the wavefunction you wrote down is the most general type you could incorporate into the

T-matrix method. It can be done and is fairly straightforward.

Schneider: If you really believe that, I think you're in for some surprises. I'm not saying it can't be incorporated; I'm saying that there are some practical questions that have to be answered before you can do this on real systems and I'm not talking about H_2.

Temkin: Do you think you can do it for H_2?

Schneider: I've done a 3-state calculation on H_2.

Hazi: Let me make a comment. Remember I once asked you whether you had the flexibility built in so that the target states, ϕ_i, could be made up from more than one configuration?

Temkin: That's not a fundamental limitation.

Hazi: But you have to make some transformations.

Rescigno: You can contract out the target state. That's not too hard.

Robb: I'd like to make the plea that we don't use the word "easy" so much. If it was so easy to do a multi-state H_2^+ calculation, it would have been done long ago and there's nothing in the literature.

Temkin: There isn't even a 2-state $e^- + H_2^+$ calculation.

Schneider: There is a 3-state $e^- + H_2$ calculation which I did and it's in the literature. Only the elastic cross sections were published, but it was a 3-state calculation.

Bardsley: Do you believe it? Was it converged?

Schneider: Certainly. It's obviously converged in the elastic channel or I wouldn't have published it.

Bardsley: Converged with respect to what?

Schneider: Well, if you look at the total cross section, it's in very reasonable agreement with experiment.

Temkin: I have another question. Admittedly there is the problem of getting the matrix elements, $H_{i\alpha,j\beta}$. The way you've written the wavefunction, you have something that looks like a scattering orbital, $\phi_{i\alpha}$. Now whether you use a traditional close-coupling approach or an L^2 - approach, is there anything fundamentally different about the kinds of problems, particularly with regard to matrix elements?

Schneider: Yes, there are differences.

Temkin: Then I would like to have them elucidated.

Schneider: There are two aspects. First, the basis states you use as the ($\phi_{i\alpha}$) are going to be very much larger than those you use to describe the target states and the psuedo-states. This is true even with multi-configuration target states. That's irrelevent and I think Tom Rescigno has said that already. We can make up the linear combination of determinants that go into the ϕ_i, then we can contract them into one function. That's not a problem.

Davidson: It's a terrible problem.

Schneider: No, I've done it, so it's not terrible.

Rescigno: It is in the general case.

Schneider: No it's not.

Rescigno: That's not true. You did a particularly easy case. If you have a very complicated open-shell case where there are lots of different spin-couplings, it's a non-trivial problem.

Schneider: Look, let me explain the philosophy. Let's say you have a CI program which can construct, as a linear combination of Slater determinants, a Hamiltonian matrix. To get the Hamiltonian matrix over the functions I wrote down just requires re-expanding in terms of the Slater determinants or spin eigenfunctions if you like and then doing a contraction in a configuration-interaction sense instead of in a one-electron sense. In other words, you use the coefficients from the N-electron diagonalization to get the contraction coefficients for the (N + 1) electron Hamiltonian.

Hazi: But what about the anti-symmetry, Barry?

Schneider: It's in there. For H_2, we used up to 16 configurations in the target wavefunction. The problem that you are going to be faced with is not the number of configurations in ϕ_i, but the very large one-electron set needed to represent the scattering orbitals. You may be able to compute the integrals, but you will have to transform two-electron integrals where two orbitals come from the inner space and two from the larger space. So you have to transform and store large numbers of two-electron integrals.

Rescigno: Let me say something. You're talking about practical difficulties. I want to point out a more fundamental difficulty and I don't think anybody appreciates it. The form of the wavefunction you've written down here is just standard close-coupling plus correlation. The only difference between this and what close-coupling people do is that the $\{\phi_{i\alpha}\}$ are a set of L^2 - functions which you're effectively using to expand continuum orbitals. So now you've got a problem because in general you're going to use the same types of functions in $\{\phi_{i\alpha}\}$ that you used to get the target functions. The point is that the channel energies in this scheme are not well defined, because in general you have approximate target wavefunctions for a many-electron target and when you use a set of L_2 - functions to expand the one-electron scattering functions, and when you make up (N + 1) electron configurations, you implicitly re-correlate the target wavefunctions.

Schneider: That's absolutely not true if you do it my way.

Rescigno: It is true. You're going to define channel energies by the diagonal matrix elements of the N-electron problem with respect to your $\{\phi_i\}$ and those are not effectively the true target functions you're using when you add another electron.

Schneider: You're saying two different things. There is a difference between re-correlation and not having exact target energies.

Rescigno: But both problems would disappear if you had exact target states.

Schneider: If you use your knowledge of the coefficients you get from the N-electron diagonalization, you can minimize the re-correlation problem. If you expand rather in terms of all spin eigenfunctions, you get a very bad re-correlation problem. I've done it in H_2 and in H_2 the extra electron got bound when I did it that way. The way I first suggested is best.

Taylor: If you do a good job on the short-range problem, then what you're adding are long-range functions and they shouldn't effect things much.

Nesbet: I have had considerable experience with this equation in its most general form for multi-channel problems, because this is exactly what's used in my electron-atom scattering variational calculations. These structural problems occur and they are soluble. They are soluble in detail in that if the $\{\phi_{i\alpha}\}$ are orthogonalized to all of the other basis functions and you're careful that you diagonalize the Hamiltonian of the matrix of the N-particle problem first and you're careful in choosing the configurations consistently; it can all be worked out in a perfectly clear way with no approximations depending on numbers being small.

Hazi: Is this true when the $\{\phi_{i\alpha}\}$ are constructed from a larger one-electron basis than for the target itself?

Nesbet: I'm saying one must orthogonalize all the $\{\phi_{i\alpha}\}$ to the other target basis functions.

Hazi: I'm sorry, but is it required that if a large number of diffuse basis functions are used in the set $\{\phi_{i\alpha}\}$, that they be used to expand the target functions as well? That's what I'm asking.

Schneider: I think no.

Temkin: He's saying exactly the opposite. He's saying they should be orthogonal to the ones in the target.

Nesbet: The Hamiltonian has to be diagonalized in the full set.

Macek: I just want to point out that in the separable approximations, I don't like to call it the T-matrix method, you are not expanding the wavefunction in an L^2 - basis.

Robb: I don't know what you mean.

Macek: You're expanding the potential in an L^2 - basis, not the wavefunction. You have some terms that put in the correct asymptotic form. In the R-matrix method you expand the wavefunction, but only inside the box. In the T-matrix method, the only thing you expand in an L^2 - basis is the potential. That's different from the other methods where you expand the scattering orbitals everywhere, including asymptotically.

Schneider: The Kohn method doesn't do that either.

Nesbet: The crucial thing is that the N-particle Hamiltonian be diagonalized in the set of functions $\{\phi_i\}$. And that must be done strictly. Moreover, the functions that are used to construct the $\{\Psi_q\}$ must be the same set as used in the $\{\phi_i\}$ but the set $\{\phi_{i\alpha}\}$ are only required to be orthogonal to this set.

Schneider: And since you have orthogonalized the $\{\phi_{i\alpha}\}$, there's one particular type of configuration that's very important to include in $\{\Psi_q\}$, namely the one that relaxes the orthogonality constriction on the $\{\phi_{i\alpha}\}$.

Bottcher: Let me say again, if you do everything by transforming to a vast basis of conventional quantum chemistry orbitals, you eventually run into linear dependency as Tom Rescigno pointed out. Now the long-range solution, and I'm not saying it's easy, is to use numerically oriented evaluation of integrals over MO's. This was done by Schaefer for $C\ell_2$.

Schneider: Look, in the LiF calculation, where we actually used the option in POLYATOM to only calculate the integrals we needed, we had 56 million octal words of integrals.

Bottcher: That's the number of primitive integrals, surely not the number over MO's.

Schneider: Of course.

Bottcher: This problem is perfectly adapted to the new generation of parallel processors.

Davidson: The real problem is that this expansion of a sine wave in Gaussians is nonsense. Let's make it perfectly clear what you're talking about. You're taking a set of Gaussians all on the same center, successively broader, and out of those your constructing something like a sine wave.

Schneider: But don't forget, in electron scattering at low energies, the wavefunctions won't oscillate rapidly on the scale of an atomic or molecular dimension. At higher energies, you do run into trouble.

Fliflet: I think that all this discussion of transformations is important and it's certainly a limitation even in static-exchange. It's also preventing the application to bigger polyatomic systems. But I think there are a number of other problems that come in when you try to go to a multi-state, multi-channel formulation. I'll bring up one, for example. If you're doing an R-matrix calculation, then your box has to be large enough to include most of the target and if you're doing a static-exchange calculation, that allows you to use a small box, may 10. au. But if you're going to do inelastic calculations where you need to include excited states which have a much bigger extent, then you have to go to a much bigger box.

Robb: But before you get to that problem, there may be a lot of low-lying excited states.

Rescigno: But closed-shell molecules don't necessarily have many low-lying valence states, they have Rydberg states.

McKoy: A lot of the triplets are low-lying.

Schneider: You can develop transformation programs specifically designed to handle the class of integrals you need. But I still think there will be practical limitations on the number of functions you can handle.

Hazi: Since we're talking about basis sets, I'd like to bring up another problem and that is that most of the calculations with very few exceptions have been restricted to the R_e for diatomic molecules. Now I believe, and this comes from my personal experience in doing the resonance width calculations as a function of internuclear distance, that even if you optimize the exponents with say ordinary Gaussians or Slaters, the one-center decomposition of the basis does change for a fixed set of orbitals as you vary the internuclear distance and therefore the basis may improve or get worse as you go to larger R.

Rescigno: It's fair to say that with the size of basis sets we're talking about, no one optimizes the orbitals.

Hazi: I'm saying that once you've chosen some orbitals, the angular properties of those molecular orbitals change as you change the internuclear distance. And so there's no guarantee that your accuracy will be equally good as you change R and nobody has computational experience in this area. I don't know the solution, but I do think people should do more calculations as a function of R to get some feeling for how the basis set adequacy holds up.

Davidson: It's true that you're going to have serious problems here if you really stretch things to large distances, because if you leave functions stuck at the middle of the molecule when there isn't anything, you'll get nonsense and if you put functions on each atom and you bring them back together, you get linear dependency very quickly. You can't use the same expansion at all distances.

Taylor: Vis-a-vis the Rydberg states and the size of the R-matrix box, there is one saving grace. You don't have to worry about exchange between the Rydberg orbital and the core.

Schneider: But you do have to worry about exchange between the Rydberg electron and the scattered electron.

Bottcher: I think the way to do this is to formally put the Rydberg state into the continuum.

Schneider: Then you're doing exactly what Fano suggested. Namely, you generalize the R-matrix method to a hyper-box containing the core plus 2 electrons outside.

Taylor: There is something we used to talk about all the time in comparing the T-matrix and R-matrix methods. The T-matrix method carries the burden that Bob Nesbet used to carry. If long-range forces are important or if nucleii are well-separated, you have to use a discrete basis all the way out. The R-matrix at least has the numerical advantage in the outer region of being able to do things without a basis set. Of course the close-coupling methods have this advantage, maybe to their detriment. How would you do LiF, with a big long-range force, in a discrete basis set?

McKoy: All you have to do is pick out the appropriate Green's function. You don't have to work with $G^{(0)}$. We can do Coulomb problems this way.

Schneider: But with real multipole problems, there is no well defined asymptotic solution. That's why we integrate numerically.

WORKSHOP ON SINGLE-CENTER TECHNIQUES

Chairman: Ron Henry

Henry: I'd like to sum up exactly where I feel we stand with the eigenfunction expansion techniques, what exactly has been done for molecules and what the future may hold for us. In particular, I think that the groups working on this - including Lee Collins, Mike Morrison, Dave Norcross and Derek Robb in some sort of loose collaboration - are using the integral equations method to solve the standard equations. This integral equation method has also been used in photoionization. I mentioned earlier that photoionization of H_2^+ and H_2 has been looked at by Ritchie's group in Alabama and also by Chapmann and Hayes. These techniques will suffer when you have a lot of exchange terms if you want to do it non-iteratively, but one of the recent advances has been that Derek Robb, Lee Collins and Mike Morrison have developed an iterative exchange type of method. They have gotten results at this point for H_2^+, H_2, N_2, CH^+ and LiH. These results have been obtained in the static-exchange approximation.

There's another large article written by Norcross and Collins dealing with the electron-polar molecule systems (PR A18, 466 (1978)). There are eight polar molecules that they have looked at using a model exchange type approximation with the model that appears to work the best, i.e., Hara's free electron gas model. Note also that Joe Dehmer and Dan Dill thought this was a good way to include exchange in their multi-center calculation, which also is a very important calculation. The polar molecules they looked at are LiCL, NaCL, LiF, NaF, KI, CsF, CsOH and KOH. Again, their interest is primarily in momentum transfer cross sections so, if you like, they're doing the elastic scattering.

Other groups are working on electron-CO, trying to do

vibrational excitation. We heard something from Dr. Choi on that and also Aaron Temkin is working on N_2 vibrational excitation. We are hoping that these methods will be fully converged at some point.

I think that where the problems lie, not just here but in the L^2 - methods as well, are in the ways to do electronic excitation. We're finding that if you just do something like elastic scattering you already have a lot of angular momentum channels, about 42 in the case of NaCL, so that if you need a similar number for each electronic channel, the methods quickly get out of hand.

One can take heart in the fact that the methods do seem to be fairly stable and can couple these large angular momenta together. One has to use special techniques when one couples an ℓ of zero with an ℓ of 42. These techniques were mentioned by Mike Morrison and have to do with stabilizing or transforming every few points in order to keep your solutions from becoming linearly dependent. The techniques are spelled out in a paper by Morrison and Collins in the work they did on H_2 and N_2. It's also spelled out in the CO_2 paper by Morrison, Lane and Collins.

This then leads me to the question - this is a session on single-center expansions - should we be using single-center expansions or should in fact one move into a multi-center approach so that you are not getting nearly as many ℓ-values coupled together? Perhaps we should be looking for a hybrid approach, using a multi-center expansion in-close and moving over at large distances to the other description.

Schneider: In that regard, I took the LiF static potential and expanded it in a Legendre series, both in spherical and prolate spheroidal coordinates. Much to my chagrin, I found out that in the important region where there is charge density, there was no big difference in how the two series converged. While it is true that the spheroidal coordinates are better as you move away from the molecule, there's no big difference in close.

Moores: That's certainly not the case in N_2. You can get away with a very small number.

Henry: In N_2, for example, Crees and Moores have a 6-term expansion, whereas Buckley and Burke have to use 14 terms.

Schneider: I'm talking about a strongly polar molecule. I would agree that with the homonuclear molecules, the prolate spheroidal expansion is much better.

Dill: I agree with you. But we're not always going to be dealing with diatomics. In our calculations we always find, incredibly sometimes, that we just don't need large expansions. Our N_2 calculation was done with 3 ℓ's around each atom and 5 ℓ's around the outer sphere.

Schneider: That does not answer the question.

Dill: I'm merely pointing out that we have a method that does not seem to require large linear systems. It may be that some sort of a hybridization might get us all to that end a lot quicker.

Hayes: I believe your comments about being able to use just 3 ℓ's, but if you approach this from a single-center expansion method, you can't get those three correctly unless you include ℓ's up to 12 or 15 in there to start with.

Dill: I agree. So don't do close-coupling with single-center expansions at short range. Do close-coupling outside if you like.

Hazi: I think it's a little misleading, because once you're using s-, p- and d- waves on centers A and B, if you re-expand onto a new center, you'll get high partial waves.

Dill: That's exactly the point, so don't attempt to re-expand.

Lane: I'd like to make a point that's obvious to everybody here but not to our colleagues in other areas, and that's the fact that the ℓ's have nothing to do with the angular momentum in close to the molecule. Outside, ℓ's have significance because outside, as has been pointed out, for low energies only a few ℓ's are important and they really are angular momenta associated with the outgoing wave. I'm only making the point because it's confusing not to us but to our experimental colleagues for instance, and sometimes to the people who are concentrating on electron-atom scattering where often ℓ conveys a different meaning.

Henry: In practice the way things are handled when you have 42 ℓ-values coupled in the interior region, you don't bring them all out to infinity. You drop all but a few as you go out because the coupling becomes extremely small.

Dehmer: I want to say that there's not a really focused opinion even within this group. It's a little sad if ℓ becomes irrelevant within the core because there are many processes like K-shell photoionization in which ℓ has a very specific meaning as the process gets started. Let's say you go from s to p in the K-shell of N_2 and then you're scattered by the anisotropic field into certain asymptotic ℓ's, but it's nice to retain ℓ's in a multi-center representation.

Lane: The ℓ's that you're talking about are those relative to the atomic centers; I'm not.

Temkin: I think we're beginning to confuse this problem with the one of doing ab initio calculations as opposed to just parametrizing. Suppose one wants to do a reasonably complicated system like a polyatomic molecule and in some sense ab initio. But by ab initio I don't mean that one is going to get precision results. One is trying to get, say, semi-accurate results but ab initio in a certain sense. Then I think that the idea of a single-center expansion, even for a polyatomic, or at least a class of polyatomics, is in fact not an unintelligent idea. It's true that to do a precision calculation you may need very many ℓ's. On the other hand, if you don't have any idea in advance, which is often the case we're faced with in astrophysical problems, the idea of using a reasonable number of ℓ's on a single center often gives a fair simulation of a polyatomic.

Nesbet: There is a technique known as the linear combination of muffin-tin orbitals which is now developing in solid state work. The point is that one uses a basis set of orbitals that are confined in the muffin-tin sense, i.e., around each atom one has a spherical region. But now you can make up a set of functions that have that structure in terms of which you can do the global integrals easily, and then use the true potential and take linear combinations of those functions. That turns out to be quite a powerful technique to get around the problem of the assumed flatness of the potential in the region between the atomic spheres. But now that you go to more complicated molecules, this becomes perhaps a desirable thing to put into a close-coupling framework for the overall scattering problem.

Dill: Does that mean that with this method you can use available integral packages and so on?

Nesbet: No, you have to develop the appropriate integral package for this kind of geometry or representation of the function.

Dill: If you're looking at a polyatomic molecule and you want to get a zero-order estimate of what's going on, we found that with CO_2 we did not need polarization to get the resonance. If we put polarization in it shifted the resonance by a few tenths of a volt. All the polarization did was to give us a grossly overestimated background near threshold. I'm just offering that as an observation. I don't understand it. But others seem to be saying that the resonance is very sensitive to the polarization. I find that with no polarization at all, the resonance is there. And with full polarization, the resonance is still there at about the same position. That needs to be understood, but if you're going to go on and make predictions, it's hard if you have a parameter that can make a resonance completely vanish.

Hazi: It might just be a cancellation of errors in your case. You're using a constant potential in the interstitial region which may be too attractive. So then you may not need the polarization, which is also attractive, to get the resonance at the correct position.

Dill: We have several cases now, N_2, CO_2, K-shells, where we get something that is observed. I suppose that if we did 100 cases and were right in 99, some would call it fortuitous. It's not fortuitous, it's an aspect of the model we must at least be aware of.

Nesbet: The point that's important here is that the polarization potential is there physically.

Dill: I agree, but I'm saying it doesn't effect the resonance, it just effects the background.

Bardsley: Can I ask, what's the radius of your outer sphere?

Dill: It's the sum of the atomic sphere radii.

Bardsley: How does it compare with the cut-off radii that people are using in the model polarization potentials?

Dill: It's about twice as big.

Bardsley: Then the second part of your statement, i.e., the resonance position is insensitive to polarization, comes about because you cut it off much further out.

Morrison: When we did the converged static calculations on CO_2 we got a spurious Σ_g resonance. We then put in exchange with the Hara potential and it went away and we did see a resonance in the proper symmetry, but it was very slight and at much too high an energy. Whether it is there in the true static-exchange model, we don't know. We then put in polarization, converged it with respect to the various parameters, and got the result everyone knows about - resonance in the right place and the rest of the cross section agrees with experiment. But Dan's question in a way leads into the question I wanted to raise. In a sense it points the finger at something we've been doing. One of the big problems is certainly convergence in ℓ. For the systems we've studied so far and hopefully for some non-linear systems that are not too grotesquely aspherical, that will be a matter of mechanics.

A more serious question to my mind right now about eigenfunction expansions addresses the question of polarization. If you choose to use an eigenfunction expansion technique and you wish to include polarization, there are a number of weaknesses inherent in the way we have introduced it. Even if you're willing to live with the approximations inherent in the semi-empirical, adiabatic, asymptotic, cut-off polarization potential, you have real problems choosing your parameter in a case where you do not have a well-established experimental structure like a resonance. So there are really two questions, at least. One is how to handle that kind of case. And the other is how to go beyond it, because when we tried to go beyond it by calculating a very accurate adiabatic polarization potential, we found that it was so accurate that we could not do scattering calculations, because it was so strong we could not cut it off realistically.

Schneider: I think you're aware of how I did that in the H_2 calculation. I did an SCF calculation in the presence of an extra charge and from that extracted an orbital, took that orbital, got rid of all the parts belonging to the Fock wavefunction because they're irrelevant, and then included that in the antisymmetrized products that I use.

Henry: So, in other words, this is an excited state.

Schneider: Well, it's not a real excited state. It's a way of computing orbitals which you can then antisymmetrize with all your other orbitals.

Temkin: What you're saying essentially is that you're using a polarized psuedo-state.

Henry: This means that you're now multiplying the number of states you have by the number of coupled channels.

Schneider: Not necessarily. You're taking the worst possible case.

Morrison: But that's not the question, Barry. What you're saying would be a reasonable way to go for smaller systems. Do people have ideas about other possible alternatives?

Bottcher: Could I ask for a clarification? Do I understand you to say that if you calculate what is called the proper polarization potential - that the potential is very strong.

Morrison: Yes, you see a much deeper potential in V_o and V_2 for example. You also see very important terms V_4 and V_6, say, coming out too large.

Lane: We're talking now of a purely adiabatic potential.

Temkin: In fact you know that there's a theorem which in a sense gives you an upper bound on the phase shift. Larry Spruch and colleagues proved a theorem which essentially says that if you use the adiabatic potential - the real adiabatic potential - then you get an upper bound on the phase shift. The phase shift is then definitely bigger than the true phase shift. That's why sometimes when you use perturbation theory for the adiabatic potential, you generally do better simply because perturbation theory is a lot weaker than the real H_2^+ problem, for example, and therefore you get a better estimate of the potential.

Morrison: In a sense, what I'm suggesting, as I tried to say yesterday, not glibly, is that we may be fortunate in that the errors that are inherent in using the semi-empirical adiabatic form make it a potential that can be easily cut off with a simple spherical cut-off with one parameter.

Temkin: I'm sure there's a certain amount of luck involved. But as molecules get more complicated one is going to get less and less precision in terms of an ab initio calculation. Therefore, the approach of, say, Dehmer and Dill is a very intelligent approach because what they're trying to do is reduce the problem to one in which there is a certain amount of art or fitting, say, but nevertheless it's methodological enough so that one can in fact make reasonable calculations and get reasonable accuracy. It seems to me we have to be a little careful in what we expect from a very complicated system.

Taylor: Dehmer and Dill's method is basically a model potential problem and falls into a category where, if you have low-lying excited states, you could be in trouble, especially if they partake in the scattering or give rise to Feshbach resonances. They can only handle valence-type resonances from the ground-state. It's a model; it's done cleverly, but it's a model potential.

Nesbet: What precludes for a molecule the use of the polarized orbital method?

Temkin: I think there's only one real polarized orbital calculation and that was for the simplest possible diatomic which is H_2^+. Nothing precludes it except that it's very difficult. In molecules you have all the additional quantum numbers and degrees of freedom. My own feeling on the matter was that even rotational close-coupling, which was calculated by Henry and Lane in H_2, included a polarization potential to get really good results and when they added polarization they were essentially going a little bit outside the frame of pure-close-coupling. So my own feeling on the matter is that one can do it but it becomes increasingly difficult as you get to many-electron systems with many degrees of freedom. Somewhere along the line you have to make compromises. The most ambitious program, of course, is the R-matrix program. I think it probably will do well, provided the molecules are not too complicated. At some point or other, we're going to have to do what Dehmer and Dill do, or it may be that their implementation won't work in the cases Howard Taylor mentioned, but maybe another formulation will be good.

Buckley: I just want to point out that as far as these local polarization potentials are concerned, my experience is that the cut-off parameter is quite a sensitive parameter. Certainly in N_2, I give a cut-off to 4 figures which is an indication of the sensitivity when I tried to fit to experiment. I think this clearly indicates that things aren't quite right with this. The other problem is that if you're going to go to different internuclear separations, how are you going to choose your cut-off?

Robb: Well, as you all know, I never even thought about looking at electron-molecule scattering until about 6 or 7 months ago. When I looked at the field I was horrified at how little had been done. Really, I thought that about fifty molecules had been studied in the static-exchange model. There are limitations in any method, otherwise people would have taken one method if it did't have limitations and just gone rampant with it. So what I decided to do was, since there's a need in my group to look at polyatomic molecules and know some basic processes there, I thought that some of the multi-center codes certainly weren't going to handle those soon, so I thought that a single-center approach was the thing to do. I realized there weren't any static-exchange single-center codes around, but that one thing you definitely could say was that the static equations in a single-center can be solved quite easily. And you can do it in a reasonable amount of computer time. So what I've come up with is the following scheme that Lee Collins and I are working on. The idea is the following. If you can get the static potential - I'm assuming there's a code that will take any molecule and, wherever the atoms are, can transform those orbitals to a single-center and can evaluate the static potential about that center. So I have that code and then I solve a set of coupled equations in the single-center sense and there might be a set of 20 by 20 equations. But this is not a problem - it takes maybe 15 seconds on a CDC 6600 to solve that per energy. Now the R-matrix codes that Barry and the Belfast group have developed are multi-center and use the Block operator which really doesn't impose the boundary conditions - it imposes the boundary conditions at the same time the basis set sees the Hamiltonian. In the atomic case, in what I've been working on for years, we impose the boundary conditions first on the basis set and then let it see the Hamiltonian. And I want to continue with that idea on molecules. So the idea would be to solve the static equations, which we can do at arbitrary energy, subject to the R-matrix boundary conditions. So I get a discrete set then of static, coupled R-matrix basis functions. But the indications are from Brian Buckley's talk that I'm putting as much physics into this as he is in his analytic orbitals on the atomic centers. So I expect that only 4 or 5 or maybe 10 of these will do per channel.

The Buttle correction is easy here because you can solve this at arbitrary energy and you can subtract out the first 5 poles and that's the Buttle correction. But the main thing is diagonalizing the Hamiltonian. That's H between two of these basis functions and H is just the kinetic energy plus the direct terms plus the exchange terms. And since the first two terms give zero when operating on a basis function, you've only got the exchange operator to diagonalize. And if you wanted to include polarization, you want to say put in an excited state of appropriate symmetry, then you just do static scattering off each state and then you have a 40 by 40 matrix. The only integrals you have to do are over short range operators. There's a lot of physics built into the basis initially and you have no problems with linear dependence since this is an orthonormal set. It's just the equivalent of the atomic problem except now most of the coupling is in the basis and it's tractable as far as computer time is concerned.

Morrison: One of the possible advantages of this is that it gives you a systematic way to check convergence of your R-matrix basis at last.

Buckley: I think one can be very systematic in testing convergence with the L^2 - methods as well.

Henry: What's the advantage in setting up the R-matrix over what you're doing right now with the iterative exchange?

Robb: With iterative exchange, sometimes you just don't converge.

Rescigno: Even if you start with the static as the first guess?

Robb: Yes, but with the static there's no trouble - you don't have to iterate to get it.

Henry: But an even better way to converge is to start with the local exchange as the first guess.

Robb: Maybe so.

Rescigno: If starting with the static won't converge in an iterative approach, doesn't that just mean that the static orbitals are not a good expansion basis when you add exchange. When you add exchange terms, you change the nodal structure of the wavefunction. So how can you guarantee that when you change the nodal structure of the wavefunction you can re-expand easily?

Temkin: He's summing over about 10 of these, not just one. He's diagonalizing the whole thing.

INDEX